Polarography of Molecules
of Biological Significance

Polarography of Molecules of Biological Significance

Edited by

W. FRANKLIN SMYTH

*Chelsea College, University of London,
Manresa Road, London, England*

1979

ACADEMIC PRESS
LONDON   NEW YORK   SAN FRANCISCO

A Subsidiary of Harcourt Brace Jovanovich, Publishers

ACADEMIC PRESS INC. (LONDON) LTD.
24/28 Oval Road
London NW1

United States Edition published by
ACADEMIC PRESS INC.
111 Fifth Avenue
New York, New York 10003

Copyright © 1979 by
ACADEMIC PRESS INC. (LONDON) LTD.

All Rights Reserved

No part of this book may be reproduced in any form by photostat, microfilm, or any other means, without written permission from the publishers

British Library Cataloguing in Publication Data
Polarography of molecules of biological significance
1. Voltammetry 2. Chemistry, Organic
I. Smyth, W. Franklin
547'.308'7   QD272.E4   78-54541
ISBN 0-12-653050-5

Printed in Great Britain by
Page Bros (Norwich) Ltd,
Mile Cross Lane, Norwich

CONTRIBUTORS

BIRCH, B. J. *Unilever (Research), Port Sunlight Laboratory, Merseyside L62 4XN, England.*

BROOKS, M. A., *Department of Biochemistry and Drug Metabolism, Hoffman-La Roche Inc., Nutley, New Jersey 07110, U.S.A.*

BROWNE, J. T., *Beecham Pharmaceuticals, Research Division, Betchworth, Surrey, England.*

CHOWDHRY, B. Z., *Department of Chemistry, Chelsea College, University of London, Manresa Road, London SW3 6LX, England.*

DAVIDSON, I. E., *Wyeth Laboratories, Huntercombe Lane South, Taplow, Maidenhead, Berkshire SL6 0PH, England.*

FLEET, B., *Department of Chemistry, Imperial College of Science and Technology, London, SW7 2AZ, England.*

FOUZDER, N., *Department of Chemistry, Imperial College of Science and Technology, London SW7 2AZ, England.*

FRANKLIN SMYTH, W., *Department of Chemistry, Chelsea College, University of London, Manresa Road, London SW3 6LX, England.*

HART, J. P., *Department of Orthopaedics, Charing Cross Hospital, Hammersmith, London, England.*

ROWE, R. R., *Murphy Chemical, Wheathampstead, St. Albans, Hertfordshire AL4 8QU, England.*

SMYTH, M. R., *Department of Microbiology, Colorado State University, Fort Collins, Colorado, U.S.A.*

WATSON, A., *Department of Chemistry, Paisley College of Technology, High Street, Paisley, Renfrewshire, Scotland.*

PREFACE

"Polarography of Molecules of Biological Significance" was conceived in the early seventies at which time a renaissance in the application of this and related techniques to the identification and determination of not only metal ions but also organic and organometallic molecules occurred. This was due to the availability of relatively inexpensive instrumentation, capable of trace analysis of a variety of materials and the realisation that structural and analytical information on, for example, the speciation of metals and degradation/metabolism of organic molecules, complementary to that obtained by chromatographic, spectroscopic and radiochemical methods, could be obtained.

In this volume, twelve young scientists from academic, industrial and medical backgrounds have selected subjects with which they are particularly cognizant and have produced reviews in which either their own research work has been interwoven into the fabric of a general voltammetric review and/or they have critically compared the application of polarographic techniques with others in a particular analytical situation. The authors have mainly concerned themselves with the literature since 1965 when "Organic Polarographic Analysis" by Petr Zuman was published.

The first two chapters are concerned with the principles of organic voltammetric analysis and are treated by consideration of the unit processes of choice of method, sampling, separation, derivitisation and determination. The remaining chapters deal with the electrochemical behaviour of many organic and organometallic molecules/species, and particularly with the resulting analytical applications of such voltammetric analysis in pharmacy, pharmacology, basic medical sciences, environmental science and agriculture.

March, 1979   W. FRANKLIN SMYTH

ACKNOWLEDGEMENT

The Editor is indebted to Dr Malcolm R. Smyth of The State University of Colorado for his editorial assistance in the preparation of this text.

FOREWORD

It is not an easy task to see your own transformation from a young, eager researcher into an eager, elder statesman of science. None of us is somehow ready to accept that years really go by and it was for me a mild shock when I realized that it was more than a quarter of a century ago when we wrote the first version of a monograph dealing with polarography in medicine, pharmacy and biochemistry. What makes it easier to accept the flow of time is the recognition that the torch is in the hands of able bearers. For me personally there is an additional pleasant feeling that two of the main contributors (B. Fleet and W. F. Smyth) were exposed to the lure of polarography during their stay in my laboratories in Prague or during my pleasant years in Birmingham, respectively.

During the last thirty years polarography and related methods have come of age too. We understand better for which compounds polarographic methods of analysis are promising and for which not, and what are the processes taking place in the course of polarographic electrolysis. Better knowledge of mechanisms of electrode processes enables development of methods less affected by other components of samples. But, above all, the range of sensitivity of polarographic methods changed. The development of differential pulse and stripping methods pushed the limits of concentration from 10^{-5}–10^{-6} M, obtainable with classical methods, to 10^{-7}–10^{-8} M or even 10^{-9} M. This increase in sensitivity can be considered predominantly responsible for the tremendous increase of interest in the use of such electroanalytical methods over the past decade, a comeback described sometimes as the "renaissance of polarography". With hindsight, the awarding of the Nobel Prize in 1959 to my esteemed teacher, Professor J. Heyrovský, for the invention of polarography, seems even more deserved than it seemed before. The reproducibility of current-voltage curves obtained with the dropping mercury electrode is still the primary advantage of polarographic methods, as it has been since the method was invented more than half a century ago. The dropping mercury electrode (d.m.e.) is still in the large majority of cases the electrode of choice when reductive processes are dealt with. Increased interest in oxidations resulted in renewed interest in the use of solid electrodes, and understanding of transport phenomena and electrode processes gained from polarography assisted greatly in the development of this area. Many useful analytical procedures have been developed involving the use of such electrodes, but it should always be realized that even when the use of such electrodes is a necessity, for instance when dealing with less easily oxidizable compounds, the degree of reproducibility is often lower than with the d.m.e.

The present volume turns attention to both electrode processes involved and analytical procedures developed from two points of view: according to the field of application (pharmacy and pharmacology, basic medical science, environmental analysis and agriculture) and according to the nature of the substance to be determined (organometallic compounds, nitrogroup containing antibiotics, oxidizable compounds). These two aspects are often complementary and their inclusion seems to make this volume attractive to a wider circle of users.

Let me express the hope that this volume will contribute to furthering the use of polarographic methods in the fields on the borderline between chemistry and biology and that it will attract new dynamic workers into a field which can offer so much satisfaction and fun.

PETR ZUMAN

CONTENTS

Contributors	v
Preface	vii
Acknowledgement	viii
Foreword	ix
Glossary of Terms	xvii
Sources of Information for Voltammetry	xix
Dedication	xx

PRINCIPLES

Chapter 1
UNIT PROCESSES IN ORGANIC VOLTAMMETRIC ANALYSIS—I. CHOICE OF METHOD, SAMPLING, SEPARATION AND DERIVATIZATION
W. Franklin Smyth and Malcolm R. Smyth

I.	Introduction	3
II.	Choice of Method	4
	A. Historical Background	4
	B. Electroactive Molecules	5
	C. Rapidity, Convenience, Accuracy and Precision	9
	D. Sensitivity	11
	E. Selectivity and Specificity	12
	F. Structure Elucidation/Identification	13
III.	Sampling and Initial Treatment of Sample	17
IV.	Separation Techniques	18
	A. General Comments	18
	B. Centrifugation/Filtration Methods	22
	C. Gel Filtration	22
	D. Ion Exchange Chromatography	23
	E. Macro Reticular Resins	23
	F. Thin Layer Chromatography	24
	G. Paper Chromatography	24
	H. High Performance Liquid Chromatography (h.p.l.c.)	24
	I. Solvent Extraction	25
	J. Other Separation Methods	28

V.	Complexation and Derivatization Procedures	28
	A. Complexation	28
	B. Nitration	29
	C. Nitrosation	30
	D. N- and S-Oxidation	31
	E. Condensation	31
	F. Hydrolysis	32
	G. Other Methods	32
References		33

Chapter 2
UNIT PROCESSES IN ORGANIC VOLTAMMETRIC ANALYSIS—II. DETERMINATION
Bernard Fleet and Nani B. Fouzder

I.	Introduction	37
II.	Preparation of Sample Solution for Voltammetric Investigation	37
	A. Choice of Solvent/Supporting Electrolyte	37
	B. Choice of pH	41
III.	Selection of a Suitable Cell and Electrodes	41
	A. Cells	41
	B. Electrodes	43
	C. The Potentiostat	46
IV.	Selection of the Optimum Polarographic or Voltammetric Technique	48
	A. D. C. Techniques	48
	B. Pulse Techniques	56
	C. A.C. Techniques	61
	D. Stripping Voltammetry	63
	E. Coulometric Techniques	66
	F. Application of Computers in Electrochemical Instrumentation	68
	G. Concluding Remarks	72
References		73

APPLICATIONS
In Pharmacy and Pharmacology

Chapter 3
THE POLAROGRAPHIC DETERMINATION OF PSYCHOTROPIC, HYPNOTIC AND SEDATIVE DRUGS
Marvin A. Brooks

I.	Introduction	79
II.	Psychotomimetic Drugs	79
III.	Psychosedative Drugs	80
	A. Antipsychotics	80
	B. Antianxiety Agents	85
IV.	Antidepressant Drugs	96
	A. Tricyclics	98
	B. Monoamine Oxidase Inhibitors	98
V.	Hypnotic or Sedative Drugs	99
	A. Miscellaneous	99
	B. Barbiturates	101
	C. 1,4-Benzodiazepines	103
VI.	Conclusion	106
References		107

Chapter 4
QUANTITATIVE ANALYSIS OF MAJOR ANTIBIOTICS CONTAINING A NITRO GROUP
J. T. Browne

I.	Polarographic Behaviour of the Nitro Group	111
II.	Applications	113
	A. Chloramphenicol	113
	B. Nitrofurans	118
	C. Nitroimidazoles	121
References		125

Chapter 5
OXIDATIVE VOLTAMMETRIC ANALYSIS OF MOLECULES OF PHARMACEUTICAL IMPORTANCE
I. E. Davidson

I.	Introduction	127
II.	Vitamins	128
	A. Vitamin C	128
	B. Vitamin E	131
	C. Other Vitamins	133
III.	Analgesics	135
IV.	Central Nervous System Agents including Tranquilizers and Sedatives	139
V.	Antituberculins	142
VI.	Naturally Occurring Compounds	142
	A. Purine and Derivatives	142

	B. Sulphur-containing Compounds	146
	C. Catechols and Catecholamines	148
VII.	Miscellaneous Compounds	150
	A. Nitrogen-containing Compounds	150
	B. Sulphur-containing Compounds	154
	C. Oxygen-containing Compounds	158
	D. Halogen-containing Compounds	159
VIII.	General Reviews on Anodic Oxidation	161
References		161

APPLICATIONS
In the Basic Medical Sciences

Chapter 6
SOME RECENT APPLICATIONS OF ORGANIC VOLTAMMETRY IN THE BASIC MEDICAL SCIENCES
Babur Z. Chowdhry

I.	Introduction	169
II.	Clinical Medicine	169
	A. Clinical Diagnosis	169
	B. Steroids, Hormones and Vitamins	173
	C. Respiratory Physiology	175
III.	Chemical Carcinogenesis	177
	A. Theoretical Aspects	177
	B. Analytical Aspects	177
IV.	Neurophysiology	179
	A. Some Direct Determinations *In Vivo*	180
	B. Some Determinations following Separation Procedures	182
V.	Virology	185
VI.	Immunology	186
VII.	Cells, Organelles and Membranes	189
VIII.	Miscellaneous Compounds	190
IX.	Conclusions	190
Acknowledgements		197
References		197

APPLICATIONS
In Environmental Science

Chapter 7
ELECTROANALYSIS OF TRACE FOREIGN ORGANIC MATERIALS IN THE AQUEOUS ENVIRONMENT
B. J. Birch and J. P. Hart

I.	Introduction	.	.	.	.	.	.	.	.	205
II.	Applications	.	.	.	.	.	.	.	.	206
	A. Carbonyl Compounds.		.	.	.	.	.	.	206
	B. Sequestering Agents	.	.	.	.	.	.	.	207
	C. Simple Aromatic Compounds		.	.	.	.	.	209
	D. Herbicides and Pesticides	.	.	.	.	.	.	210
	E. Polymeric Materials	.	.	.	.	.	.	.	214
	F. Surfactants	.	.	.	.	.	.	.	215
	G. Miscellaneous Molecules	.	.	.	.	.	.	219
III.	Conclusions	.	.	.	.	.	.	.	.	223
References		.	.	.	.	.	.	.	.	224

APPLICATIONS
In Agriculture

Chapter 8
ELECTROANALYSIS OF AGROCHEMICALS
R. R. Rowe and Malcolm R. Smyth

I.	Introduction	.	.	.	.	.	.	.	.	229
II.	Organochlorine Compounds	.	.	.	.	.	.	230
III.	Organophosphorus Compounds	.	.	.	.	.	234
IV.	Nitro Compounds	.	.	.	.	.	.	.	239
V.	Sulphur-Containing Compounds	.	.	.	.	.	244
VI.	Carbamates and Ureas	.	.	.	.	.	.	249
VII.	Heterocyclic Nitrogen Compounds	.	.	.	.	.	251
VIII.	Organometallic Compounds	.	.	.	.	.	.	253
IX.	Miscellaneous Compounds	.	.	.	.	.	.	255
X.	Conclusions	.	.	.	.	.	.	.	.	257
References		.	.	.	.	.	.	.	.	257

Chapter 9
THE ELECTROCHEMICAL BEHAVIOUR AND ANALYSIS OF ORGANOMETALLIC COMPOUNDS OF MERCURY, TIN, LEAD AND GERMANIUM
Nani B. Fouzder and Bernard Fleet

I.	Introduction		.	.	.	.	.	.	.	261
II.	Mercury	.	.	.	.	.	.	.	.	262
	A. Introduction	.	.	.	.	.	.	.	262
	B. General Polarographic Behaviour		.	.	.	.	262
	C. Polarographic Analysis of Organomercurials	.	.	.	266
III.	Tin	.	.	.	.	.	.	.	.	.	268

	A. Introduction	268
	B. General Polarographic Behaviour	268
	C. Polarographic Analysis	276
IV.	Lead	280
	A. Introduction	280
	B. General Polarographic Behaviour	280
	C. Polarographic Analysis	284
V.	Germanium	286
	A. Introduction	286
	B. General Polarographic Behaviour	287
	C. Polarographic Analysis	289
VI.	Conclusions	290
References		291

Chapter 10
POLAROGRAPHIC BEHAVIOUR AND ANALYSIS OF THE ORGANIC COMPOUNDS OF ARSENIC
A. Watson

I.	Introduction	295
II.	The Arsonic Acids	297
III.	Phenyl Arsenoxide	303
IV.	Diphenyl Arsinic Acid	305
V.	Diphenyl Arsine Oxide	307
VI.	Triphenyl Arsine Oxide	308
VII.	Quaternary Arsonium Salts	310
VIII.	Organic Arsenic Halides	310
IX.	The Electroinactive Compounds	311
X.	Analytical Applications	311
References		318

SUBJECT INDEX 321

GLOSSARY OF TERMS

a.c. polarography	alternating current polarography
AN	acetonitrile
a.s.v.	anodic stripping voltammetry
BR buffer	Britton Robinson buffer
c.s.v.	cathodic stripping voltammetry
c.v.	cyclic voltammetry
d.c. polarography	direct current polarography
d.m.e.	dropping mercury electrode
DMF	dimethylformamide
DMSO	dimethylsulphoxide
d.p.a.s.v.	differential pulse anodic stripping voltammetry
d.p.c.s.v.	differential pulse cathodic stripping voltammetry
d.p.p.	differential pulse polarography
$E_{\frac{1}{2}}$	half-wave potential
E_p	peak potential
EDTA	ethylenediamine tetraacetic acid
e.s.r.	electron spin resonance spectroscopy
g.l.c.	gas liquid chromatography
g.l.c.–m.s.	gas liquid chromatography—mass spectroscopy
g.l.c.–t.e.a.	gas liquid chromatography—thermal energy analysis
h.m.d.e.	hanging mercury drop electrode
h.p.l.c.	high performance liquid chromatography
i_{lim}	limiting current
i_p	peak current
l.s.v.	linear sweep voltammetry
m.f.e.	mercury film electrode
m.o.	molecular orbital
m.s.	mass spectroscopy
n.p.p.	normal pulse polarography
NAD/NADH	nicotinamide adenine dinucleotide
p.g.e.	pyrolitic graphite electrode

r.g.e.	rotating graphite electrode
s.c.e.	standard calomel electrode
TBAB	tetrabutylammonium bromide
TBAI	tetrabutylammonium iodide
TEAI	tetraethylammonium iodide
TEAOH	tetraethylammonium hydroxide
TEAP	tetraethylammonium perchlorate
THF	tetrahydrofuran
TMAB	tetramethylammonium bromide
t.l.c.	thin layer chromatography
u.v. spectroscopy	ultra violet spectroscopy

SOURCES OF INFORMATION FOR VOLTAMMETRY

(1) Ascatopics (E.D.7)
(2) Current Contents (all sections; chemical sciences, biological sciences, weekly subject index).
(3) UKCIS (United Kingdom Chemical Information Service)
(4) Chemical Abstracts
(5) Biological Abstracts
(6) Analytical Abstracts
(7) Electrochemical Abstracts*
(8) Chemical titles
(9) Int. Abstr. Biol. Sciences
(10) Bioresearch Index
(11) Pharmaceutical Abstracts
(12) Referativny-I Zhurnal (Russian)
(13) Index Medicus
(14) Specialized Journals of an analytical, electrochemical and biological nature.
(15) Excerpta Medica (all sections)
(16) Specialist Periodical Reports in electrochemistry.†
(17) Specialist Computer Retrieval systems at various universities also provide reference lists

---

* Best general source on theory and practice of electrochemical techniques.
† Published by Chemical Society, London.

DEDICATION

This book is dedicated to three beautiful girls, my wife Kate and daughters Clare and Frances who had to suffer my moods during its production. May some measure of peace now return to the family.

PRINCIPLES

Chapter 1

UNIT PROCESSES IN ORGANIC VOLTAMMETRIC ANALYSIS—I. CHOICE OF METHOD, SAMPLING, SEPARATION AND DERIVATIZATION

W. FRANKLIN SMYTH

Department of Chemistry, Chelsea College, Manresa Road, London, England.

and

MALCOLM R. SMYTH*

Department of Microbiology, Colorado State University, Fort Collins, Colorado, U.S.A.

I. INTRODUCTION

The most systematic way of discussing an analytical method is in terms of constituent unit processes; choice of method, sampling and initial treatment of sample, separation, derivatization and determination. It is this course that is adopted in the first two chapters of this volume for the voltammetric analysis of molecules of biological significance. This chapter will concern itself with the first four of these processes and the second with the determination step. It is envisaged that this methodological approach should be particularly useful for recent devotees to the subject, as well as gathering together the recent literature in a somewhat unique form.

---

* *Present address*: Institut fur Chemie der Kernforschungsanlage, Jülich, West Germany.

II. CHOICE OF METHOD

A. HISTORICAL BACKGROUND

The polarographic method has had a somewhat chequered career since its inception by Heyrovsky (1922, 1924) as the first instrumental method for trace metal analysis. It was extremely popular for some 25 years and a large file of both theoretical and practical data was assembled by laboratories all over the world. With reference to molecules of biological significance one can quote the authoritative works of Brezina and Zuman (1958), Milner (1962), Volke (1964) and Zuman (1964) who surveyed the polarographic behaviour and collected together analytical methods for many such molecules. During the late 1950s polarography suffered a decline in popularity for the solution of analytical problems, despite the impetus created by the award of the Nobel Prize to Jaroslav Heyrovsky in 1959. This decline can be traced to several factors, the main ones being the limited sensitivity of d.c. polarography, difficulties in the interpretation of polarographic waves and also the growing competition from chromatographic and spectroscopic methods.

During the mid-1950s, however, the first of the modern variants of classical polarography started to appear. Cathode ray polarography (or fast linear sweep voltammetry) was the first of these (Randles, 1947; Reynolds and Davis, 1953; Davis and Seaborn, 1953) followed by the development of square wave polarography and pulse polarography by Barker and co-workers at Harwell (Barker and Jenkins, 1952; Barker, 1958; Barker and Gardner, 1960). At about the same time, improved electronic instrumentation became available, in particular low cost operational amplifiers which led to the use of a wider range of input waveforms and signal processing techniques.

During the last decade a renewal of interest has been detected in the literature and this can be traced to two principal factors (i) the availability of relatively inexpensive instrumentation (e.g. pulse polarographs) capable of rapid and sensitive analysis of both inorganic and organic molecules and (ii) the need for information complementary to that provided by spectroscopic and chromatographic methods, e.g. on the speciation of metals and on the metabolism of organic molecules in biological samples.

In attempts to decrease the background current in pulse polarographic techniques, Vassos and Osteryoung (1973) have developed a low noise digital pulse polarograph whereas Abel *et al.* (1976) have modified a commercial instrument, the PAR Model 174, to permit variation of instrumental parameters. Christie *et al.* (1976) have also developed a new technique called alternate drop pulse polarography which is effective at discriminating against the capacitive current. These improvements in pulse polarographic instrumentation have come about mainly through an increased understanding of the input and output waveforms involved, a subject which has recently been reviewed by Osteryoung and Hasebe (1976).

Another significant trend in the last few years has been the advent of microprocessor controlled polarographs. Computers have been used for several years to carry out data processing from rapid d.c. and linear sweep voltammetry and the logical conclusion of this has been seen in the recent introduction of a microprocessor controlled multimode polarograph (Princeton Applied Research, 1976). In addition to blank subtraction and rapid print-out of concentration currently performed by these polarographs, future developments in this area are likely to be concerned with identification of unknown peaks, which poses many problems in trace organic analysis. They can also carry out automatic analyses when connected to a sample turntable and this should prove of great value in formulation assays.

The development of electrochemical detectors for high performance liquid chromatography (h.p.l.c.) is seen as a major breakthrough in the determination of trace amounts of naturally occurring molecules in biological materials, based on the oxidation processes these molecules undergo at solid electrodes.

During this period there has also been a rapid expansion in the general area of electrotechnology, i.e. in the development of industrial electrochemical synthesis, electrochemical energy conversion and electrochemical techniques for effluent control. Interest has also been shown in the development of selective voltammetric detectors responsive to a variety of gases, e.g. oxygen (Clark, 1956) and carbon monoxide (Bergman and Windle, 1972), and enzyme-substrate systems (Hicks and Updike, 1967). Although beyond the scope of this volume, developments of this nature are dependent on fundamental studies of the electrode processes in which polarography and other voltammetric techniques often play a vital role.

B. ELECTROACTIVE MOLECULES

Molecules of biological significance whose electroactivity has been used in the development of analytical methods may be classified as follows:

(i) Those undergoing reduction processes. These are monitored principally at the d.m.e. although solid indicator electrodes have also been used e.g. glassy carbon (see p. 25).

The commonly encountered electroreducible bonds or groupings along with their mechanisms of reduction are listed in Table I. The analytical applications of these processes are discussed in subsequent chapters of this book. There are also a number of heterocyclic compounds, e.g. phthalimides,

TABLE I. *Electroreducible bonds/groupings*

Electroreducible bond/grouping	Reduction mechanism	Comments and examples
(a) Carbon–Carbon Bond (i) Double bond	$2e^-$ process resulting in saturation of bond	Occurs at high negative potentials and can give analytically usable waves in tetraalkylammonium supporting electrolytes dissolved in organic solvents. The reduction potential is made less negative if the bond contains electron-withdrawing groups (e.g. as in cephalosporins) or the degree of conjugation is increased by the presence of say a keto group (as in 3-keto $\Delta_{4,5}$-steroids)
(ii) Triple bond	Reducible if conjugated to an aromatic ring	An unconjugated triple bond can be rendered electroactive as in the case of lynestrenol by heating with 60% H_2SO_4/CH_3OH to produce an electroreducible group (Van Bennekom et al., 1975)
(b) Carbon–Halogen Bond	Carbon–halogen bond is cleaved	In general, polyhalogenated compounds give rise to well defined waves/peaks, e.g. hexachlorophane can be determined with a sensitivity of 100 ng ml^{-1} in plasma (Jacobsen and Rojahn, 1972)
(c) Carbon–Oxygen Bond	$2e^-$ process generally for aldehydes and ketones	Conjugated carbonyl compounds, e.g. quinones and benzophenones reduce at less negative and more analytically usable potentials than non-conjugated ones (e.g. formaldehyde and acetaldehyde). Sugars such as ketoses and aldoses are reduced in kinetically controlled processes at relatively negative potentials and give ill defined peaks on application of d.p.p.

TABLE I—continued

Electroreducible bond/grouping	Reduction mechanism	Comments and examples
(d) Carbon–Nitrogen Bond		
(i) Single bond	Reductive splitting occurs in quaternary phenyl and alkyl ammonium salts	Reduction occurs at high negative potentials; an exception being 1-benzoyl,5,5'-diethylbarbituric acid (Kahl and Pasek, 1970)
(ii) Double bond	$2e^-$ process resulting in saturation of bond	Well defined waves/peaks produced when group is conjugated to aromatic nucleii e.g. 1,4-benzodiazepines (Brooks, this volume, Chapter 3)
(e) Nitrogen–Nitrogen Bond		
(i) Single bond	Reductive cleavage of bond consuming $2e^-$	For example, in benzhydryl piperazine derivatives, this process occurs together with saturation of adjacent azomethine group (Smyth et al., 1976)
(ii) Double bond	Saturation of bond, in some cases followed by cleavage to form a mixture of amines	The latter process occurs when the group is conjugated to an aromatic one possessing activating substituents, e.g. 4-hydroxyazobenzene (Florence, 1974)
(f) Nitrogen–Oxygen Bond	Reduction process is dependent on oxidation state of nitrogen atom	Aromatic nitro compounds can reduce in $2e^-$, $4e^-$, or $6e^-$ steps depending on the substituents whereas aliphatic counterparts are reduced in $4e^-$ steps to the hydroxylamine. Waves/peaks for the aromatic molecules are well defined Aliphatic nitroso-containing molecules generally produce ill defined waves/peaks whereas aromatic nitroso ones have been observed to give well defined peaks on application of d.p.p. at less negative potentials (Franklin Smyth et al., 1975)

TABLE I—continued

Electroreducible bond/grouping	Reduction mechanism	Comments and examples
		Aliphatic N-oxides are reduced at relatively negative potentials. In the absence of extensive conjugation, so are heterocyclic N-oxides whose potentials can be unaffected by substituents (e.g., 4-nitroquinoline N-oxide and its 4-amino and 4-hydroxyamino metabolites)
		Hydroxylamines are only reduced in the protonated state and generally give ill defined peaks on application of d.p.p. Aryl hydroxylamines have been reported to give well defined waves in alkaline media (Iversen and Lund, 1969)
		Some aliphatic oximes give relatively small reduction peaks on application of d.p.p. (Beckett et al., 1977)
(g) Carbon–Sulphur Bond		
(i) Double bond	Can be reduced in 2e$^-$ step	For example, some 5,5' disubstituted thiobarbiturates (Smyth et al., 1970a)
(h) Sulphur Oxides	Mechanism dependent on oxidation state of sulphur atom	Aromatic S-oxides give well defined 2e$^-$ peaks (Beckett et al., 1974)
(i) Sulphur–Sulphur Bond	2e$^-$ process resulting in saturation of bond	Discussed further in Chapters 5 and 6

anthocyanins, pyrimidines etc., and organometallic compounds which can be reduced at the d.m.e. For a more detailed account of the electrochemical behaviour of these compounds and those mentioned in Table I, standard texts such as Milner (1962), Zuman (1964), Zuman and Perrin (1969), Baizer (1973) and Chapters 9 and 10 of this volume should be consulted.

(ii) Those undergoing oxidation processes at the d.m.e. and other indicator electrodes such as Pt, glassy C, etc. Examples of such molecules are phenols, aromatic amines and many heterocyclic compounds. The electrochemical behaviour of these compounds has been dealt with by Adams (1969) and resulting applications are reviewed by Davidson (Chapter 5, this volume).

(iii) Those that give rise to catalytic processes such as catalytic hydrogen evolution. Examples are heterocyclic substances containing nitrogen in unbuffered acidic or buffered solutions, and naturally occurring molecules such as cysteine and some proteins which show catalytic waves in presence of some heavy metals, e.g. Co or Ni (Mairanovskii, 1968). Some recent applications of this latter process are discussed by Chowdhry in Chapter 6 of this volume. In addition, catalytic pre-waves of metal ions have been used for analytical purposes, e.g. the determination of thiamine using a cobalt(II) catalytic pre-wave (Sanz Pedrero and Lopez Fonseca, 1972).

(iv) Those that possess none of the above properties but which can adsorb on the electrode surface thus giving rise to tensammetric waves. Examples are given in the chapter by Birch and Hart (Chapter 7).

(v) Electroinactive molecules that are determined by production of an electroactive species/molecule. This is discussed in greater detail later in this chapter.

C. RAPIDITY, CONVENIENCE, ACCURACY AND PRECISION

Voltammetric methods are found particularly rapid, convenient, accurate and precise in formulation analysis where the reducible or oxidizable active ingredient can frequently be determined in the presence of generally electroinactive excipients, irrespective of whether the latter are soluble or not in the chosen supporting electrolyte. The method of choice is to grind up the formulation in a mortar, mix it with a solvent (methanol or dimethylformamide for free bases and others, and water for the corresponding salts) in which the active constituent is soluble and then to dilute an aliquot of the supernatant with appropriate supporting electrolyte prior to electroanalysis. This procedure eliminates both time-consuming solvent extraction steps and calculations of recovery common to photometric and chromatographic methods, while the resulting accuracy and precision are at least comparable if not better than the above mentioned methods. Errors of the order of 1–2% and similar coefficients of variation are not uncommon. Voltammetric methods can also be used as rapid means for stability testing of formulations, provided that the degradation products are electroactive at different potentials or, alternatively, electroinactive.

In vitro dissolution rate measurements can also be rapidly followed by voltammetric sensors placed directly in a sample loop with a filter system to remove suspended particulate materials. Continuous sampling is thus possible and the percentage of substance dissolved can be measured with time. This eliminates the need for serial sampling, extraction and tedious calculations often needed to perform dissolution rate determinations by photometric or chromatographic methods (M. A. Brooks, Personal communication).

Automation of quality control procedures has been reported by Silvestri (1972), Bargagna Prunai et al. (1972), Jacobsen and Jacobsen (1971), Cullen et al. (1973), Lund and Opheim (1975, 1976) and, as mentioned previously, there is now instrumentation available (PAR Model 374) which employs a pressurized d.m.e. in conjunction with an automatic sampling system. This instrument also contains a microprocessor unit which is capable of storing polarographic curves in its memory and of subtracting blank currents. This instrument could ideally be applied to on-line quality control analysis and is discussed in greater detail in Chapter 2 by Fleet and Fouzder.

Direct voltammetric analysis of organic molecules in biological fluids is rapid and can be of great importance in *in vivo* biological monitoring. These methods have been applied to some aspects of clinical diagnosis (Chowdhry, Chapter 6) and could be more extensively applied to the determination of foreign substances which exist in biological fluids at relatively high trace concentrations (as is the case in many toxicological investigations). C.s.v. has recently been applied to the direct determination of some foreign organic compounds, i.e. thioamides in biological fluids resulting in a method that was both rapid and sensitive with a relative standard deviation of between 2–4% (Davidson and Franklin Smyth, 1977).

When the molecule of interest has to be separated from the biological matrix then not only does the method become more time consuming but the accuracy and precision of the overall method is affected. The relative standard deviation of most methods involving voltammetric determination of organic compounds following separation from biological fluids is rarely better than 3–4% (at the 1 μg ml^{-1} level). When derivatization is also required then the relative standard deviation can be of the order of 5–10% (see Section V).

In the early stages of testing the toxicity and distribution of potential pharmaceutical compounds in experimental animals, the analytical method of choice usually involves the use of a radiochemical technique. Bearing in mind current awareness of the dangers of the use of radioactive compounds and the need for rapid and complementary "cold" methods of bioassay, sensitive polarographic methods could be more extensively applied at this stage to those compounds that are electroactive.

D. SENSITIVITY

Voltammetric methods of analysis have sensitivities varying from 10^{-3}M (d.c. polarography) to 10^{-9}–10^{-10}M (stripping voltammetry). Voltammetry at solid electrodes (generally employed for oxidative measurements) has a sensitivity of the order of 10^{-6}M. This can be improved upon by using semi-micro solid electrodes in conjunction with flowing streams of supporting electrolytes as in electrochemical detectors for liquid chromatography. The highest sensitivities are obtained using the d.m.e. and h.m.d.e. in certain polarographic modes, e.g. linear sweep, differential pulse, anodic and cathodic stripping modes. Linear sweep and differential pulse polarography have sensitivities greater than that of spectrophotometry and equivalent to those of spectrofluorimetry, gas–liquid chromatography with a flame ionization detector and liquid chromatography with a u.v. detector. Only the stripping methods can attain the sensitivity of either gas–liquid chromatography with electron capture or mass spectrometric detection, or radioimmunoassay; all of which have demonstrated sensitivities in the low nanogram to picogram ml^{-1} range.

The more sensitive voltammetric methods have found application in pharmaceutical analysis for single tablet assays and in the determination of small amounts of electroactive degradation products. In d.p.p., given adequate separation of E_p values, these degradation products can be determined irrespective of whether the impurity peak occurs at a more or less negative potential than the parent drug.

For the trace determination of drugs and other organic molecules in biological fluids, the overall method generally employs clean-up, separation and concentration steps similar to those required by spectrophotometric or chromatographic procedures. The concentration step is particularly important since many clinically effective drugs are administered in very low doses. It is not at all uncommon to obtain peak blood levels of the parent compound and its metabolites in the range of 25–50 ng ml^{-1} following 5 to 10 mg doses. Therefore, analytical procedures of high sensitivity are required for the simultaneous determination of the parent compound and/or its metabolic products in blood and plasma for periods of up to 48–96 hours after administration. Differential pulse polarography and fast linear sweep polarography are the voltammetric techniques currently used for the routine determination of drugs in blood or plasma at these low concentrations. The analysis of urinary excretion products is more easily performed since the urine concentrations usually exceed 1 µg ml^{-1}. In addition, polarographic methods are ideally suited for the analysis of urinary metabolites since they are usually very polar compounds. This is in contrast to gas–chromatographic methods which often require derivatization to reduce the polarity of the compounds to yield volatile derivatives suitable for analysis.

The application of stripping voltammetry to the determination of organic molecules of biological significance has received scant attention in the analytical literature considering that its sensitivity lies in the nanogram to picogram ml^{-1} range. Such organic molecules can be determined by stripping methods in basically three ways:

(a) pre-concentration either by adsorption of the species to be determined or by adsorption of a product of the electrode reaction, followed by stripping of the adsorbed species [e.g. as in the determinations of methylene blue (Perone and Oyster, 1964) and of some chlorinated pesticide derivatives (Supin and Budnikov, 1973)],
(b) anodic pre-concentration as sparingly soluble salts or complexes with mercury ions, followed by cathodic stripping [e.g. the determination of some thioamides directly in plasma in *in vivo* samples (Davidson and Franklin Smyth, 1977)];
(c) indirect determination using stripping peaks of metals which form stable compounds or complexes with the substance to be determined [(e.g. the determination of bromazepam, a 1,4-benzodiazepine which complexes with Cu(II) ions (Smyth *et al.*, 1977a)].

E. SELECTIVITY AND SPECIFICITY*

In general, voltammetric methods on their own are not particularly selective although some elegant examples to the contrary exist in literature; for example, Bieder and Brunel (1971) have applied rapid a.c. polarography to the determination of the urinary metabolites of ethionamide and prothionamide in man. In the potential region -0.7 to -1.3 V, the parent compound and the three metabolites (the sulphoxide, the 2-carbamoyl-4-pyridine and the 2-isonicotinic acid derivatives) were simultaneously determined.

When metabolites or breakdown products are reduced or oxidized at the same potential as the parent compounds thin layer or liquid chromatographic separation of the initial extract from the biological fluid is usually employed to improve the selectivity of the assay.

Voltammetric methods can show exceptional specificity when the electroactive parent compound produces electroinactive metabolites or degradation

---

* Specificity is here defined as the ability of the electroanalytical technique to only give a response to a particular molecule/species in a mixture.

products. This is aptly illustrated in the case of sulphur containing molecules that both degrade and metabolize to non-sulphur containing species, e.g. thioamides (Davidson and Franklin Smyth, 1977).

F. STRUCTURE ELUCIDATION/IDENTIFICATION

For voltammetry to become more widely accepted as a technique suitable for organic trace analysis, it must not only offer some measure of selectivity, specificity and sensitivity comparable to other trace methods of analysis, but should also offer structural information on the nature of unknown metabolites/degradation products.

Those metabolic reactions that give rise to a change in potential of the parent compound or transform an electroinactive molecule into an electroactive one are discussed in a recent article by Franklin Smyth and Smyth (1976). These are listed in Table II.

Voltammetric methods also have the ability to discriminate between oxidation states of metals contained in some organometallic species, which is important in a study of their metabolic reactions in environmental situations. This is illustrated more fully in the chapters by Fouzder and Fleet (Chapter 9) and Watson (Chapter 10).

If the parent compound and its possible metabolites are closely related structurally so that similar electrochemical mechanisms are likely to operate for all members of the series then the $E_{\frac{1}{2}}/E_p$ value can be related to the Hammett polar function in a linear fashion (Zuman, 1967). This has been attempted for substituted 1,4-benzodiazepines by Brooks *et al.* (1975). Alternatively, one can relate the energy of the lowest vacant molecular orbital to the $E_{\frac{1}{2}}/E_p$ value. This has been studied for various groups of compounds, e.g. heterocycles (Bergman, 1954), substituted benzyl-*p*-chloranilines (Lalithambika *et al* 1975) and substituted nitrobenzenes (Smyth, 1976). M. o. calculations have also been used to predict the reduction pathway of organochlorine compounds (Beland, 1975).

On a more practical level, information on the structure of the molecule giving rise to an "unknown" peak can be obtained by the experienced researcher by a study of the variation of its $i_{\lim}/i_p$ and $E_{\frac{1}{2}}/E_p$ values with pH*, $i_{\lim}/i_p$ values with the height of the mercury column†, etc.

Some of these structural elucidations are currently hampered by shifts in the E_p values that can occur following extraction from some biological

---

* Some N-oxides will only reduce in the protonated form; the N–NO group is reduced in 4e⁻ in acidic media and by half that value in alkaline media.
† It should be remembered that the "unknown" peak may not be due to reduction or oxidation processes say, a catalytic process produced by an N-heterocyclic compound.

TABLE II. *Metabolic reactions accessible to polarographic investigation**

Metabolic reaction	Notes
Dealkylation	A significant change in electroactivity can be observed if this occurs on a hetero atom e.g. $-S-CH_3^a \xrightarrow{M} -SH^{a'}$ $\underset{}{\bigcirc}N^+-CH_3^r \xrightarrow{M} \underset{}{\bigcirc}N^c$ (Smyth et al., 1970b) 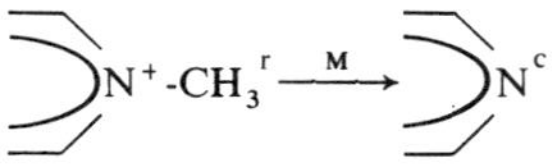 (Zuman and Perrin, 1969) or adjacent to an electroactive group permitting tautomerization to a species with different electrochemical behaviour. 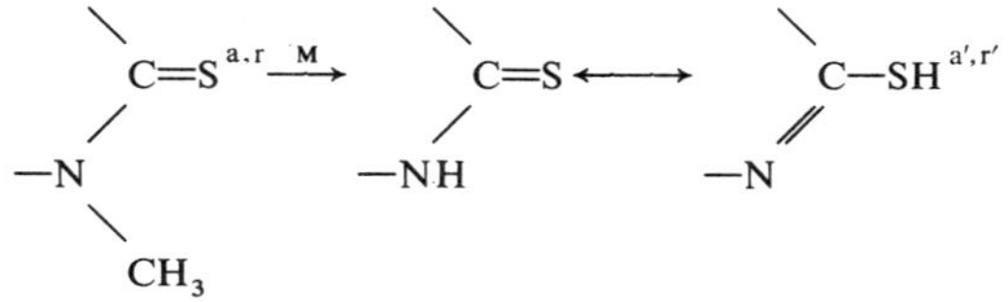 $C=S^{a,r} \xrightarrow{M} C=S \longleftrightarrow C-SH^{a',r'}$ (Smyth et al., 1970a) $-N \quad -NH \quad -N$ CH_3 C-dealkylation has been observed to shift the potential of a catalytic process due to the pyridine entity in a benzhydryl piperazine derivative. 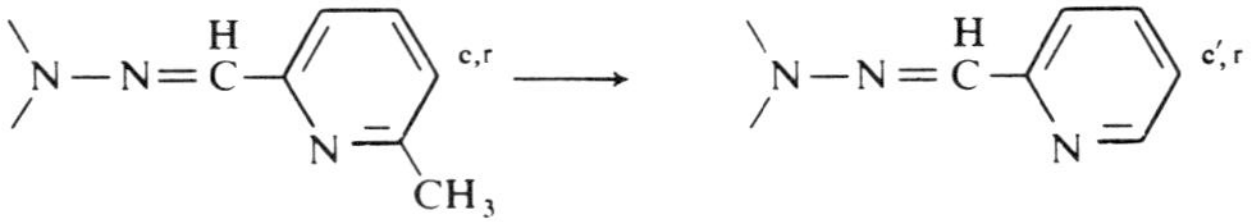 $\underset{N}{\bigcirc}\!\!\!\underset{CH_3}{}\!\!\!N=C\!-\!H \xrightarrow{c,r} \underset{N}{\bigcirc}N=N=C\!-\!H^{c,r}$ (Smyth, 1976)
Hydroxylation	A significant change in electroactivity will be observed on aromatic hydroxylation if the electroactive group is conjugated to the aromatic entity. 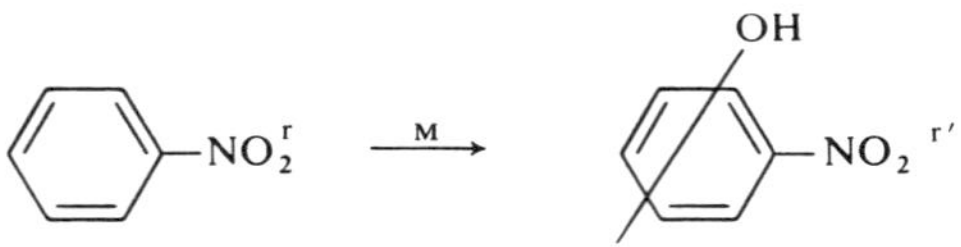 $\bigcirc\!-\!NO_2^r \xrightarrow{M} \underset{OH}{\bigcirc}\!-\!NO_2^{r'}$ (Milner, 1962) In aliphatic hydroxylation, the process must occur on or adjacent to the electroactive group for a shift in potential to occur.

TABLE II—continued

Metabolic reaction	Notes
Deamination	This can result in the production of an electro-reducible molecule, e.g. $$C_6H_5.CH_2.CH(CH_3).NH_2 \xrightarrow{M} C_6H_5.CH_2.CO.CH_3$$ amphetamine — benzylmethyl ketone
N-oxidation and S-oxidation	These processes can result in the production of electroreducible molecules, e.g. chlorpromazine $\xrightarrow{M}$ chlorpromazine N-oxide[r] and S-oxide[r] (Beckett et al., 1974)
Desulphuration	This metabolic reaction is frequently encountered in the conversion of $R_1(R_2)C=S^{a,r}$ to $R_1(R_2)C=O^{a,r}$, both of which can be oxidized or reduced at the d.m.e. depending on the nature of R_1 and R_2, (Smyth et al., 1970a).
Nitrogroup reduction	$R-NO_2^r \xrightarrow{M} R-NO^{r'} \xrightarrow{M} R-NHOH^{a,r'''} \xrightarrow{M} R-NH_2^{a'}$ (Beckett et al., 1977) $\updownarrow M$ $R'CH=NOH^{r''}$ (note that the species R—NO could be an N-oxide[r'''] or a nitroxide[r'''']).
Azogroup reduction	Two metabolic reactions can be followed...

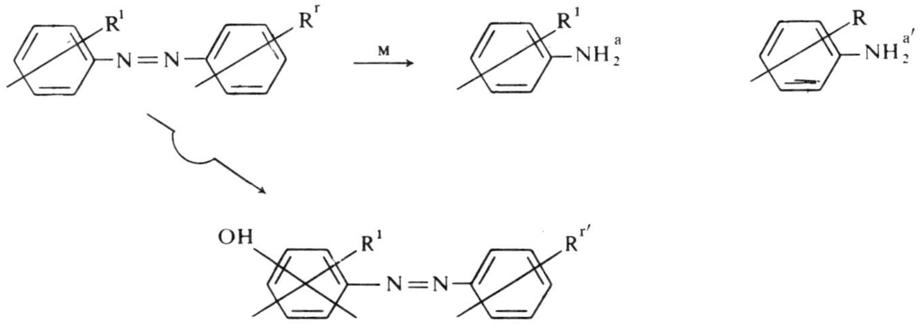

TABLE II—continued

Metabolic reaction	Notes
Hydrolysis	Polarographic methods have been used when the starting compound and hydrolysis product(s) are all electroreducible, as has been found with some 1,4-benzodiazepines.

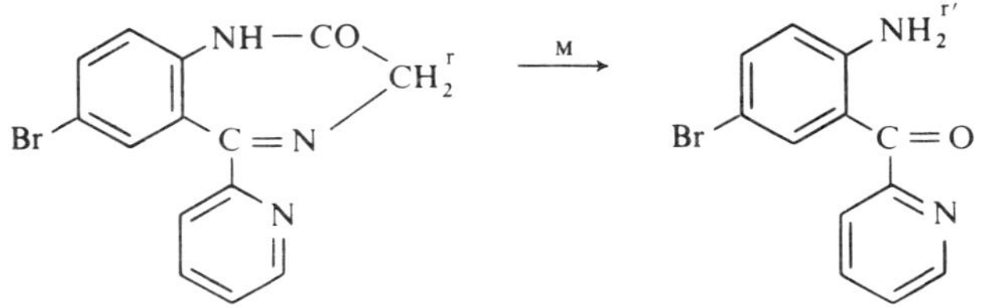

[Structures: starting 1,4-benzodiazepine with NH–CO–CH$_2^r$, C=N, Br, pyridine ring $\xrightarrow{M}$ hydrolysis product with NH$_2^{r'}$, C=O, Br, pyridine ring]

(de Silva et al., 1974)

C-Nitrosation and N-Nitrosation	Creatinine $\xrightarrow{M}$ [structure with CH$_3$, =NOH^r, HN, C=O ring]
	Creatine $\xrightarrow{M}$ CH$_3$–N–CH$_2$COOH$^{r'}$
	N=O

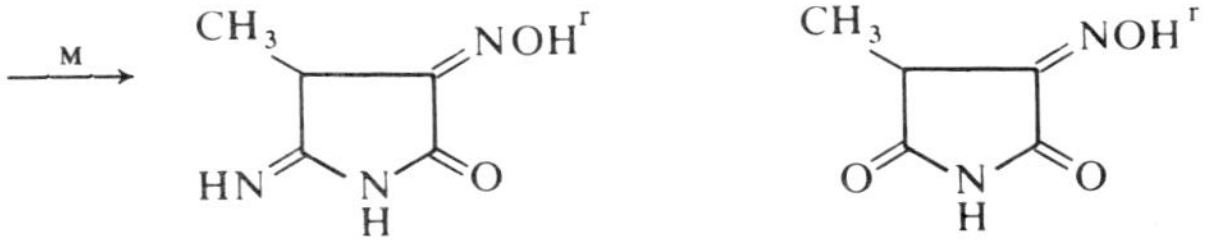

(Velisek et al., 1974)

Conjugation Reactions	Methylation, Acetylation, formation of glucuronides generally only give small shifts in potential.

* $\xrightarrow{M}$ signifies a metabolic reaction; [a] group gives anodic waves/peaks; [a'] group gives anodic waves/peaks at different potential; [c] group gives catalytic waves; [c'] group gives catalytic waves at different potential; [r] group gives reduction waves; [r'] group gives reduction waves at different potential.

fluids, e.g. urine, believed to be caused by as yet unidentified co-extractable interferences. Confirmatory evidence on the nature of an unknown metabolite should always be obtained using mass spectroscopy and other techniques.

Burmicz (1979) has recently investigated d.p.p. as an identification tool in forensic drug analysis. He has found, in spite of the large number of drugs that are encountered in forensic science, that relatively few "acidic and neutral" drugs are reducible. Resolution of mixtures of these reducible drugs, e.g. methaqualone, haloperidol, persedon, etc., is possible by variation of pH. Identification of 1,4-benzodiazepines by a combination of solvent extraction, hydrolysis and polarographic examination has recently been proposed by Franklin Smyth et al. (1978).

III. SAMPLING AND INITIAL TREATMENT OF SAMPLE

Many liquid and solid samples have been analysed for organic molecules by voltammetric methods. The liquid samples that have been mainly encountered are body fluids, samples from the aqueous environment (see Birch and Hart, Chapter 7) and some liquid formulations.

The body fluids which have been most analysed using voltammetry are blood and urine, although cerebrospinal fluid, vitrous humour, breast milk, sweat and saliva have also been investigated. In the case of blood a volume of about 3 ml is usually required in order to obtain detection limits (using d.p.p.) of about 20–30 nanograms of material. After a sample has been taken, it should be centrifuged to separate the red cells from the serum (addition of anticoagulant to the serum gives plasma). It is imperative at this stage to prevent any haemolysis of the red blood cells (caused by excess shaking or standing for a long period at room temperature) since traces of haemoglobin or its degradation products (e.g. haem) in the serum (or plasma) can cause interference in the subsequent voltammetric analysis. There are special problems associated with sampling urine, since it is often difficult to obtain precise measurements of volume which relate to specific sampling times. In addition, variations can occur in the electroactive composition of urine (which is greater than that found in plasma) which are dependent on the physiological state of the animal at the time of sampling.

Unless facilities are available to analyse samples immediately, it is advisable to freeze them prior to analysis in order to prevent: (i) further metabolism of the drug species in the body fluid—this is especially true in blood where there are small concentrations of metabolizing enzymes (e.g. hydroxylases) in both the serum and in the red cell membranes; (ii) chemical degradation of the drug species, e.g. hydrolysis; (iii) further binding (or conjugation) of the drug species to protein (or glucuronic acid, SO_4^{2-}, etc.) if an equilibrium has not been attained in the fluid one is analysing and; (iv) chemical modifications of naturally occurring electroactive constituents which would lead to changes in the "blank".

After allowing a frozen sample to return to room temperature, it should then be placed in a water bath at 37°C to redissolve any suspended material that may be present. The use of manufactured devices to aid separation of the substance of interest from clot-like substances in the serum has been criticized by Reid (1976) in his review on sample preparation techniques in drug analysis.

Solid samples such as drug and agrochemical formulations, and materials of biological importance such as crops, foods and tissues have all been analysed by voltammetric methods. Initial treatment of drug formulations is discussed earlier in this chapter; Rowe and Smyth (Chapter 8) have surveyed the application of these methods to a variety of crop formulations. Crop residues, foods and tissues are generally prepared for the voltammetric analysis of trace organic and organometallic species by homogenization of the sample followed by extraction of the species into a suitable solvent. Methanol, acetonitrile, chloroform and ethyl acetate have been most commonly used. In certain assays for organometallic species, the sample is digested with HCl prior to solvent extraction and determination of the metal entity (see Fouzder and Fleet, Chapter 9 and Watson, Chapter 10).

IV. SEPARATION TECHNIQUES

A. GENERAL COMMENTS

As mentioned previously, there are several reported cases where voltammetric determinations of trace amounts of foreign organic compounds have been carried out directly in biological fluids (particularly plasma). These methods are generally only applicable, however, to those compounds that are amenable to stripping analysis or those that contain readily reducible moieties, such as the $-NO_2$ and $>C=N-$ groups, and which can adsorb strongly on the Hg drop. Examples of the latter compounds include metronidazole (Kane, 1961) dinitro-o-cresol (Mikolajek, 1969) and some 1,4-benzodiazepines (Halvorsen and Jacobsen, 1972; Fidelus et al., 1972; Jacobsen et al., 1973). Other indicator electrodes have also been used for direct determinations in biological fluids; Mason and Sandman (1976), for example, have determined nitrofurantoin directly in urine using a rotating Pt electrode. However, these direct assays suffer from several drawbacks:

(i) Interferences caused by the reduction or oxidation of naturally occurring substances in biological fluids. In the case of blood, plasma or serum, the compounds most likely to interfere with the analysis of "foreign" organic compounds are given below. Proteins, for instance, can catalyse hydrogen evolution in the presence or absence of cobalt ions. In the absence of cobalt ions, amino acids (such as lysine) containing NH_2 groups within their molecular structure act as protein donors in the following manner:

$$-NH_3^+ + e^- \rightarrow -NH_2 + H\cdot$$
$$-NH_2 + DH \rightleftarrows -NH_3^+ + D^-$$

This behaviour gives rise to a double peak* (in d.p.p.) having E_p values of -0.79 and -1.05 V vs s.c.e. (in phosphate buffer, pH 7.4) and thus limits the potential range which can be investigated using direct methods of analysis. The polarographic behaviour of proteins is more fully dealt with by Homolka (1971), and that of the haem proteins is discussed by Mairanovskii (1968).

The presence of metal species in the biological fluid can also affect the voltammetric behaviour of the molecule/species under study since apart from being electroactive themselves, they can also complex with various organic constituents in the matrix. For example, amino acids do not give rise to a polarographic wave unless complexed to Cu(II) or Ni(II) ions (Blaedel and Todd, 1960); bilirubin forms complexes with various metal species (Ryan, 1975) and malonic acid (and other constituents of urine) forms complexes with lead (Maheswari et al., 1974; A. Clatworthy, personal communication). Compounds containing sulphur either in the form of —S—S— or —SH can also interfere in the determination of foreign organic compounds. In the cysteine–cystine (Milner, 1962; Mairesse-Ducarmois et al., 1974) and reduced–oxidized glutathione (Milner, 1962; Mairesse-Ducarmois et al., 1975) systems, the reduced forms (i.e. —SH) can form mercury salts and thus interfere with polarographic or c.s.v. methods for the determination of other molecules, which give a response in that potential region. This is exemplified by the work of Tomana (1966) who showed that glutathione interfered with the oxidation wave obtained for ascorbic acid in tissue homogenates. The oxidized forms of these compounds contain the —S—S— linkage, which is also present in some thiuram pesticides, and can thus interfere with their determination. Sugars such as glucose, fructose, and ribose give rise to kinetic waves (Heyrovsky and Zuman, 1968) at very negative potentials ($-1.5 \rightarrow -1.7$ V) and will interfere with the polarographic determination of compounds that reduce at a similar potential. Other compounds which are found naturally in body fluids and which can interfere with the voltammetric analysis of foreign organic compounds at the trace level include:

* For serum albumin.

(a) vitamins, e.g. ascorbic acid, pyridoxine, nicotinamide and folic acid (Heyrovsky and Zuman, 1968; Lindquist and Farroha, 1975; Gulaid, 1976; de Silva et al., 1973);

(b) co-factors, e.g. NAD/NADH (Knevel, 1968; Blaedel and Jenkins, 1975), and their complexes with metal ions, e.g. pyridoxine (Chaturvedi and Gupta, 1974) and isonicotinic acid (Korneva et al., 1972);

(c) urinary metabolic products, e.g. uric acid (Dryhurst, 1972) and creatinine (Blass and Thibert, 1974); and

(d) the naturally occurring neurotransmitters (Radejova and Sharka, 1974).

(ii) Interferences caused primarily by the effect that macromolecules (such as proteins and polysaccharides in body fluids; tannins in plant material and humic and fulvic acids in natural waters) have on the diffusion controlled processes of the smaller organic molecules which are analytically of interest. Proteins, for instance, adsorb strongly on the Hg drop and can affect a non-linear relationship between i_d or i_p and concentration in the lower concentration ranges, as has been demonstrated for the determination of nitrazepam at concentrations lower than 1 µg ml^{-1} in serum (Halvorsen and Jacobsen, 1972). In addition, these macromolecules can cause distortions in the shape of the wave/peak being measured.

(iii) Interferences caused by de-aeration of plasma or urine which results in foaming and is eliminated either by the use of a maximum suppressor, addition of methanol or of a higher alcohol, e.g. n-octanol. These additions can result, however, in a diminution of the height of the wave/peak and in some cases a shift in $E_{\frac{1}{2}}/E_p$ value thus yielding erroneous results.

The most severe restriction on these direct methods of analysis is, however, that they are inherently non-selective. The presence of structurally related compounds (e.g. metabolites) can greatly interfere with the analysis and sometimes extraneous background material in the biological specimen can overlap and sometimes obscure the analytical wave of interest.

Separation procedures are employed in voltammetric methodology, therefore, both to remove these electroactive interferences and to improve the selectivity of the method for compounds of related structure and similar electrochemical behaviour.

If dissociation of the protein bound fraction of the compound in plasma or serum is desired it can often be achieved by a five-fold dilution of the sample (Trevor et al., 1972). Another method of achieving this is to employ another compound which binds more strongly to the protein thus releasing the compound of interest; for example, butazolidin has been used to displace salicyclazosulphapyridine from its binding site on protein prior to polarographic analysis (Nygard et al., 1966). Dyes (such as methyl orange) have been used for the same purpose as in the case of chlorpromazine

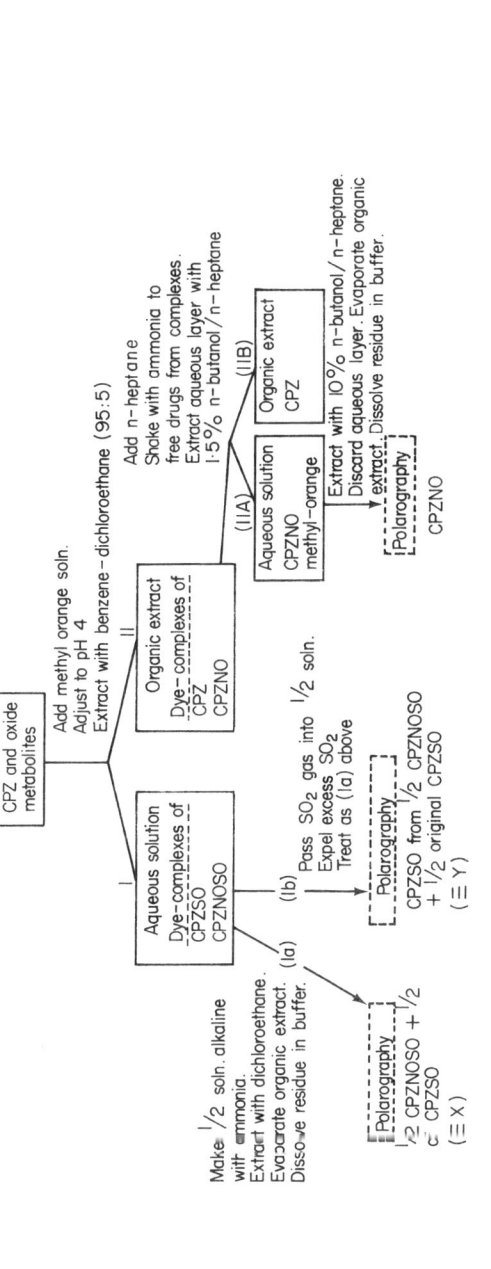

FIG. 1. Scheme for the determination of the N- and S-oxygenated metabolic products of chlorpromazine (after Beckett et al., 1974). Note CPZNOSO is reduced by SO_2 to CPZSO. Polarographic reduction peak height of CPZNOSO = $2 \times$ peak height of CPZSO. $X - Y$ = polarographic value for the NO of $\frac{1}{2}$ CPZNOSO, = $\frac{1}{4}$ of total CPZNOSO. $4(X - Y)$ = total CPZNOSO. $2X - 4(X - Y)$ = total sulphoxide. $2(2Y - X)$ = total sulphoxide.

(Beckett et al., 1974). Figure 1 shows the scheme developed for the polarographic analysis of this major tranquilizer and its three major metabolites—the N-oxide, the S-oxide and the N-oxide S-oxide. Alternatively proteins can be precipitated using many agents [e.g. tungstic acid, $(NH_4)_2SO_4$, trichloroacetic, perchloric and other acids]; in the case of perchloric acid (5% w/v), the precipitant can subsequently be removed as its insoluble potassium salt (Weil-Malherde, 1971). In the case of urine, conjugates can be broken down either by acid hydrolysis [this can give rise, however, to a mixture of products with resulting poor t.l.c. separation (Weil-Malherde, 1971; Kullberg and Gorodetzky, 1976)] or by the use of an enzyme preparation, e.g. glusulase which contains both glucuronidase and sulphatase fractions. In the case of glucuronides, periodate may sometimes be used (A. Clatworthy, personal communication).

The separation techniques most commonly employed in voltammetric methodology are given below.

B. CENTRIFUGATION/FILTRATION METHODS

These methods are generally employed to remove the interference caused by molecules of relatively high MW (e.g. proteins) or to remove the products of precipitation reactions mentioned previously. A notable example of their use is in the determination of penicillins and cephalosporins in plasma (Benner, 1970) in which protein was removed by filtering through a UM-2 filter at a pressure of 50 p.s.i. Celite 54T has also been used to remove interferences in the polarographic determination of saccharin in beverages and food (Dungen, 1976). These methods suffer, however, from the inability to remove other electroactive interferences from the sample.

C. GEL FILTRATION

This has been little employed in voltammetric methodology because it is generally time consuming (depending on the length of column and flow rate needed to effect separation of the compounds of interest) and results in a dilution of the original sample. It can be useful, however, in those cases where absolute sensitivity is not required and where a great deal of electroactive interference is encountered in the biological matrix. Sephadex LH-20 has found particular application in the determination of various N-nitrosamines in bacon (Hasebe and Osteryoung, 1976) and in synthetic cutting fluids (Smyth et al., 1977b). Very good separations of N-nitrosamines on this column have also been reported by Eisenbrand et al. (1970).

Other examples where gel filtration has been employed in voltammetric methodology include the determination of meprobamate in serum following separation on Sephadex G-25 (Hynie and Prokes, 1964) and in the determination of metation and its analogues following separation on silica gel layers (Seifert and Davidek, 1971).

D. ION-EXCHANGE CHROMATOGRAPHY

Ion-exchange chromatographic methods have found most application in the removal of electroactive interferences from urine (particularly organic acids), e.g. in the determinations of chlorpromazine (Porter and Beresford, 1966) and morphine (Orlov et al., 1964). They have also been used to effect quantitative recoveries from plasma. Amberlite IR-120(H^+) resin has been used to separate methyldopa from various body fluids with varying degrees of recovery—89% from urine, 80% from whole blood and 100% from plasma (Stewart et al., 1974). Ion exchange chromatography has also been used to separate surfactants in water (Linhart, 1972). This is discussed in greater detail by Birch and Hart (Chapter 8).

Lasheen (1961) has used the cation exchange resin Dowex 50 in the H^+ form to remove interfering species such as oxalate and acetate anions prior to the polarographic determination of saccharin in plant extracts. Williams (1973) also used a cation exchange column (CG-120) to separate bis (tributyl) and dibutyltin oxides prior to a polarographic determination.

E. MACRO RETICULAR RESINS

Macro reticular resins have been widely used to concentrate and separate both charged and uncharged organic molecules in sea water samples. They are synthetic polymers, in bead form, possessing high adsorption capacity for organic materials. They are available commercially with a variety of surface polarities, surface areas and porosities which can considerably modify their sorption behaviour. The usual method of use is to employ the beads in column operations. The (dilute) test solution is slowly passed through about 2 g of resin held in a column about 5 cm long. The organic material adsorbed onto the column is then eluted, usually with portions of diethyl ether, which may then be concentrated (by evaporation) by factors of up to 1000.

The method has been extensively tested by Riley (personal communication) on a wide variety of materials yielding excellent recoveries and appears to be the most reliable and universally applicable concentration method yet devised. This procedure could, therefore, be used prior to voltammetry of those molecules that are electroactive or in conjunction with derivatization of inactive molecules (e.g. nitration of alkyl benzene sulphonates—see Birch and Hart, Chapter 7).

F. THIN LAYER CHROMATOGRAPHY

This chromatographic technique has been widely applied in voltammetric methodology both to separate structurally related substances and, to a lesser extent, to remove electroactive interferences. It is usually carried out following a suitable solvent extraction from the body fluid and concentration of the extract. The extract can then be spotted on the plate (silica or Al_2O_3) and the mixture run in a suitable solvent system in order to achieve adequate resolution of the compounds of interest. These can then be scraped from the plate, taken up in methanol or DMF (the latter solvent has been recommended to prevent readsorption on the support material—R. G. Cooper, personal communication) and diluted with the appropriate supporting electrolyte. Interference caused by metal species in the support material can be removed by complexation with EDTA (Oelschlager et al., 1976). Silica plates have been used for the determination of some corticosteroids (Hake, 1966), tocopheronolactone (Schmandke and Crohlke, 1965) and the metabolites of flurazepam (de Silva et al., 1974), trimethoprim (Brooks et al., 1973a) and 2-hydroxynicotinic acid (de Silva, et al., 1973) in urine. An Al_2O_3 support has been used in the determination of brucine and strychnine (Ch'en et al., 1966). T.l.c. has also been used to remove interference caused by ballast materials in organic solvents prior to nitration procedures (Fidelus and Zietek, 1970).

G. PAPER CHROMATOGRAPHY

The application of paper chromatography to polarographic methodology has been discussed by Pazdera et al. (1957). It has found most application in certain toxicological cases, e.g. in the determination of barbiturates (Prokes and Vorel, 1960), meprobamate (Hynie and Prokes, 1964) and of some morphine analogues (Mithers, 1961). In the case of barbiturates, however, the method was reported as being only semiquantitative. More recently, Engst and Schnaak (1974) have combined paper chromatography and polarography to determine ethylene thiourea residues in foods, following nitrosation. A sensitivity of 0.05 µg ml^{-1} with a recovery of 70% was reported.

H. HIGH PERFORMANCE LIQUID CHROMATOGRAPHY (h.p.l.c.)

Although h.p.l.c. can be used as an efficient and rapid means of separating structurally related compounds, most interest in this area has been centred on the development of electrochemical detectors for use with this chromatographic technique. Although a d.m.e. can be used to monitor the eluant in a flowing stream (Little, 1974) it is more common to employ solid electrodes (such as glassy carbon) which can be built into systems such as the wall-jet cell (Yamada and Matsuda, 1973; Fleet and Little, 1974). These electrodes considerably increase the potential range available for the detection of compounds undergoing oxidation processes, e.g. ascorbic acid (Thrivikraman et al., 1974), and can also be used for the determination of reducible compounds such as the 1,4-benzodiazepines, (Fig. 2; Franklin Smyth and Smyth, unpublished results). The application of h.p.l.c. (with electrochemical detection) to the determination of trace organics in body fluids has, however, been mainly centred around those compounds which give rise to oxidation processes at carbon electrodes. This method of analysis has been pioneered by Kissinger (1977) and specific applications of the technique are mentioned in the chapters by Davidson (Chapter 5) and Chowdhry (Chapter 6).

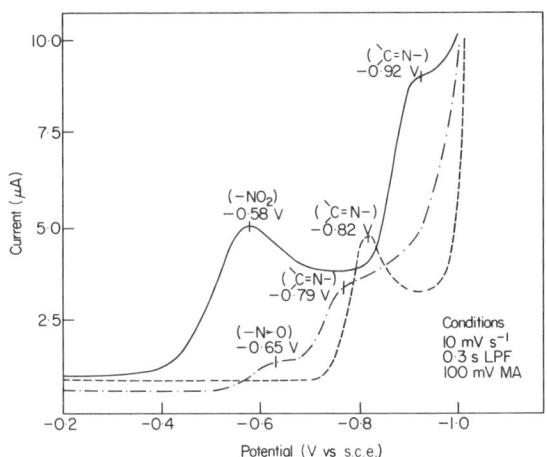

FIG. 2. D.c. voltammograms obtained for 10^{-4} M solutions of nitrazepam (———) chlordiazepoxide (–·–·–) and diazepam (----) at a glassy carbon electrode in 0.1M HCl/50% MeOH.

I. SOLVENT EXTRACTION

This is by far the most universally employed separation technique in voltammetric methodology, since it can remove a large quantity of electroactive interferences, offer a degree of resolution between compounds of related structures and also serve as a preconcentration step.

The two main factors which should be considered in the development of an extraction scheme for an organic compound in a biological fluid are (i) choice of solvent, and (ii) choice of pH. Concerning (i), it is generally advisable to employ a solvent whose polarity is as low as will effect quantitative extraction of the compound of interest. For example, in extractions from plasma it has been shown that non-polar solvents (such as benzene and petroleum ether) co-extract much less electroactive interference than some polar solvents (Smyth, 1976). If one is forced to use a polar solvent such as ethyl acetate or ether then one should reduce both the time and strength of shaking (extractions from body fluids should be carried out on mixers which allow gentle equilibration of the phases) consistent with quantitative extraction. This is particularly the case with ethyl acetate which when shaken with acidic fractions hydrolyses to form ethanol and acetic acid, thus resulting in a merging of the phases. Another problem arises with the use of ether which as well as extracting various metal species from the body fluid also contains small concentrations of peroxide which can effect the production of N- and S-oxide products, similar to those formed on metabolism (thus decreasing the specificity of the assay).

Concerning (ii), it has been shown (Smyth, 1976) that although the available potential range is extended following an alkaline extraction, "cleaner" blanks may be obtained for solvent extractions carried out on plasma samples buffered at acidic pH. Excluding any chromatographic "clean-up", there are other methods which can be employed to remove the interferences remaining after the preliminary extraction step, e.g. back-extraction techniques [acid extraction into ether will remove undesirable keto compounds and organic acids whereas an alkaline extraction into benzene will remove amines and other neutral species (Essien, 1974)]. The addition of anhydrous salts (e.g. Na_2SO_4) to the organic extract will serve to remove polar compounds that are adsorbed onto their surface (Zietek and Fidelus, 1968).

Solvent extraction can also be used to separate compounds of similar structure by variation of solvent composition and the composition and pH of the aqueous phase. The main factors which affect the selectivity of this method are: (i) the relative polarity of the compounds to be separated and (ii) their state of ionization across the pH range. Concerning (i), it is a general rule that the more polar the compound, the more polar the solvent should be to extract it from the body fluid. It is the converse of this rule, however, which has been of real importance in obtaining selective solvent extraction procedures. This is aptly illustrated by the work of Groves (1976) on the extraction characteristics of flurazepam and its metabolites into a variety of solvents of different polarity (Fig. 3). He showed that one could achieve greater differentiation between the parent compound and its metabolites (metabolic processes are designed to make the compound more polar (e.g. by dealkylation and hydroxylation reactions, so that it can be more easily excreted in the urine) by using a non-polar solvent such as petroleum ether. The effect of increasing chain length (and hence increased lipophilicity) on solvent extraction profiles has been well illustrated by Whelpton and Curry (1976) for some long-chain ester analogues of fluphenazine.

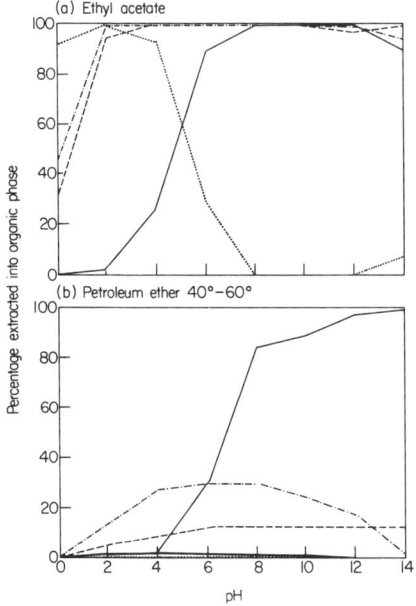

FIG. 3. Distribution of flurazepam and its metabolites between organic solvents and aqueous buffered phases. Flurazepam ———; NH ————; CH_2COOH ······; CH_2CH_2OH ————; MH,3-OH ———.

The selectivity of solvent extraction procedures can further be improved by a study of the acid-base behaviour of the compounds of interest since only molecules with an overall neutral charge can be extracted into the organic solvent. In the case of fluphenazine (FPZ), whose major metabolites are 7-hydroxy fluphenazine (7-OH FPZ) and fluphenazine sulphoxide (FPZSO), differentiation of the three compounds (using toluene) is possible since only FPZ is extracted at pH 4.0, 7-OH FPZ is not extracted from alkaline solution (due to deprotonation of the phenolic moiety) and FPZSO shows two S-shaped portions in the percent extraction–pH curve due to protonation of one of the N atoms and the S→O species respectively. In the case of flurazepam and its metabolites one is also able to attain this selectivity because the products of metabolism all contain different ionizable groups, i.e. —COOH, —NH and —OH, which undergo their respective acid-base equilibria at pH values well separated from each other (Groves, 1976). In addition, the parent compound contains a triethylamino side chain which has a pK_a of the order of 9–10. At pH values lower than this, however, it was found that the compound could still be extracted into ethyl acetate. This was attributed to an ion-pairing between the protonated species and acetate ions in the buffer since at a pH below 5.0 the extractability of flurazepam decreased to zero (in solution pH <2.5) corresponding to the association of acetate ions with protons. This behaviour has been utilized in the determination of this compound and its major blood metabolites in the plasma of test animals (Clifford et al., 1974). The use of ion pairing behaviour for increasing the selectivity of solvent extraction techniques has been further dealt with by Schill (1976).

J. OTHER SEPARATION METHODS

Other separation methods, such as evaporation of volatile organic molecules from water or toxicological samples, the adsorption of organic molecules onto activated charcoal, or the use of foam fractionation have not been widely used in conjunction with voltammetric techniques for the determination of organic molecules.

V. COMPLEXATION AND DERIVATIZATION PROCEDURES

If a molecule does not have inherent polarographic activity or if it exhibits behaviour that is of little analytical value, then there are a variety of complexation and derivatization procedures which can be employed to achieve an indirect method of analysis.

A. COMPLEXATION

Complexation procedures have been employed in polarographic methodology for a variety of reasons: (i) to impart polarographic activity to the molecule; most amino acids are polarographically inactive unless complexed to Cu(II) or Ni(II) ions (Blaedel and Todd, 1960; Davis and Bordelon, 1970)—this process gives rise to single waves due to metal ion reduction. Proteins in the presence of Co(II) ions give rise to characteristic double waves (i.e. Brdicka waves, see Chowdhry, Chapter 6). In the case of nitrilotriacetic acid (NTA), which is also polarographically inactive, complexation with Cd(II) ions permitted its trace determination (Asplund and Wanninen, 1971) whereas the alkaloids solamidine and solamine could best be determined as their Cu(II) complexes (Pierzchalski, 1963); (ii) if the resulting complex is electroinactive then an indirect method exists for the determination of the molecule based on the decrease in wave height of free metal ion in solution containing the organic molecule, as in the complexation of metronidazole with Cu(II) ions for example (Chien et al., 1975); (iii) complexation can sometimes result in an improvement in definition of an existing wave; in the case of the dithiocarbamate disulfiram, for example, complexation of the disulphide moiety with Cu(II) ions resulted in the enhancement of the wave (Porter and Williams, 1972); (iv) complexation can also serve to remove an interfering wave, as in the case of lignin sulphonic acids where a complexation procedure was used to remove an interfering catalytic wave (Erlebach et al., 1970).

B. NITRATION

This is by far the most common derivatization procedure employed in polarographic methodology since it can be carried out with relative ease and efficiency and gives rise to well-defined analytically useful waves. This subject has been discussed in several reviews, notably by Skora-Zietek (1965) and by Nurnberg and Wolff (1966).

The procedure has mainly been carried out on aromatic nucleii. In

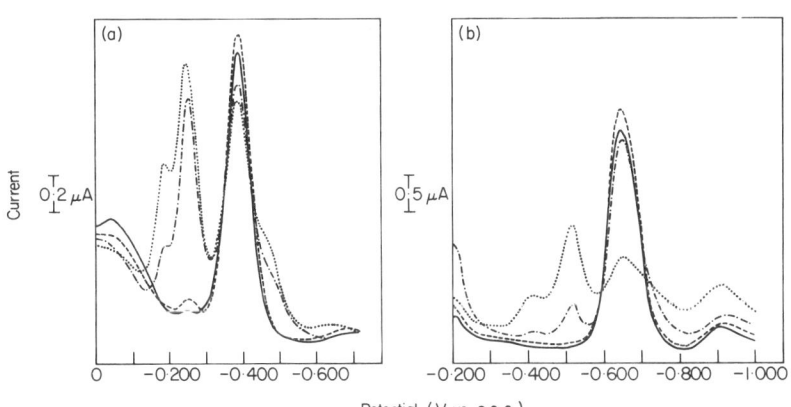

FIG. 4. Effect of nitration temperature on (a) phenobarbital and (b) diphenylhydantoin. (———) 0°C; (– – – –) 25°; (– · – · –) 65°; (·······) 105°). (After Brooks et al., 1973b).

general, conditions are employed which give quantitative recovery of the mono-nitrated derivative by the use of mild conditions, e.g. $NaNO_2/HCl$ or HNO_3/CH_3COOH. The effect of temperature is also an important factor in the development of an analytical nitration procedure. This is well illustrated by the work of Brooks et al. (1973b) on the determination of phenobarbital and diphenylhydantoin in serum (Fig. 4), where it can be seen that the best peak for analytical purposes is obtained for a low reaction temperature of 0–25°C.

The nitration of microquantities of materials leads, however, to much larger errors than methods not requiring a derivatization procedure (generally ± 5–10% for 1 µg material). These errors arise mainly from: (i) impurities in the nitration mixture, e.g. inorganic nitrates/nitrites, which can interfere with the subsequent analysis; (ii) nitration of traces of organic solvent, e.g. benzene and toluene, which remain following separation of the compound of interest from the biological matrix; (iii) nitration of naturally occurring substances which remain following the preliminary separation procedure(s), e.g. phenyl-containing compounds and sugars [the latter compounds have been shown to interfere with the nitration of 4-hydroxy-benzoic acid (Tammilehto and Perala, 1971)]; (iv) side reactions which occur in the nitrating mixture, e.g. acid hydrolysis.

Because of these interferences, it is sometimes better to employ a further solvent extraction or chromatographic step rather than to analyse the derivative directly in the (un)diluted nitration mixture. In the case of 2,4-dichlorophenoxyacetic acid, interfering ballast materials in the chloroform (used to extract it from the biological matrix) were removed by t.l.c. resulting in better reproducibility for the nitration procedure (Fidelus and Zietek, 1970). Nitration has also been carried out with some success on eluates from chromatographic columns (Sachweh et al., 1966).

C. NITROSATION

Nitrosation is also quite commonly employed in polarographic methodology and is of most importance in the determination of molecules possessing centres of high electron density, e.g. phenols and secondary amines. The most common reagent used is $NaNO_2/HCl$, which can, however, also result in the formation of nitro derivatives. It is imperative, therefore, to identify the products of the derivatization procedure (by using m.s., for example) before going on to determine the compound of interest in the biological matrix. The factors affecting nitrosation procedures have been reviewed by Pasciak and Lewandowska (1972). Recently Pasciak and Gajewska (1976) have reported that the nitrosation process for phenol is markedly affected by the presence of alcohol in the nitrosating mixture, since when it is present

p-nitrosophenol is the main product whereas when it is absent the corresponding p-nitro derivative is formed. They have also suggested that ethylnitrite is important in the nitrosation reaction. For analytical purposes the interfering influence of nitrate can be removed using ammonium amidosulphonate whereas degassing in acid solution is sufficient to remove nitrite. For kinetic reasons KBr can sometimes be added to speed up the reaction (Bronstad and Friestad, 1976). The other main source of interference in this procedure comes from the nitrosation of naturally occurring substances which remain following the initial separation stage, e.g. phenolic compounds (including metabolites) and amines such as creatine, creatinine (Velisek et al., 1974) and biotin (Davidek, 1961). Bronstäd and Friestad (1976) have separated interferences in natural waters by anion exchange chromatography prior to the determination of glyphosphate residues by nitrosation. With the increasing awareness concerning possible carcinogenic side effects, it can be foreseen, however, that this derivatization procedure may fall from favour in the near future.

D. N- AND S-OXIDATION

These derivatization techniques have mostly been employed for pharmaceutical quality control analysis and are carried out using either H_2O_2 (about 30%) or chloroperbenzoic acid (in ethyl acetate). These methods are not recommended for trace analysis in body fluids since the reactions involved are also common metabolic processes and hence their use would decrease the specificity of the assay (unless some form of chromatographic separation had been employed prior to the derivatization procedure).

E. CONDENSATION

The most common condensation reactions that are employed in polarographic methodology are those involving the production of semicarbazone (Fleet, 1966; Pribyl and Nedbalkova, 1969; Afghan et al., 1975) or hydrazone (Turyan and Tolstikova, 1974) derivatives of carbonyl compounds. The formation of semicarbazone derivatives proceeds according to:

$$RR^1C=O + H^+ \rightleftarrows RR^1C=\overset{+}{O}H$$

$$R^{11}NH_2 + H^+ \rightleftarrows R^{11}\overset{+}{N}H_3$$

$$RR^1C=\overset{+}{O}H + R^{11}NH_2 \rightleftarrows RR^1\overset{|}{C}-\overset{+}{N}H_2R^{11} \quad (OH)$$

$$RR^1\overset{|}{\underset{OH}{C}}-\overset{+}{N}H_2R^{11} \xrightarrow[-H_2O]{-H^+} RR^1C=NR^{11}$$

The fastest rates of reaction usually occur when the concentration of protonated carbonyl compound is high and that of the protonated amine is low. In general, a pH of about 4–6 will yield optimum recovery of the semicarbazone derivative. Semicarbazide is a particularly useful agent to carry out this reaction since it is not polarographically active itself and can thus be added in a large excess to ensure that the reaction has gone to completion and also to minimize hydrolysis (Fleet, 1966). The reduction of aldehyde and ketone semicarbazones occur essentially by a 4-electron process:

$$RR^1C{=}N{-}NHCONH_2 + 2e^- + 2H^+ \xrightarrow{E_1} RR^1C{=}\overset{+}{N}H + H_2NCONH_2$$

$$RR^1C{=}\overset{+}{N}H + 2e^- + 2H^+ \xrightarrow{E_2} RR^1CHNH_2$$

and gives rise to well defined waves for analytical purposes.

Steroids can similarily be determined as their hydrazone derivatives.

F. HYDROLYSIS

Hydrolytic procedures that have been employed in voltammetric methodology can be divided into two categories. (i) Those that attack a side chain in the molecular structure to convert an electroinactive compound into an electroactive one; Franke and Buecher (1965) have converted 17-keto steroids into their β-vinylhydrazone derivatives by heating in 15% HCl at 100°C for 10 minutes. Van Bennekom et al. (1975) have used a similar procedure to determine the artificial steroid lynestrenol in pharmaceutical preparations. (ii) Those that release a product which is itself electroactive; hydrolysis of some benzophenoxazine dyes results in formation of electroreducible 5-keto products (Kotoucek and Zavadilova, 1972) or may be easily converted to one, e.g. in the case of the production of CS_2 (from disulfiram) which was subsequently determined as copper diethydithiocarbamate (Brown et al., 1974).

G. OTHER METHODS

Other derivatization methods that have been employed in voltammetric methodology include pyrolysis, e.g. of cellulose and its esters, (Bezugly et al., 1972), oxidation, e.g. of adrenaline to iodoadrenochrome (Dezelic et al., 1967), and distillation, e.g. of mandelic acid to benzaldehyde (Bardodij et al., 1964).

Sodium cyclamate has been determined by its reaction with acetic acid/sodium nitrite, production of sulphate, reaction of sulphate with excess Pb(II) and voltammetric determination of the remaining Pb(II) (Kurayuki et al., 1966). Similarly, several sulphur-containing pesticides have been determined following liberation of S^{2-} and reaction with Pb(II) (Kosmatyi and Kavetskii, 1973).

Organic molecules have also been determined voltammetrically by following their reaction with oxygen and estimation of the oxygen remaining using its reduction wave (see Chapter 6).

REFERENCES

Abel, R. H., Christie, J. H., Jackson, L. L., Osteryoung, J. and Osteryoung, R. A. (1976) *Chem. Instrum.* **7**, 123.
Adams, R. N. (1969). "Electrochemistry at Solid Electrodes". Marcel Dekker, New York.
Afghan, B. K., Kulkarni, A. V. and Ryan, J. F. (1975) *Anal. Chem.* **47**, 488.
Asplund, J. and Wanninen, E. (1971) *Anal. Letters*, **4**, 267.
Baizer, M. M. (1973). "Organic Electrochemistry". Marcel Dekker, New York.
Bardodij, Z., Fiserova-Bergerova, V. and Lederer, E. (1964). *Pracovni Lek* **16**, 414.
Bargagna Prunai, P., Cinci, A. and Silvestri, S. (1972). *Farmaco, Ed. Prat.* **27**, 89.
Barker, G. C. (1958). *Anal. Chim. Acta.* **18**, 118.
Barker, G. C. and Gardner, A. W. (1960). *Z. Anal. Chem.* **173**, 79.
Barker, G. C. and Jenkins, I. L. (1952). *Analyst*, **77**, 685.
Beckett, A. H., Essien, E.E. and Franklin Smyth, W. (1974). *J. Pharm. Pharmacol.* **26**, 399.
Beckett, A. H., Rahman, R. and Franklin Smyth, W. (1977). *Anal. Chim. Acta.* **92**, 353.
Beland, F. A. (1975). *Diss. Abst. Inst. B.* **35**(9), 4385.
Benner, E. J. (1970). *Antimicrob. Agents Chemotherapy*, **10**, 201.
Bergman, I. (1954). *Trans. Faraday Soc.* **50**, 829.
Bergman, I. and Windle, D. (1972). *Ann. Occup. Hyg.* **15**, 329.
Bezugly, V. B., Shtal's, S., Bmitrieva, I. N. and Kononenko, L. V. (1972). *Zav. Lab.* **38**, 1067.
Bieder, A. and Brunel, P. (1971). *Ann. Pharm. Franc.* **29**, 461.
Blaedel, W. J. and Jenkins, R. A. (1975). *Anal. Chem.* **47**, 1137.
Blaedel, W. J. and Todd, J. W. (1960). *Anal. Chem.* **32**, 1018.
Blass, K. G. and Thibert, R. J. (1974). *Microchem. J.* **19**, 1.
Brezina, M. and Zuman, P. (1958). "Polarography in Medicine, Biochemistry and Pharmacy". Interscience, New York.
Bronstäd, J. O. and Friestad, H. O. (1976). *Analyst*, **101**, 820.
Brooks, M. A., de Silva, J. A. F. and d'Arconte, L. (1973a). *J. Pharm. Sci.* **62**, 1395.
Brooks, M. A., de Silva, J. A. F. and Hackman, M. R. (1973b). *Anal. Chim. Acta*, **64**, 165.
Brooks, M. A., BelBruno, J. J., de Silva, J. A. F. and Hackman, M. R. (1975). *Anal. Chim. Acta*, **74**, 165.
Brown, D. R., Porter, G. S. and Williams, A. E. (1974). *J. Pharm. Pharmacol.* **26**, 95p.
Burmicz, J. S. (1979). Ph.D. thesis, University of London.
Chaturvedi, D. N. and Gupta, C. M. (1974). *J. Inorg. Nucl. Chem.* **36**, 2155.
Ch'en, C. C. *et al.* (1966). *Yao Hsueh Hsueh Pao*, **13**, 131.
Chien, Y. W., Lambert, H. J. and Sanvordeker, D. R. (1975). *J. Pharm. Sci.* **64**, 957.
Christie, J. H., Jackson, L. L. and Osteryoung, R. A. (1976). *Anal. Chem.* **48**, 242.
Clark, L. C., Jr (1956). *Trans. Amer. Soc., Artif. Internal Organs*, **41**, 2.
Clifford, J. M., Smyth, M. R. and Franklin Smyth, W. (1974). *Z. Anal. Chem.* **272**, 198.
Cullen, L. F., Brindle, M. P. and Papariello, G. J. (1973). *J. Pharm. Sci.* **62**, 1708.
Davidek, J. (1961). *Naturwissenschaften*, **48**, 403.

Davidson, I. E. and Franklin Smyth, W. (1977). *Anal. Chem.* **49**, 1195.
Davis, D. G. and Bordelon, W. R. (1970). *Anal. Letters*, **3**, 449.
Davis, H. M. and Seaborn, J. E. (1953). *Electronic Eng.* **25**, 314.
de Silva, J. A. F., Strojny, N. and Munno, N. (1973). *Anal. Chim. Acta*, **66**, 23.
de Silva, J. A. F., Puglisi, C. V., Brooks, M. A. and Hackman, M. R. (1974). *J. Chromat.* **99**, 461.
Dezelic, M., Trknovnik, M., Popovic, R. and Dimitrijevic, D. (1967). *Acta. Pharm. Jugoslav.* **17**, 81.
Dryhurst, G. (1972). *Anal. Chim. Acta*, **58**, 183.
Dungen, P. W. L. M. Van den (1976). *Z. Lebensm. Unters-Forsch*, I, 61.
Eisenbrand, G., Spaczynski, K. and Preussman, R. (1970). *J. Chromat.* **47**, 304.
Engst, R. and Schnaak, W. (1974). *Die Nahrung*, **18**, 597.
Erlebach, J., Lischke, P. and Kucera, Z. (1970). *Chem. Listy*, **64**, 984.
Essien, E. (1974) Private communication.
Fidelus, J. and Zietek, M. (1970). *Mikrochim. Acta.* No. 5, 1010.
Fidelus, J., Zietek, M., Mikolajek, A. and Gruchowska, Z. (1972). *Mikrochim. Acta*, No. 1, 84.
Fleet, B. (1966). *Anal. Chim. Acta*, **36**, 304.
Fleet, B. and Little, C. L. (1974). *J. Chromat. Sci.* **12**, 747.
Florence, T. M. (1974). *J. Electroanal. Chem.* **52**, 115.
Franke, R. and Buecher, M. (1965). *Acta. Biol. Med. Germ.* **14**, 1.
Franklin Smyth, W., Watkiss, P., Burmicz, J. S. and Hanley, H. O. (1975). *Anal. Chim. Acta* **78**, 81.
Franklin Smyth, W. and Smyth, M. R. (1976). *Proc. Anal. Div. Chem. Soc.* **13**, 223.
Franklin Smyth, W., Smyth, M. R., Groves, J. A. and Tan, S. B. (1978). *Analyst*, **103**, 497.
Groves, J. A. (1976). Ph.D. Thesis, University of London.
Gulaid, A. (1976). M.Sc. Dissertation, Chelsea College, University of London.
Hake, J. (1966). *J. Electroanal. Chem.* **11**, 31.
Halvorsen, S. and Jacobsen, E. (1972). *Anal. Chim. Acta*, **59**, 127.
Hasebe, K. and Osteryoung, J. G. (1976). Unpublished results.
Heyrovsky, J. (1922). *Chem. Listy*, **16**, 259. [See also Brdicka, R. (1950). *Coll. Czech. Chem. Commun.* **15**, 691 (biography)].
Heyrovsky, J. (1924). *Trans. Faraday Soc.* **19**, 785.
Heyrovsky, J. and Zuman, P. (1968). *In* "Practical Polarography". Plenum Press, New York.
Hicks, G. P. and Updike, S. J. (1967). *Nature, Lond.* **214**, 986.
Homolka, J. (1971). *Methods Biochem. Anal.* **19**, 435.
Hynie, I. and Prokes, J. (1964). *Chem. Zvesti.* **18**, 425.
Iversen, P. E. and Lund, H. (1969). *Anal. Chem.* **41**, 1322.
Jacobsen, E. and Jacobsen, T. V. (1971). *Anal. Chim. Acta*, **55**, 293.
Jacobsen, E. and Rojahn, T. (1972). *Anal. Chim. Acta*, **61**, 320.
Jacobsen, E., Jacobsen, T. V. and Rojahn, T. (1973). *Anal. Chim. Acta*, **64**, 473.
Kahl, W. and Pasek, W. (1970). *Rocz. Chem.* **44**, 2425.
Kane, P. O. (1961). *J. Polarog. Soc.* **VII**, 58.
Kissinger, P. T. (1977). *Anal. Chem.* **49**, 447A.
Knevel, C. R. (1968). *Anal. Biochem.* **22**, 179.
Korneva, L. E., Tuichiev, E. T. and Murtazaev, A. M. (1972). Mater Yubileinoi Resp. Nauchn. Knof. Farm. Posryashch. 50-Letiyn. Obraz. SSSR. p. 158.
Kosmatyi, E. S. and Kavetskii, V. N. (1973). *Zh. Anal. Khim.* **28**, 1028.
Kotoucek, M. and Zavadilova, J. (1972). *Coll. Czech. Chem. Commun.* **37**, 3212.
Kullberg, M. P. and Gorodetzky, C. W. (1976). *Clin. Chem.* **20**, 177.
Kurayuki, Y., Miznoya, Y. and Kojima, H. (1966). *J. Pharm. Sci. (Japan)*, **86**, 890.
Lalithambika, M., Sinha, L. and Sudha, K. (1975). *Z. Phys. Chem. (Frankfurt-am-Main)*, **94**, 127.
Lasheen, A. M. (1961). *Proc. Amer. Soc. Hart. Sci.* **77**, 135.

Lindquist, J. and Farroha, S. M. (1975). *Analyst*, **100**, 377.
Linhart, K. (1972). *Tenside*, **9**, 241.
Little, C. L. (1974). Ph.D. Thesis, University of London.
Lund, W. and Opheim, L. N. (1975). *Anal. Chim. Acta*, **79**, 35.
Lund, W. and Opheim, L. (1976). *Anal. Chim. Acta*, **88**, 275.
Maheswari, A. K., Jain, D. S. and Saraswat, H. C. (1974). *J. Electrochem. Soc. India*, **23**, 159.
Mairanovskii, S. G. (1968). "Catalytic and Kinetic Waves in Polarography". Plenum Press, New York.
Mairesse-Ducarmois, C. A., Patriarche, G. J. and Vandenbalck, J. L. (1974). *Anal. Chim. Acta*, **71**, 165.
Mairesse-Ducarmois, C. A., Patriarche, G. J. and Vandenbalck, J. L. (1975). *Anal. Chim. Acta*, **76**, 299.
Mason, W. D. and Sandman, B. (1976). *J. Pharm. Sci.* **65**, 599.
Mikolajek, A. (1969). *Mikrochim. Acta*, No. 6, 1229.
Milner, G. W. C. (1962). "The Principles and Applications of Polarography and Other Electroanalytical Processes", 3rd Edn. Longmans,
Mithers, K. (1961). *Acta. Pharm. Tox. KBH*, **18**, 199.
Nurnberg, H. W. and Wolff, G. (1966). *Z. Anal. Chem.* **216**, 169.
Nygard, B., Olofsson, J. and Sandberg, M. (1966). *Acta. Pharm. Suec.* **3**, 343.
Oelschlager, H., Lumbantoruan, S., Volke, J. and Kraft, G. (1976). *Z. Anal. Chem.* **279**, 257.
Orlov, Y. E., Ignatov, Y. L. and Shostenico, Y. V. (1964). *Med. Prom. SSSR*, **18**, 44.
Osteryoung, J. G. and Hasebe, K. (1976). *Rev. Polarog. Japan*, **22**, 1.
Pasciak, J. and Gajewska, T. (1976). Proc. 2nd National Conference on Analytical Chemistry, Varna (Bulgaria), p. 209.
Pasciak, J. and Lewandowska, T. (1972). *Chemia. Analit.* **17**, 919.
Pazdera, H. J., McMullen, W. H., Ciaccio, L. C., Missan, M. R. and Grenfell, T. C. (1957). *Anal. Chem.* **29**, 1649.
Pierzchalski, T. (1963). *Chem. Anal. (Warsaw)*, **8**, 443.
Perone, S. P. and Oyster, T. J. (1964). *Anal. Chem.* **36**, 235.
Porter, G. S. and Beresford, J. (1966). *J. Pharm. Pharmacol.* **18**, 223.
Porter, G. S. and Williams, A. (1972). *J. Pharm. Pharmacol.* **24** (Suppl), 144
Pribyl, M. and Nedbalkova, J. (1969). *Z. Anal. Chem.* **224**, 244.
Princetown Applied Research Corp. (1976). Technical Data, Model 374 Polarograph. P.O. Box 2565, Princeton, New Jersey, U.S.A..
Prokes, J. and Vorel, F. (1960). *Chem. Zvesti.* **14**, 818.
Radejova, E. and Skarka, B. (1974). *Farmaceuticky.* **43**, 119.
Randles, J. E. B. (1947). *Analyst*, **72**, 301.
Reid, E. (1976). *Analyst*, **101**, 1.
Reynolds, G. F. and Davis, H. M. (1953). *Analyst*, **78**, 314.
Ryan, M. D. (1975). *Anal. Chem.* **47**, 1717.
Sachweh, H., Seidenglanz, G. and Richter, J. (1966). *Arzneimittelstandardisierung*, **7**, 697.
Sanz Pedrero, P. and Lopez Fonseca, J. M. (1972). *Analyst*, **97**, 81.
Schill, G. (1976). *In* "Methodological Developments in Biochemistry" (F. Reid, ed.), Vol. 5, p. 87. North-Holland, Amsterdam.
Schmandke, H. and Crohlke, H. (1965). *Clin. Chim. Acta*, **11**, 491.
Seifert, J. and Davidek, J. (1971). *J. Chromat.* **59**, 446.
Silvestri, S. (1972). *Pharm. Acta. Helv.* **22**, 209.
Skora-Zietek, M. (1965). *Dissert. Pharm.* **17**, 301.
Smyth, M. R. (1976). Ph.D. Thesis, University of London.
Smyth, M. R. and Franklyn Smyth, W. (1978). *Analyst*, **103**, 529.

Smyth, M. R., Franklin Smyth, W., Palmer, R. F. and Clifford, J. M. (1976). *Anal. Chim. Acta,* **86**, 185.
Smyth, M. R., Franklin Smyth, W. and Beng Tan, A. (1977a) *Anal. Chim. Acta.* **92**, 129.
Smyth, M. R., Rowley, P. G. and Osteryoung, J. G. (1977b). Proceedings 174th ACS Meeting, Chicago.
Smyth, W. F., Svehla, G. and Zuman, P. (1970a). *Anal. Chim. Acta,* **51**, 463.
Smyth, W. F., Svehla, G. and Zuman, P. (1970b). *Anal. Chim. Acta,* **52**, 129.
Stewart, J. T., Lo, H. C. and Mason, W. D. (1974). *J. Pharm. Sci.* **63**, 954.
Supin, G. S. and Budnikov, G. K. (1973). *Zh. Anal. Chim.* **28**, 1303.
Tammilehto, S. and Perala, M. (1971). *Pharm. Acta. Helv.* **46**, 351.
Thrivikraman, K. V., Refshange, C. and Adams, R. N. (1974). *Life Sci.* **15**, 1335.
Tomana, M. (1966). *Coll. Czech. Chem. Commun.* **31**, 4728.
Trevor, A., Rowland, M. and Way, E. L. (1972). *In* "Fundamentals of Drug Metabolism and Drug Disposition" (B. N. LaDu, H. G. Mandell and E. L. Way, eds), p. 369. Williams and Williams, Baltimore.
Turyan, Y. I. and Tolstikova, O. A. (1974). *Tr. Krasmoda, Politechn. Inst.* **63**, 95.
Van Bennekom, W. P., Reeuwijk, H. J. E. M. and Schute, J. B. (1975). *Anal. Chim. Acta* **74**, 387.
Vassos, B. and Osteryoung, R. A. (1973). *Chem. Instrum.* **5**, 257.
Velisek J., Davidek, J. and Klein, S. (1974). *Z. Lebensn-Unters Forsch.* **155**, 203.
Volke, J. (1974). "Die Polarographie in der Chemotherapie, Biochemie und Biologie". Abhandlung der DAW, Berlin.
Weil-Malherde, H. (1971). *In* "Methods of Biochemical Analysis-Supplementary Volume. Analysis of Biogenic Amines and their Related Enzymes" (D. Glick, ed.), p. 119. Interscience, New York.
Whelpton, R. and Curry, S. H. (1976). *In* "Methodological Developments in Biochemistry" (E. Reid, ed.), Vol. 5, p. 115, North-Holland, Amsterdam.
Williams, A. I. (1973). *Analyst,* **98**, 233.
Yamada, J. and Matsuda, H. (1973). *J. Electroanal. Chem.* **44**, 189.
Zietek, M. and Fidelus, J. (1968). *Acta. Polon. Pharm.* **15**, 77.
Zuman, P. (1964). "Organic Polarographic Analysis". Pergamon Press, Oxford.
Zuman, P. (1967). "Substituent Effects in Organic Polarography". Plenum Press, New York.
Zuman, P. and Perrin, C. L. (1969). *In* "Organic Polarography", 4th Edn. Interscience, New York and London.

Chapter 2

UNIT PROCESSES IN ORGANIC VOLTAMMETRIC ANALYSIS—II. DETERMINATION

B. FLEET and N. B. FOUZDER

*Chemistry Department, Imperial College,
South Kensington, London, England*

I. INTRODUCTION

The determination step in organic voltammetric analysis can be conveniently divided into three separate operations:

(1) Preparation of the sample solution for voltammetric investigation following the unit processes discussed in Chapter 1. The choice of solvent, supporting electrolyte and pH are particularly of importance and are fully discussed in the text.

(2) Selection of a suitable cell and electrodes. The cells and electrodes used for analytical voltammetry are reviewed in this subsection.

(3) Selection and application of the optimum polarographic technique. A brief survey of the principles and major areas of application of the different techniques are presented in this section.

II. PREPARATION OF SAMPLE SOLUTION FOR VOLTAMMETRIC INVESTIGATION

A. CHOICE OF SOLVENT/SUPPORTING ELECTROLYTE

Voltammetric investigations can be carried out in both organic and aqueous solvents. The organic solvents commonly used include acetonitrile (AN), dimethylformamide (DMF), dimethylsulphoxide (DMSO), cellosolves and benzene–methanol mixtures. These solvents are made conducting by the addition of salts such as lithium perchlorate, tetraalkylammonium salts such as perchlorate, or fluoroborate. The main advantage of carrying out voltammetric investigations in such media is that one can study electrode processes in the absence of proton involvement. For organic molecules, this can lead to different electrode mechanisms (e.g. barbiturates are not reduced in aqueous media but give well defined waves in R_4NClO_4/DMSO supporting electrolytes). A serious limitation to the widespread use of voltammetry in aprotic media is the problem of solvent purity; for this reason most applications have been reported in formulation analysis where higher concentrations ($> 10^{-5}$M) are usually encountered. A notable exception in the literature is the assay of the coccidiostat, nicarbazin (an equivalent mixture of 4,4′dinitrocarbanilide and 2-hydroxy,4,6dimethylpyrimidine) in chicken tissue. Following homogenization of the tissue, extraction with ethyl acetate and evaporation of the solvent, the residue was dissolved in 0.1M $(C_2H_5)_4$ $NClO_4$/DMSO and subjected to polarography in the aprotic medium. This medium (purified by shaking the DMSO with Al_2O_3 plus centrifugation) was chosen for solubility considerations and also because a sharp peak was produced following addition of 2×10^{-3}M benzoic acid (Michielli and Downing, 1974). Mann (1969) has reviewed the various non-aqueous solvents for electrochemical use, including methods of purification, choice of reference electrode and selection and purification of supporting electrolytes.

The majority of polarographic trace analyses are carried out, therefore, in aqueous or aqueous/alcoholic solvents. Common electrolytes that are used in voltammetric analysis include simple salt solutions (e.g. KCl, LiCl), the common acids/alkalis (e.g. HCl, H_2SO_4, KOH, NaOH) or simple buffered systems (e.g. acetate, formate, phosphate buffers). The simple buffer solutions are only effective, however, over small pH ranges, and therefore it is common to employ mixed buffer systems, e.g. Britton–Robinson (B.R.) buffers, in the investigation of the voltammetric behaviour of an organic compound. These can be used over a wide pH range with only slight changes in the ionic strength so that pH can be studied as the variable. Their use, however, increases the possibility of interaction between an ion in the electrolyte and the substance to be determined. This can sometimes lead to shifts in the potential of the electrode process which if not identified can lead to erroneous results. In those cases where an interaction is observed, however, the possibility exists to resolve structurally related compounds. For example chlortetracycline interacts with borate ions and this results in a shift in potential which permits its identification in a mixture containing oxytetracycline (Telupilova and Masinova, 1953). In order to minimize the effect of the migration current the ratio of the concentration of supporting electrolyte to that of the substance to be determined (i.e. the depolarizer) should be at least 10^3. For trace analysis, anomalies can occur if the concentration of the supporting electrolyte is not maintained at a level of about 0.01M or greater. This is particularly true in d.p.p. where a decrease in background current is obtained using high salt concentrations (Osteryoung and Hasebe, 1976).

For reduction processes occurring at relatively negative potentials ($<$ ca -1.5 V) analyses can often be carried out in electrolyte solutions containing tetraalkylammonium salts, e.g. tetraethylammonium perchlorate (TEAP), which extend the cathodic potential range to about -2.5 V and permit the determination of compounds such as 3-keto steroids (with unsaturation at positions 4–5) which reduce at approximately -1.8 V (vs s.c.e.) in TEAP/ 50% CH_3OH. These salts can also be added to other supporting electrolytes to decrease background interference. This is due to their strong adsorptive properties at the d.m.e. and has proved useful in the d.p.p. determination of some carcinogenic N-nitrosamines using H_2SO_4 as the supporting electrolyte (Hasebe and Osteryoung, 1975).

The waves obtained for most reduction processes occur at negative potentials and are unaffected by the oxidation of Cl^- and OH^- ions in the supporting electrolyte. This is not the case, however, for most oxidation processes which occur at potentials near and usually more positive than zero volts. Since the working range for the d.m.e. is restricted to potentials more negative than $+0.4$ V (vs s.c.e.), care should be taken not to employ Cl^- or OH^- ions in the supporting electrolyte which would limit the working potential range even further (this is not the case with a.c. methods where up to M Cl^- can be tolerated—Gruendler and Choschzich, 1972).

In certain cases one can employ a supporting electrolyte which obviates the need to purge the solution with an inert gas (in order to remove the interfering influence of O_2 in cathodic processes); for example, sodium sulphite can be used for this purpose (when in alkaline solution) as has been demonstrated by Iversen and Lund (1969) for the determination of some aliphatic hydroxylamines. Ascorbic acid also has the ability to remove O_2 from solution (Florence and Farrar, 1973) but this electrolyte should not be used for the determination of compounds that undergo oxidation processes. These reagents often contain significant amounts of metal ions which precludes their use for trace analysis.

If a wave exhibits maxima, these can often be removed by the addition of a maximum suppressor (e.g. gelatin, or surfactants such as Triton X-100 are most commonly employed for this purpose). The concentration of these agents should, however, be maintained at the lowest possible level which effects the removal of the maximum and improves the wave shape (Hasebe and Osteryoung, 1975). If they are present in too high a concentration, they can often affect the electrode process e.g. in the case of transazobenzene, the addition of Na dodecylsulphate or cetylpyridinium chloride to the supporting electrolyte (citric acid/Na_2HPO_4 pH 5.05) affected the $E_{\frac{1}{2}}$, i_{lim}, wave shape and reversibility of the electrode process (Westmoreland et al., 1972).

The use of aqueous/alcohol (methanol or ethanol) mixtures has been widely employed in voltammetric methodology, mostly to ensure solubility of the compound under test and to improve the wave shape of compounds undergoing complicated electrode processes. In general it is advisable to maintain the concentration of alcohol in the mixture as low as possible although concentrations of up to 50% have been tolerated under certain circumstances. The main effects of using aqueous/alcohol mixtures have been discussed by Mairanovskii (1965) and are summarized below: (i) shifts can be observed in the $E_{\frac{1}{2}}$, e.g. in the case of benzophenone the $E_{\frac{1}{2}}$ of the first process becomes more positive on the addition of ethanol; (ii) changes can occur in the overall electrode process, e.g. addition of ethanol to 2-nitrofuran in neutral solution causes a 4-electron process to change to a one-electron process and a new wave to appear at more negative potentials (Stradyn et al., 1965); (iii) addition of organic solvents can minimize problems of adsorption at the electrode and can also cause decreases in i_{lim} values; (iv) addition of organic solvents can increase the rate of protonation of compounds in the bulk of the solution, e.g. for pyridine and 2,6-lutidine (Mairanovskii and Gultyai, 1965) but can also decrease the rate of the electrode process when antecedent surface protonation occurs.

With regard to the choice of supporting electrolyte, most analytical grade reagents are of adequate purity; in many cases the supporting electrolyte also acts as a buffer system. For trace analysis, however, many ultra-pure reagents still contain significant amounts of heavy metals. Two main methods for the purification of supporting electrolytes are recommended (PAR, 1974b). The first is based on controlled potential electrolysis at a large area mercury pool cathode for approximately 24 hours. After deaeration of the solution the potential is set to the maximum cathodic value that the system will stand, i.e. at least -1.5 V, if zinc, a major trace contaminant, is to be removed. The second method, known as isothermal distillation, has been used to prepare ultra-pure hydrochloric acid, ammonium hydroxide and ammonium chloride. For example, a 500 ml beaker filled with hydrochloric acid and a 500 ml beaker half filled with distilled water are placed in a dessicator, the lid closed and the system allowed to stand for 7–10 days. At the end of this period the beaker of water will have come to equilibrium with that containing the acid. A similar system is employed for ammonium hydroxide and combining the two solutions will give ultra-pure ammonium chloride.

One further method worth noting is crystal adsorption. It has been observed that if a saturated solution of potassium chloride is allowed to stand for a long period of time in contact with solid crystalline potassium chloride then traces of heavy metal ions are preferentially adsorbed onto the solid phase leaving the supernatant liquor in a very pure state. This method requires at least 6–12 months for effective purification.

B. CHOICE OF pH

In protogenic media protons are involved in the electrode processes of most organic molecules and consequently the pH of the medium must be carefully controlled. The effect of pH on the voltammetric behaviour, particularly the limiting current of the compound under investigation, must also be studied in order to select the optimum pH for analytical measurement. The stability of the compound as a function of pH must also be checked; this is most simply carried out by running sequential polarograms on the same test solution.

III. SELECTION OF A SUITABLE CELL AND ELECTRODES

A. CELLS

To an electrochemist a cell design is a very personal thing and the literature contains a large number of tributes to the glass-blowing art! However, apart from some special designs of cells for techniques such as coulometry or spectroelectrochemistry many cells have been described which are entirely adequate (i.e. they give support for the electrodes, prevent atmospheric oxygen from entering the test solution and provide an approach to ideal geometry).

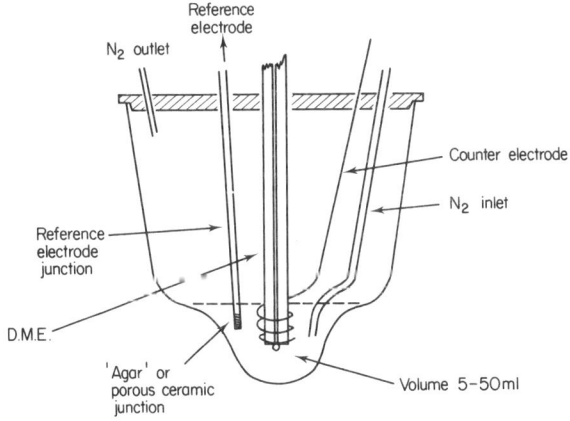

Fig. 1. Polarographic cell.

Modern electrochemical methods employ a three-electrode principle of operation in which minimization of the ohmic drop is achieved by a reference electrode probe being situated as close as possible to the working electrode. A detailed discussion of the problems of cell design has been given by Harrar and Shain (1966) and Harrar (1975). For most analytical and mechanistic studies a conventional cell (Fig. 1) is adequate. This type of cell accepts sample volumes from 5 to 50 ml and may also be thermostatted.

For the analysis of trace amounts of drugs and their metabolites in body fluids, Brooks and Hackman (1975) have described a micro-cell of volume 0.5 ml containing a 3-electrode system having a d.m.e. as the working electrode, platinum as the counter electrode and Ag/AgCl as the reference electrode (Fig. 2).

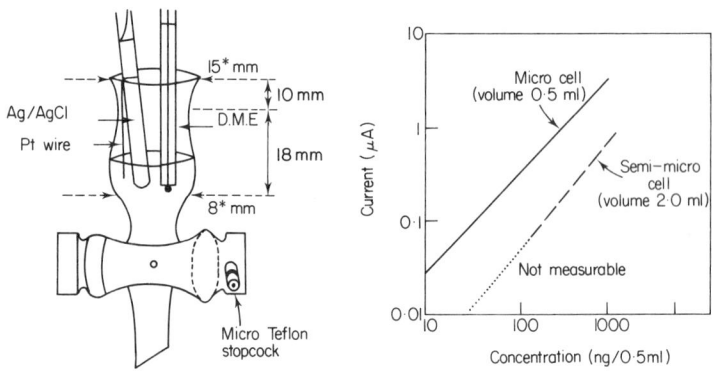

FIG. 2. (a) Schematic diagram of polarographic micro cell and (b) comparison of sensitivities obtained with a micro cell and semi-micro cell for the reduction of N-desalkylflurazepam (Brooks and Hackman, 1975).

In many cases a s.c.e. is used in preference to the Ag/AgCl reference electrode since most electrode potentials are quoted versus s.c.e. in the literature. Cells of larger volume (2–5 ml) offer greater manageability and, provided that one is not working at the limit of detection (see Fig. 2), these are usually sufficient for most analyses. Where possible one should isolate the two auxiliary electrodes from the bulk of the solution (by means of sintered glass etc.) so that in the case of the counter electrode one minimizes the interference caused by products of the electrolysis (or by the adsorption of surface active organics present at trace levels) and in the case of the reference electrode one minimizes the interference caused by Cl^- or OH^- leaching out into the supporting electrolyte. In cases where one employs a silver wire as a pseudo reference electrode it is necessary, however, to employ halide ions in the supporting electrolyte.

Some more specialized cell designs include those for stripping voltammetry (Stulik and Stulikova, 1973), microcoulometry using the dropping mercury electrode (Manousek, personal communication) (see also section IV, E, i), macro scale coulometry, rapid anodic stripping voltammetry and coulometry by use of a rotating cell (Clem et al., 1973), and continuous monitoring of electrolysis products by spectroscopic techniques (Janata and Mark, 1967). Cell designs for continuous flow voltammetry are rather specialized (Tenygl 1978) and include specific voltammetric detector systems (Fleet and Little, 1974).

For trace analysis it is essential that a source of pure nitrogen or argon for de-aeration be available since even the residual ppm levels of oxygen in high purity nitrogen can cause problems. Two main methods of carrier gas purification have been described (PAR, 1974a).

The first method involves chemical scrubbing in which the carrier gas is passed through an absorbing solution consisting of a vanadous chloride solution prepared by dissolving 2 g ammonium metavanadate in 25 ml concentrated hydrochloric acid with heating and dilution to 250 ml. A few g of amalgamated zinc is then added to the scrubbing tower and nitrogen gas passed through. A clear violet colour indicates that all the vanadium is in the +2 state. Regeneration of the solution after prolonged use can be achieved by addition of more zinc amalgam which is easily prepared by addition of mercury to granular or powdered zinc in distilled water which has been acidified slightly by addition of a few drops of concentrated hydrochloric acid.

The second method consists of a copper based catalyst which is first of all conditioned by heating to 120–140°C in a stream of hydrogen gas. This treatment must be repeated for periodic renewal of the catalyst surface, generally when the nitrogen cylinder is changed. An alternative, if more expensive version of this system is the Oxisorb™ disposable cartridge which is utilizable for the lifetime of one gas cylinder.

B. ELECTRODES

As with cell designs a vast variety of electrode configurations for polarography and voltammetry have been described. These have been fully reviewed by Tenygl (1978). For the purpose of this review only the more useful types of dropping mercury and solid electrodes will be mentioned.

(i) *Mercury electrodes*
The dropping mercury electrode (d.m.e.). The classical d.c. polarographic

technique employs a dropping mercury electrode—generally a capillary of 0.02–0.08 mm internal diameter and 10–20 cm long which has a natural drop time of ca -10 s. Modern techniques have tended to use capillaries with very short drop times (ca 100 ms) in contrast to more commonly used drop times in the range 1–5 s controlled by mechanical drop detachment. Mercury outflow is caused either by having a head of mercury of some 20–100 cm or alternatively by applying a gas pressure to the mercury reservoir.

The hanging mercury drop electrode (h.m.d.e.). This electrode employs a stationary drop which is extruded from a capillary via a micrometer driven piston in the mercury reservoir. Alternatively the Hg drop may be attached to micro platinum or gold electrodes although the latter design is tedious to assemble and can lead to problems of intermetallic compound formation. Recently Princeton Applied Research have introduced a universal electrode assembly which consists of a gas pressure operated dropping mercury electrode which can also be operated in the stationary hanging drop mode (PAR, 1976a).

The dropping mercury electrode in continuous monitoring. Practical problems and design modifications for the use of the d.m.e. in continuous monitoring have been reviewed by Novak (1962), Strafelda and Dolezal (1967) and Tenygl (1978). One of these involves treatment of the capillary with silicone oil to prevent build up of electrolysis products inside the capillary orifice.

(ii) *Non-mercury electrodes*

Despite the unique advantages of the dropping mercury electrode it does have some serious limitations, principally its mechanical instability and negligible anodic range. The search for suitable electrode materials for solid electrode voltammetry has been prompted by electrooxidation studies, the search for suitable mercury film substrates in stripping voltammetry and the design of voltammetric detectors for on-line or continuous monitoring. A wide range of electrode materials have been investigated (Alder *et al.*, 1971) and Adams (1969) provides a good survey both of practical problems and techniques.

Traditionally, the noble metals, platinum, gold and silver, have been the most widely used electrode materials while other metals such as lead and nickel have found application in the anodic oxidation of organic compounds.

Recently various forms of carbon including carbon paste (Adams, 1969; Lindquist, 1973), vitreous (glassy) carbon (Zittel and Miller, 1965; Yoshimori *et al.*, 1965; Florence, 1970) and impregnated graphites (Matson *et al.*, 1965; Beilby, 1964) have been widely used in electroanalytical work and have

2. UNIT PROCESSES IN ORGANIC VOLTAMMETRIC ANALYSIS (II)

largely displaced metal electrodes both for anodic voltammetry and stripping voltammetry.

For use in stripping voltammetry two types of mercury plating procedure have been used on glassy carbon or impregnated graphite substrates: (i) Mercury film plating prior to pre-electrolysis step (Hume and Carter, 1972; Copeland *et al.*, 1973). (ii) Simultaneous co-deposition of mercury film and trace metal of interest (Florence, 1970; PAR, 1972).

The latter procedure, devised by Florence (1970), has become widely used in recent years. Both techniques rely on a clean electrode surface to ensure a uniform film formation and the Hg(II) used for the deposition process should be free from trace heavy metals. Griepink (personal communication) recommends an electrolytic purification of the Hg(II) used for plating. Deposition of a known amount of Hg(O) on a silver foil electrode is first carried out; this is then rinsed and transferred to the electrolysis vessel used for the mercury film plating or the reservoir for Hg(II) and electrolytically redissolved.

Several groups of workers (Florence, 1970; Stulikova 1973; Berger and Fleet, 1977) have observed that the first stripping peak on thin film a.s.v. always gives erroneous results. The stripping peaks on second and subsequent scans always show much better reproducibility.

The relative merits of glassy carbon versus wax impregnated graphite as a substrate for mercury film formation is a subject of much controversy within the electrochemical community (Interface, 1974). On balance glassy carbon appears to be the most popular provided that a good grade of glassy carbon such as that produced by Tokai* is used. Several workers, most notably Environmental Science Associates (ESA, 1974) and Clem and Sciamanna (1975), have strongly recommended wax or polystyrene impregnated graphite.

Some newer electrode materials such as boron carbide have been described but appear inferior to glassy or impregnated carbon. One interesting material which shows considerable promise is carbon fibre; this material has a very high surface-to-weight ratio of 10^4–10^5 cm^2 g^{-1}. Short lengths of a few fibres also show promise as a microelectrode for anodic voltammetry (Fleet and das Gupta 1976). The major practical problem of working with solid electrodes is the maintenance of a uniform working electrode surface. This problem is discussed in Section IV, A, iv.

Most solid electrodes, Pt, Au and the various types of carbon, show similar anodic potential ranges ($+1.2 \to +1.6$ V in aqueous media; up to $+2.0$ V in organic solvents) whereas the cathodic limits show considerable variation. The various carbon electrodes (glassy, pyrolitic and wax impregnated) show a good cathodic range (up to -1.0 V) while platinum (-0.4 V) and gold (-0.7 V) have a smaller range.

---

* Tokai Electrode Co. Ltd., 2–3 Kita-Aoyama 1 Chome, Minatoku, Tokyo, Japan. European agents: EDT Research, 65 Ivy Crescent, London W4, England.

(iii) *Reference electrodes*

The role of the reference electrode is often neglected in voltammetric measurements and as in potentiometry is a frequent source of problems. For two-electrode operation a large area (ca 2–3 cm^2) calomel electrode is recommended, i.e. a mercury pool with saturated KCl above the pool and separated from the test solution by a ceramic frit, agar or "Vycor" (Corning Glassworks, Corning, New York, U.S.A.) porous glass junction. For three-electrode operation a conventional potentiometric calomel or silver chloride electrode is adequate. In polarography in non-aqueous media the choice of the reference electrode is more complex (Mann, 1969) but a simple solution is to use an anodized silver wire in a 0.1M solution of tetramethylammonium chloride which provides a "pseudo" reference electrode (Ryan *et al.*, 1974).

C. THE POTENTIOSTAT

Early designs of d.c. polarographs used a motor driven potentiometer to apply the working potential to a two-electrode cell. By using a large area reference anode such as a mercury pool the reference electrode potential remained essentially constant while the working microelectrode was polarized. In this arrangement the polarizing potential was applied to the d.m.e. through the reference electrode and across the cell and the overall cell potential could be defined as

$$E_{cell} = E_{reference} - E_{working} + iR$$

where the *iR* term, known as the ohmic drop, is the potential required to overcome the resistance of the experimental system, mainly of the cell. At low current levels in highly conducting media this system works well; unfortunately at higher current levels it is possible to polarize the reference electrode. A further complication is that in poorly conducting media such as organic solvents the resistance across the cell is included in the measurement and causes distortion of the polarogram.

The first solution to this problem was the development of the three-electrode manual potentiostat. In this system the cell voltage was applied between the working and reference electrodes while the true working potential was measured (with a second reference electrode) in a separate circuit with a high impedance voltmeter. This device provided a means for controlling the potential between the electrodes but had the serious disadvantage that it required constant adjustment of the control voltage during the measurement of a polarogram.

The first automatic electronic potentiostat was devised by Hickling (1942) and has provided the basis for all present day analogue and digital potentiostats.

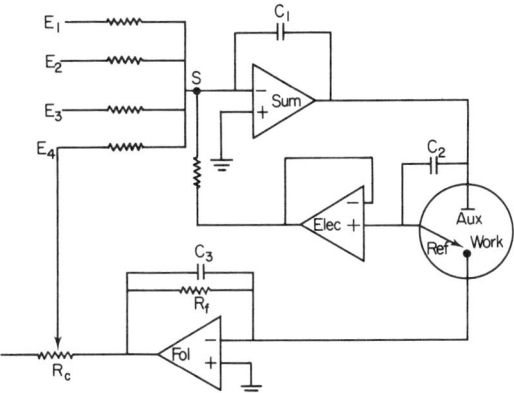

FIG. 3. Schematic circuit for the potentiostat.

A schematic diagram of a typical modern electronic potentiostat is shown in Fig. 3.

In this scheme Sum amplifier is the control or summing amplifier which supplies the input signal (d.c. level, voltage ramp, etc.) to the cell, Elec is a high impedance, unity gain, non-inverting voltage follower, while Fol is a variable gain current follower which serves to maintain the working electrode at zero volts (virtual earth) and provide an output proportional to the polarographic current. The operation of this circuit can best be visualized by realizing that an operational amplifier reacts in a manner required to maintain zero potential difference between its inputs.

Therefore the stable condition for the loop containing the Sum and Elec amplifiers is when the output from Elec is equal in magnitude but opposite in polarity to the sum of the input voltages. This means that the voltage at point S, the summing junction, will always be at zero volts. Because the reference electrode is maintained at a potential given by the negative sum of potential inputs $-(E_1 + E_2 + E_3)$ and the working electrode at zero volts, it is apparent that the working electrode potential relative to the reference is always maintained at $(E_1 + E_2 + E_3)$, i.e. at the input potential.

When an input voltage is applied to the Sum amplifier it outputs a voltage to the auxiliary electrode which causes a current to flow between the auxiliary and working electrode. This current polarizes the working electrode until it reaches the potential defined by the input to Sum and the current flow remains sufficient to maintain this state. The capacitors C_1-C_3 are inserted to prevent the tendency to oscillation which is a feature of this type of control circuit. Modern potentiostats often employ a device known as "positive feedback". The output from the FoI amplifier under experimental conditions in the absence of the electroactive species provides an additional potential input E_4 which allows complete compensation of residual ohmic potential drop.

The potentiostat is the heart of modern polarographic instrumentation both for d.c. and for the more rapid pulse and a.c. techniques. Its principal features are that it can maintain a set potential to within an accuracy of a few millivolts and causes only a minimal current, of the order of a few picoamps, to be drawn through the reference electrode. It also has a very rapid response time typically of the order of a few microseconds when driving a resistive load. This feature is of extreme importance in modern pulse and a.c. techniques where the system must follow the rapidly varying input signal.

IV. SELECTION OF THE OPTIMUM POLAROGRAPHIC OR VOLTAMMETRIC TECHNIQUE

A. D.C. TECHNIQUES

(i) *Polarography with the dropping mercury electrode (d.m.e.).*

The polarographic wave. D.c. voltammetry at the dropping mercury electrode or classical polarography involves the measurement of current voltage curves at a dropping mercury electrode when a slow linear voltage ramp input (typically 100 mV min^{-1}) is applied to the cell.

In an electrolysis cell three types of mass transport are possible—migration, diffusion and convection. In polarography the experimental conditions employed (quiet solution, excess of inert background electrolyte) eliminate contributions from convection and migration so that the current arises solely from diffusion. The unique advantages of the dropping mercury electrode (renewable, reproducible electrode surface, large negative potential range) are supplemented by the fact that well defined S shaped current voltage curves are obtained for each electrode process.

The analytical parameters of interest are as follows. (i) The current on the plateau of the wave, i_{lim}, which is linearly related to concentration, c, by the Ilkovic equation (1) when the current is diffusion controlled:

$$i_d = 0.607\, nFc\, D^{\frac{1}{2}}\, m^{\frac{2}{3}}\, t^{\frac{1}{6}} \tag{1}$$

where n is the number of electrons taking part in the electrode process, F is the Faraday, D is the diffusion coefficient and m and t are the characteristics of the mercury outflow and electrode drop time respectively. (ii) The half-wave potential $E_{\frac{1}{2}}$ which is the potential where the current equals one half of the limiting value. (iii) The slope of the current potential curve.

Both these latter parameters can be used for the identification of the molecule/species that is undergoing the electrochemical reaction.

Equation (1) is uncorrected for the value of the charging current. This charging current, of primary importance in polarography, is due to the adsorption of anions or cations at the electrode surface to form a double layer. This double layer has a finite capacitance and therefore a significant current is required to charge the electrode–solution interface to the required potential. It is the existence of this charging current, which effectively limits the sensitivity of the d.c. method to ca 5×10^{-5}M.

This capacitive current is very large at the start of the drop lifetime but decreases with the growth of the drop. The faradaic current on the other hand, defined by the Ilkovic equation, increases during the drop growth (Fig. 4). This important difference between the faradaic and capacitive currents is the basis on which most modern polarographic techniques have been developed.

Most reductions of metal ions and some organic and organometallic electrochemical reactions are "reversible" (i.e. the rate of electron transfer is fast in comparison with the mass transport process). The equation which

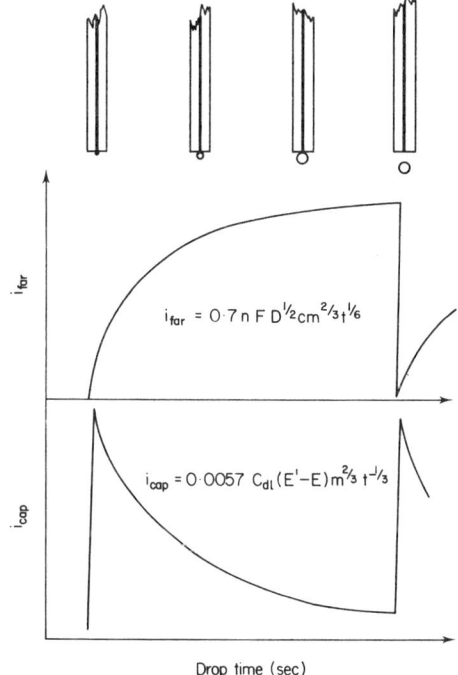

FIG. 4. Time dependence of Faradaic and capacitive currents.

describes the shape of the current potential curve has been defined (Heyrovsky and Kuta, 1965):

$$E = E° + \frac{RT}{nF} \log\left(\frac{i_d - i}{i}\right)\left(\frac{D_R}{D_O}\right)^{\frac{1}{2}} \qquad (2)$$

where i is the current at potential E, i_d is the diffusion limited current, $E°$ is the standard redox potential for the system and D_R and D_O are the diffusion coefficients of reduced and oxidized forms of electroactive species respectively. Thus if the values of D are known, or are assumed to be equal, equation (2) above can be used to calculate the value of n for the reversible electrode process.

However, most electrochemical reactions of the molecules/species considered in this volume are "irreversible", i.e. the kinetics of the reaction must be considered as well as the thermodynamics in order to deduce the relationship between current and potential.

The current potential relationship for an irreversible process is given by:

$$E = E° + \frac{RT}{\alpha nF} \log\left(\frac{i_d - i}{i}\right)\left(\frac{D_R}{D_O}\right)^{\frac{1}{2}} \qquad (3)$$

where α, the transfer coefficient represents the fraction of an increase in $E_{applied}$ which will favour the reduction process and consequently $1 - \alpha$ the fraction which hinders the oxidation process. For most electrode processes α values lie in the range 0.3 to 0.7.

Protons are also involved in many organic electrode processes, e.g. for the reaction $O + mH^+ + ne^- \rightarrow R$

$$E_{\frac{1}{2}} = E°_{\frac{1}{2}} - 0.059 \, m/n . \, pH \qquad (4)$$

where m = number of protons and n = number of electrons involved in a reversible reaction.

This illustrates the importance of adequate buffering so that the electrode/solution interface does not undergo any appreciable pH change. It also provides the basis for the use of the $E_{\frac{1}{2}}$–pH dependence as a means for characterization of the electrode process. A more detailed discussion of these topics can be found in Delahay (1954) and Zuman and Perrin (1969).

Types of limiting current. Although the limiting current for most electrode reactions is controlled by the rate of diffusion of the electroactive species there are several other types of current that have been observed (Heyrovsky and Kuta, 1965). Recent applications of catalytic processes are discussed in the chapter by Chowdhry (Chapter 6). Adsorption currents, caused by adsorption of the reactant or product of electrolysis on the electrode surface, generally complicate polarographic analysis when using d.c. techniques but are an advantage in trace analysis where the electroactive species can form a surface monolayer (Booth and Fleet, 1970a, b).

Analytical applications. The analytical applications of d.c. polarography at the d.m.e. for organic substances of biological significance have been reviewed by many authors, principally Zuman (1964), Zuman and Perrin (1969) and recently Brezina and Volke (1975).

(ii) *Linear sweep voltammetry*

Linear potential sweep chronoamperometry was one of the first of the modern variants of the classical polarographic method. The original method was often referred to as cathode-ray polarography and was based on the application of a rapid voltage sweep to the electrode during the last part of the mercury drop lifetime. Nowadays potential sweep techniques are also employed with stationary mercury electrodes, i.e. h.m.d.e. and various solid electrodes. In linear sweep methods a peak response is obtained due to the combined effects of high mass transfer on the non-steady state followed by the progressive depletion of the reactant concentration in the diffusion layer.

Linear sweep techniques are not widely used in modern polarographic analysis since the sensitivity is severely limited by the fact that the double layer charging current is also dependent on sweep rate as is the peak potential.

Thus attempts to improve sensitivity of the method by using increased sweep rates simply result in an increasingly sloping background current contribution. Subtractive techniques using matched twin cells have been employed in an attempt to overcome this problem but the difficulties in obtaining matched cells have generally been too great. Linear sweep chrono-amperometry is, however, still widely used for the measurement step in stripping voltammetry.

Analytical applications. A considerable body of literature exists on the applications of cathode-ray polarography mainly in inorganic analysis (Rooney, 1962). The annual reviews of polarographic applications published by the Czechoslovak Academy of Sciences and the Bibliografica Polarografica published by the Italia Crns provide a useful source of reference to this early work.

(iii) *Cyclic voltammetry*

An important variant of linear sweep voltammetry is the technique of cyclic voltammetry where the imposed signal consists of a triangular waveform with the first sweep being followed by a reverse sweep back to the initial potential. This gives a peak of opposite polarity on the reverse sweep. Important mechanistic information is obtained from the peak potential separation, the ratio of cathodic to anodic peak currents and the effect on these two parameters caused by variations in the voltage sweep rate.

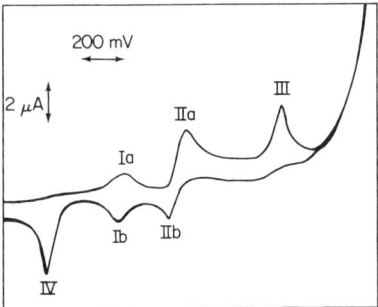

FIG. 5. Cyclic voltammogram of triphenyltin acetate 5×10^{-4}M in pH 7.3 buffer, 50% ethanol, start potential -0.1 V, scan rate 200 mVS^{-1}; peaks 1a and 1b correspond to the adsorption/desorption of free radical Ph$_3$Sn.; peaks IIa and IIb to the same process in bulk solution; peak III is a one-electron reduction of radical to form triphenyltin anion Ph$_3$Sn$^-$ and peak IV is the 2-electron oxidation of triphenyltin hydride to form the triphenyltin anion Ph$_3$Sn$^-$ (Booth and Fleet, 1970a).

Analytical applications. An example where cyclic voltammetry has been used to establish the optimum conditions for a polarographic analytical procedure is in the reduction of organotin compounds (Booth and Fleet, 1970a, b; Fleet and Fouzder, 1975a; Fig. 5) which are widely used as herbicides. This is detailed in the chapter by Fleet and Fouzder (Chapter 9, p. 261).

(iv) *Voltammetry at solid electrodes*

The theoretical principles governing linear potential sweep chronoamperometry at the dropping mercury and hanging mercury drop electrodes apply also to solid electrodes of fixed area. The principal reasons for using solid electrodes have been referred to earlier, namely in electro-oxidation studies, in the design of voltammetric detectors and as substrates for mercury deposition in anodic stripping voltammetry. Solid electrodes have also found widespread use in amperometric titrations (Stock, 1976 and personal communication). Peak shaped current voltage curves are obtained for solid electrodes in stationary solution.

When a solid electrode is used in a stirred solution, e.g. as a titration endpoint sensor, then a current voltage curve similar to a d.c. polarogram is obtained. In this case the electrolysis current is controlled both by diffusion and convection. The equation for the current is given by:

$$i_{\text{lim}} = \frac{n\text{FADC}_\text{o}}{\delta} \quad (5)$$

where δ is the diffusion layer thickness and A is the surface area of the electrode. Since the thickness of the diffusion layer is controlled by the stirring rate, equation (5) can be rewritten as follows

$$i_{\lim} = K' n\text{FADC}_o \qquad (6)$$

where K' is a mass transport coefficient which depends on several factors including electrode and cell geometry, rate of stirring or rotation and is consequently difficult to define.

The major practical problem with solid electrode voltammetry is the maintenance of a uniform working electrode surface. Contamination of the electrode surface can be caused either by adsorption of the products of the electrode reaction or by surface film formation caused by the working electrode potential exceeding the anodic or cathodic limits for the solvent system under investigation. The practical consequences of these phenomena are either the decrease in peak currents with successive voltage sweeps or the occurrence of spurious peaks caused by re-oxidation or re-reduction of the surface films. Thus the effective use of solid electrode voltammetry is crucially dependent both on the preliminary electrode preparation and the maintenance of a reproducible working surface. For the preparation of glassy carbon or impregnated graphite electrodes the electrode is first polished with either silica or diamond abrasive powder of 1 μ particle size, followed by a final polishing with 0.1 μ powder. Occasional repolishing may be required, especially if the electrode has been polarized beyond the potential limits. Maintenance of a reproducible electrode surface is a complex problem and a variety of approaches have been suggested to the problem (Tenygl, 1978). The techniques employed have been based either on mechanical abrasion or electrochemical potential cycling. In the simplest case the electrode may be removed from the cell and cleaned between measurements. Various automated approaches to the problems have been described. A rotating drum of abrasive material such as carborundum or even ground glass which intermittently contacts the working electrode has been described by Stock (personal communication). Addition of carborundum chips, mechanically agitated by a gas stream, to the cell has also been suggested.

Potential cycling methods where the working electrode is intermittently polarized to a potential of opposite polarity to the working potential have been described by numerous workers and are effective in many cases. One problem which becomes apparent when potential cycling is used is the fact that solid electrodes exhibit far longer time constants for the decay of the double layer charging current than are observed on mercury electrodes (Blaedel and Jenkins, 1974; Berger and Fleet, 1977). This effect is also of importance when pulsed measurement techniques are used. The reason for this anomaly in the case of carbon electrodes is the involvement of surface chemical groupings which undergo finite equilibrium processes; consequently slow chemical reactions occur when the working potential is changed.

Anodic voltammetry. Up to the present time voltammetry at anodic potentials has been a neglected area of research. This situation is due entirely to the problems of using solid electrode voltammetry, especially the problems of surface adsorption referred to earlier.

The electro-oxidation of organic compounds is a particularly promising area for study since it opens up a whole new range of compounds which are not reducible, and many of these are of importance in the biomedical and pharmaceutical sciences.

The fact that a wide variety of electrode materials have been used for anodic voltammetry is indicative of the importance of the catalytic nature of the electrode surface material on the electrode mechanism and consequently the products of the reaction. In most cases the electrode is regarded as an electron sink for the supply or removal of electrons although electron transfer may be preceded or followed by adsorption of reactants, intermediates or products. The majority of electro-oxidations of organic species have tended to favour platinum or carbon electrodes (Ross *et al.*, 1975). For oxidations at nickel and lead anodes there is evidence that the electrode process occurs via a reduction of the type:

$$NiO(OH) + org. \rightarrow Ni(OH)_2 + product$$

and not by direct electron transfer from the compound to the electrode (Fleischmann *et al.*, 1971).

The increasing interest and commercial viability of large scale industrial electro-organic synthesis has also led to renewed activity in the area of solid electrode voltammetry since in many cases a detailed study of the electrode process at a microelectrode is used to define the optimum conditions for the large scale process.

Analytical applications of anodic voltammetry. A survey of the literature will show that there is an increasing number of applications, particularly for the determination of naturally occurring molecules in biological media (see Chapter 6). Adams (1969) has surveyed most of the early work in this field.

Analytical applications of cathodic voltammetry. There are few reported cases of applications of cathodic voltammetry in the literature.

Mason and Sandman (1976) have determined the drug nitrofurantoin (see Browne, Chapter 4) directly in the urine of clinical patients using the rotating platinum electrode. It is also possible to differentiate between various 1,4-benzodiazepines (Chapter 1).

(v) *Hydrodynamic voltammetry*

Techniques in which the working electrode operates in a stirred or flowing solution where mass transport is due to both diffusion and convection are becoming increasingly important. Originally as in the case of the rotating disc electrode the method was used for mechanistic studies where the controlled mass transport conditions enabled the study of rates of electron transfer or intermediate chemical steps in the electrode process to be made. More recently the theoretical basis of this technique has assumed considerable importance due to the increasing applications of voltammetric detectors in continuous monitoring applications in flowing streams. A useful concept for understanding the contribution of convection to the overall electrode process is that of the Nernst diffusion layer (Fig. 6), which assumes that even under rapidly stirred conditions there is still a thin motionless layer (ca 0.1 mm) next to the electrode across which mass transfer takes

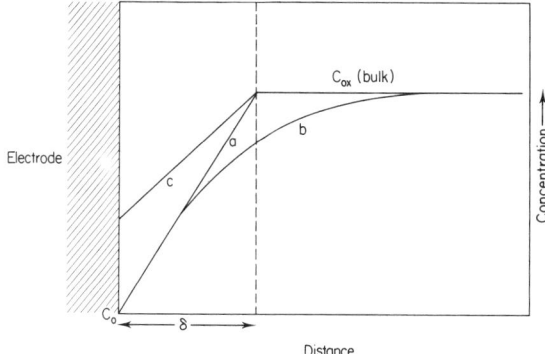

FIG. 6. Nernst diffusion layer, δ = diffusion layer thickness, concentration-distance profile under limiting current conditions (a) theoretical and (b) observed. Curve c shows concentration-distance profile at potentials below the limiting current on the ascending portion of the d.c. wave.

place solely by diffusion. Further refinements in the theoretical treatment of hydrodynamic voltammetry with various electrode configurations have been carried out by Levich (1962) and co-workers while the principles of current distribution and mass transport in flowing systems have been reviewed by Newman (1973).

Continuous flow sensors. The increasing trend towards automation has been reflected in the wide range of electrode designs which have been described for continuous flow voltammetry (Tenygl, 1978). These devices may be voltammetric in origin; for example, tubular electrodes, wire or point designs, or thin layer cells. Alternatively an equally wide range of devices operating on coulometric principles have been reported. In these the electrode usually consists of a packed bed of conducting particles or fibres; in some cases porous electrodes (e.g., fuel cells) have been used (Fleet et al., 1972a, b).

Two designs of flow cell are worth noting. Tubular electrode configurations have been extensively studied by Oesterling and Olsen (1967a, b) and by Blaedel and Boyer (1971). These authors have also used a tubular mercury electrode with a platinum tube as the substrate for the mercury film.

Another promising design, developed in the authors' laboratory, is the wall-jet cell (Fleet and Little, 1974; Fleet et al., 1977). This device is a thin layer cell where the sample solution enters the cell through a fine nozzle and impinges on the centre of a planar glassy carbon disc electrode. Reference and auxiliary electrodes are located downstream of the working electrode. The features of this cell design are the high sensitivity due to enhanced mass transport at the point of solution impact, ultra-low dead volume of the order of 0.5–5 µl, and the wide operating potential range of the glassy carbon electrode. It has found application for on-line monitoring of metals, continuous flow anodic stripping voltammetry (See section D) and as a detector for high performance liquid chromatography (EDT Research, 1976).

B. PULSE TECHNIQUES

(i) *Normal pulse polarography*

All of the modern variants of the classical polarographic method have sought to improve sensitivity by eliminating the contribution from the capacitive current due to the double layer charging of the drop. While

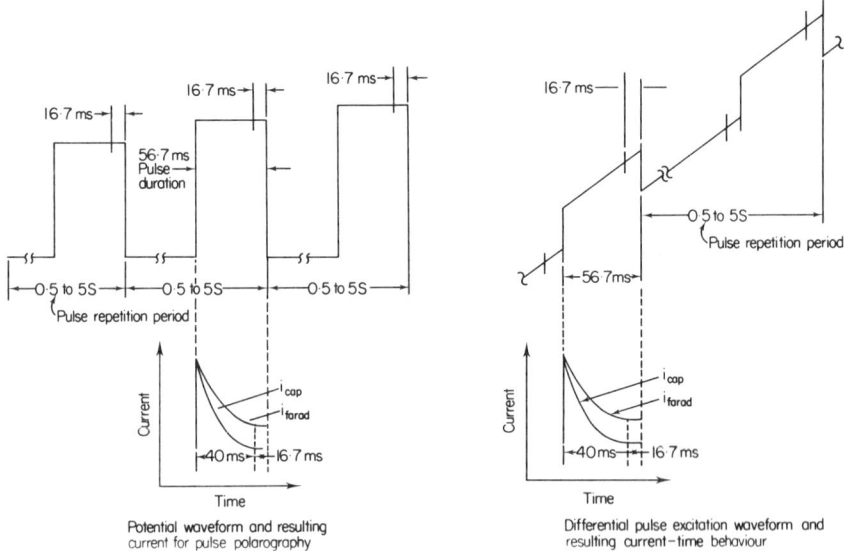

FIG. 7. Excitation waveform and current time behaviour for (a) normal pulse and (b) differential pulse polarography.

there is a vast array of techniques available, only a very few have found routine acceptance. Pulse polarography was invented by Barker and co-workers at Harwell in the late 1950s (Barker and Gardner, 1960). In Barker's original technique a series of increasing amplitude voltage pulses were imposed on successive drops; the current was sampled at the initial potential and after a selected time following the pulse application. Subsequently various modifications to the original technique have been made and nowadays pulse techniques are the most widely used in analytical polarography/voltammetry.

Utilizing the differing time dependences of faradaic and charging currents, the pulse method imposes a series of pulses of increasing amplitude to successive drops at a preselected time near the end of each drop lifetime. The input signal waveform and current response obtained are shown in Fig. 7. The initially high charging current decays away very rapidly and the residual faradaic current is sampled during the final part of the 50–60 millisecond time pulse.

When a potential pulse is applied to an electrode of constant area the capacitive current charges rapidly with time (See Fig. 4). At a stationary electrode, i_c decays as an exponential function

$$i_c = \frac{\Delta E}{R} e^{-(t/RC_{dl})} \tag{7}$$

where ΔE is the pulse amplitude, R is the uncompensated resistance (dependent on cell and solution characteristics), t is the time elapsed since application of the pulse and C_{dl} the double layer capacitance. The faradaic current at a planar electrode, however, is defined by the Cottrell equation which indicates that the current decays as a function of $t^{-\frac{1}{2}}$. By sampling the current for a short period (usually for a period equal to one cycle of the mains supply), when the capacitive component has decayed to a steady state, a favourable measurement of the faradaic current over the capacitative current can be achieved.

It is interesting to compare the diffusion limited currents for pulse and d.c. polarography as follows:

$$\frac{i_{\text{pulse}}}{i_{d.c.}} = \left(\frac{3t_p}{7t_u}\right)^{\frac{1}{2}} \simeq 5\text{--}10 \tag{8}$$

where t_p is the time after application of the pulse and t_d is the drop time in the d.c. mode. While this ratio predicts that normal pulse will only be a factor of 5–10 more sensitive than d.c. polarography, in practice the ratio is nearly two orders of magnitude. The additional increase is due to the ability of the technique to discriminate against the capacitive component.

Pulse polarography is extremely useful in analytical applications since it can respond to both reversible and irreversible processes. A further feature of the pulse technique is that as the measurement pulse is only applied for a small fraction of the total drop time (0.5–5 seconds) only a small amount of material is deposited on the electrode. One implication of this is that the pulse method is far less affected by problems of adsorption than is the d.c. technique.

(ii) Differential pulse polarography

While normal pulse polarography gives a marked improvement in sensitivity over the d.c. method, it still gives a similar S shaped polarogram. A much more useful variant, which is nowadays by far the most widely used polarographic technique for analysis, is differential pulse polarography. In this mode, small amplitude (10–100 mV) pulses of ca 60 m sec duration, superimposed on a conventional d.c. ramp voltage, are applied to the d.m.e. near to the end of the drop lifetime.

The current output is sampled at two time intervals (Fig. 7): immediately on the ramp prior to the imposition of the pulse and then again at the end of the pulse (after 40 milliseconds) when the capacitance current has decayed. It is the difference in these two currents that is displayed. As the greatest increase in current for a given voltage increment will occur at the half-wave potential the i–E curve in d.p.p. will have a peak shape.

The theoretical relationship between the peak current, i_p, and pulse modulation amplitude, ΔE, has been derived by Parry and Osteryoung (1965). The maximum peak current when the pulse modulation amplitude is less than the value of RT/nF is defined as:

$$i_p = \frac{n^2 F^2 A C_0}{4RT} \left(\frac{D}{\pi t}\right)^{\frac{1}{2}} \Delta E \qquad (9)$$

For a differential pulse polarogram with very small values of ΔE then the peak current potential coincides with the half-wave potential, $E_{\frac{1}{2}}$. With larger values of modulation amplitude, however the peak current potential is no longer coincident with $E_{\frac{1}{2}}$ and is given by

$$E_p = E_{\frac{1}{2}} - \Delta E/2 \qquad (10)$$

Consequently it is clear that maximum sensitivity in differential pulse polarography is obtainable for large values of ΔE. Increases in ΔE, however, also result in increased peak broadening with consequent loss of resolution. These definitions only apply to reversible systems; for irreversible systems i_p values are generally lower and peak widths broader than predicted reversible values (Parry and Osteryoung, 1965).

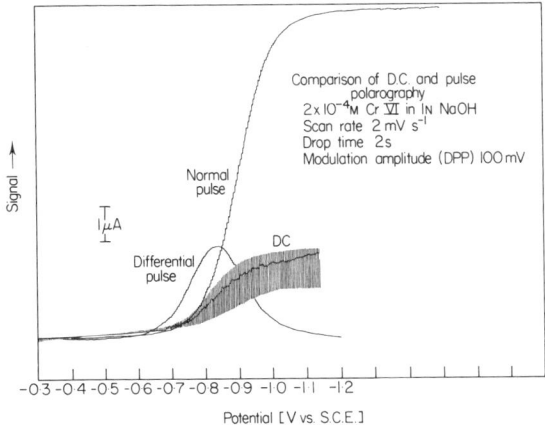

FIG. 8. Comparison of d.c., pulse and differential pulse polarograms.

From a consideration of equation (9) it would appear that differential pulse polarography is inherently less sensitive than the normal pulse procedure but in practice it shows a slightly improved sensitivity due to the better resolution of the current–voltage curves at very low concentrations. Figure 8 shows the pulse response to be larger than the differential pulse or d.c. responses at relatively high concentrations. In contrast, at low levels the favourable signal-to-noise ratio of differential pulse polarography gives well-defined polarograms where no d.c. response is obtained.

(iii) *Other pulse techniques*

One uncompensated source of error in pulse polarography is due to the slight growth of the mercury drop during the current sampling interval. Christie and Osteryoung (1974) and Christie et al. (1976) have devised a method known as "alternate drop pulse polarography" which allows complete compensation of the capacitive background due to drop expansion. In their approach only every other drop is pulsed from the base line potential E_1 to the measurement potential E_2°, the intervening drop remains at E_2 throughout its drop life. The current output is the difference in currents measured at the same time and same potential for pulsed and non-pulsed drops. Although this technique considerably improves the slope of the background current over normal mode pulse it is still not entirely flat and the authors attribute this fact to the influence of capillary noise. This capillary response current results from the slow relaxation of the potential in a solution layer trapped between the mercury thread and the capillary wall and in the cracks in the capillary. It is interesting to note that Smoler (1954) carried out a microscopic examination of a large number of capillaries prepared by breaking a longer length of capillary tube and nearly all were found to have one or more small cracks running radially from the orifice.

Lane and Hubbard (1976) have used a technique described as differential double pulse voltammetry (d.d.p.v.) for the analytical determination of catecholamines in physiological media.

Electro-oxidation of the parent catecholamine yields the open-chain o-quinone (II).

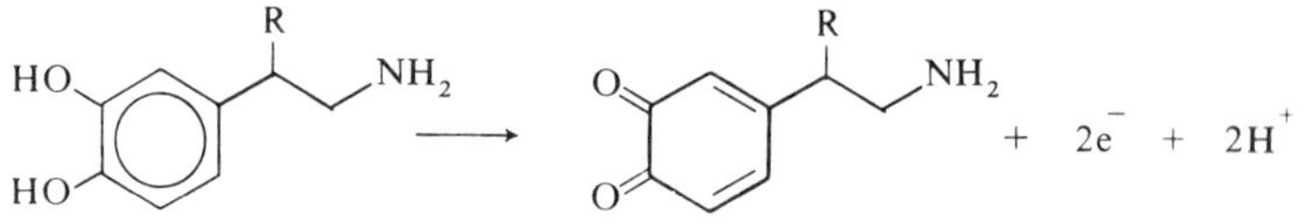

$$\text{(I)} \longrightarrow \text{(II)} + 2e^- + 2H^+$$

$$R = \begin{cases} H, \text{dopamine, DA} \\ OH, \text{norepinephrine, NE} \end{cases}$$

Intracyclization of the o-quinone proceeds within seconds, producing the 5,6-dihydroxyindoline (III).

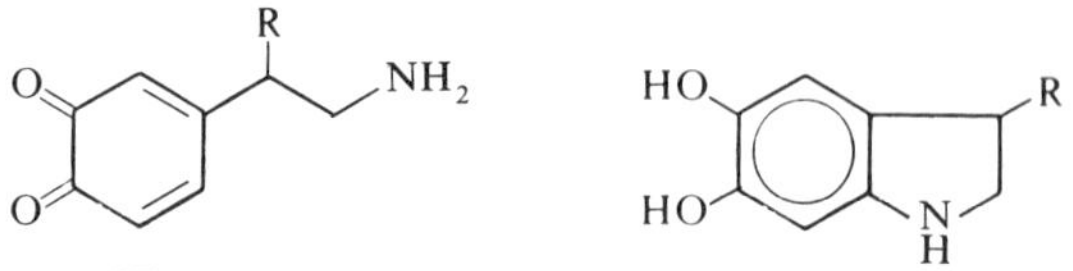

The cyclic intermediate (III) is more easily oxidizable than the parent catecholamine (I) and accordingly is rapidly oxidized by the o-quinone (II) to produce an aminochrome (IV).

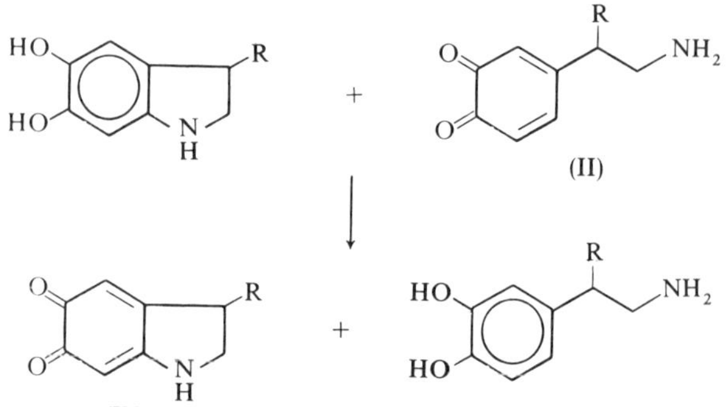

(IV) polymerizes readily to melanin-like products. The rates of these reactions are such that they can occur to an appreciable extent on the time scales employed in the conventional differential pulse experiment. It was concluded that progressive build-up of an insulating film on the electrode surface is a direct consequence of the slowly increasing d.c. ramp inherent in the d.p.p. potential-time waveform. In the operation of the technique two unsymmetrical square wave input signals are employed; the first for measurement of the catecholamine and the second to cause the reduction of the o-quinone back to the parent compound. Each pulse is applied for a comparatively short time (typically $\leqslant$ 20 ms) and since the rates of formation of the 5,6-dihydroxyindoline derivatives of DA and NE at physiological pH are $0.26\,\text{s}^{-1}$ and $0.59\,\text{s}^{-1}$ respectively, it was found that the reduction of the o-quinones back to the parent catecholamines occurred prior to appreciable intracyclization.

Analytical applications. Many analytical applications of pulse and differential pulse polarography have been reported (PAR, 1976b; Brooks et al., 1973). Linear calibration curves have been reported for a large range of compounds with a limit of detection of the order of several ppb (Fleet and Fouzder, 1975a, b, c).

C. A.C. TECHNIQUES

(i) *Fundamental harmonic a.c. techniques*

In fundamental a.c. polarography a small amplitude sinusoidal potential of fixed frequency and amplitude is superimposed on the conventional d.c. ramp applied to the cell (Breyer and Gutmann, 1946; Breyer and Bauer, 1963). The current flowing through the cell will contain both direct current and alternating current components but the plot of a.c. current versus applied potential will give a symmetrical peak shaped response (Fig. 9). This response arises because an alternating faradaic current can only occur when the small amplitude perturbation in the applied signal can cause a significant change in the ratio of the interfacial concentration of the reduced and oxidized form of the electroactive species.

Fundamental harmonic a.c. polarography has a similar sensitivity to d.c. polarography and would have little utility as a technique if it were not for the possibility of using various sophisticated electronic techniques to minimize charging current contributions. Generally these methods are based on the differences in the phase angle between charging current and faradaic current. The capacitive current shows a phase angle of 90° relative to the applied sinusoidal potential while the faradaic current has a phase angle approaching 45° for a diffusion controlled process and approaching 0° for charge transfer control. By employing phase sensitive detectors one can effectively discriminate against the undesired charging current.

The full capabilities of the a.c. method can therefore only be realized when both the amplitude and phase of the cell current are measured. While direct phase angle measurements of electrode processes are of considerable mechanistic and analytical importance (McAllister and Dryhurst, 1972), the primary analytical aim is rejection of the 90° or quadrative component of the current. Several instrumental approaches have been employed for the solution of this problem; operational amplifier circuits for phase selective sampling have been reviewed by Smith (1966) while lock-in amplifier systems have been reviewed by PAR (1975), and Bard and Flego (1975). Smith (1972) has also described the use of on-line digital computers for a.c. measurements including phase sensitive detection.

The use of higher harmonics and intermodulation techniques have also been investigated (Smith, 1966; Neeb, 1965; Jee, 1971; Bond, 1972).

(ii) *Tensammetric methods*

This group of methods are based on non-faradaic admittance and depend on the fact that capacity of the electrical double layer is a function of the concentrations and nature of all electroactive and electroinactive species in the solution. While most ions have relatively little influence on the double layer, certain species, such as surface active agents and some large organic molecules which are preferentially adsorbed at the mercury interface, can markedly affect the double layer capacity. The C_{dl} is much lower in the presence of the surfactant than under normal conditions since the large surfactant ions exclude the normal electrolyte ions from the interface region. At sufficiently large positive or negative potentials the normal electrolyte forces between electrode and electrolyte anions or cations become so strong that the adsorbed species is repelled from the electrode and a sudden change in the double layer capacity occurs. These so called "tensammetric" waves are useful in that they offer the possibility of determining many compounds which are not normally electroactive.

Analytical applications. The analytical capabilities of a.c. polarography have been examined by a number of workers, notably Bond and Canterford (1971a, b, 1972; Bond, 1972) and Fleet and Jee (1976) and their findings may be summarized as follows:

(a) The technique can detect reversible reductions in the presence of irreversible ones. This has application in organic analysis in aprotic media where normally irreversibly reduced molecules can be reversibly reduced e.g., to radical anions (Woodson and Smith, 1970). Fleet and Jee (1969) have determined olefins in protogenic media (aqueous methanolic solvent) via reduction of the Hg(II) olefin addition compounds using a.c. polarography.
(b) The effect of oxygen on the peak current of a more negatively reduced compound can only be ignored in highly acidic media (e.g., 5M HCl).

(c) Sensitivity of a.c. polarography is less than pulse and differential pulse methods.
(d) Selectivity of the technique is comparable if not better than other voltammetric methods.
(e) Tensammetric methods are of limited use for the identification and determination of organic molecules.

D. STRIPPING VOLTAMMETRY

Inherently one of the most sensitive analytical techniques, stripping voltammetry has received a great deal of attention in recent years (Barendrecht, 1967; Vydra *et al.*, 1976; Kissinger, 1976; PAR, 1974b) especially for environmental monitoring. It is extremely simple in concept, the method consisting of a two stage process; the first, a pre-concentration step, consists of controlled deposition of the species of interest onto a stationary electrode. This is followed by the measurement step which consists of electrolytically stripping the deposited species back into solution. This is illustrated in Fig. 9 with the input waveforms and output signals of other techniques, already discussed. In practical terms the pre-electrolysis step involves either the use of a stirred solution or a rotated electrode under carefully controlled hydrodynamic and electrolytic conditions while the stripping step involves the imposition of a potential step or any of the conventional polarographic input waveforms such as pulse, a.c. or voltage ramp. The magnitude of the stripping signal can be related to the concentration of the species of interest.

The pre-electrolysis step. A variety of techniques have been employed for the deposition step. The most popular has been based on the use of a stationary electrode in a stirred solution although rotated electrodes, continuous flow tubular or wall-jet electrodes, packed particle bed electrodes and several other designs have been described.

For deposition with a stationary electrode in a stirred solution it is essential that the hydrodynamic parameters be rigidly controlled. In practice this means a controlled stirring rate, electrolysis time and reproducible location of the working electrode in the electrolysis cell. Mathematical treatment has shown that the mercury film electrode is capable of more sensitive measurements than the hanging mercury drop electrode (h.m.d.e.) under a given set of conditions. In fact, it is recommended that for trace analysis at levels down to ca 1–10 pbb the h.m.d.e. be used and that the m.f.e. is used for levels below this.

The stripping step. After pre-electrolysis the accumulated species must be removed from the electrode in the stripping step. A variety of electrochemical and chemical procedures have been used for this step. The most common

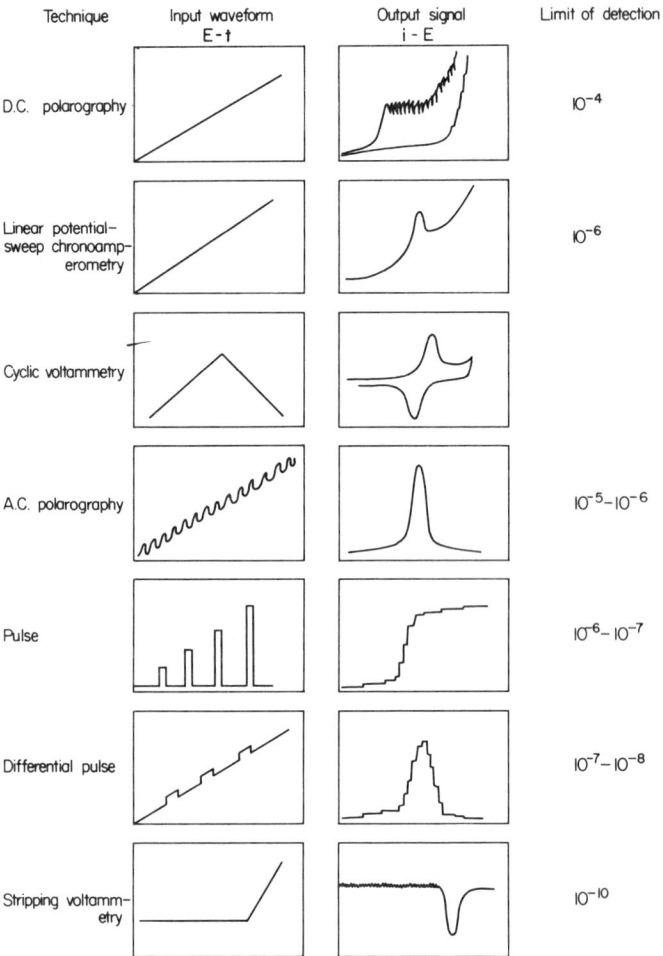

FIG. 9. Summary of signal input waveforms and current-potential response for a range of voltammetric techniques.

has been the use of a linear voltage sweep, i.e. the technique of anodic stripping voltammetry (a.s.v.) generally refers to stripping with a fast (ca 100–200 mV s^{-1}) voltage sweep from a cathodic to an anodic potential. More recently the differential pulse technique has become more popular for the stripping process since it offers a much more effective discrimination against the background charging current and hence an increased sensitivity (Copeland et al., 1973).

Analytical applications of anodic stripping voltammetry (a.s.v.). The main application of stripping voltammetry to date has been in the anodic stripping of trace metals, particularly toxic heavy metals such as lead, cadmium, zinc and mercury in a variety of sample media from natural waters to biological fluids such as whole blood (Siegerman and O'Dom 1972). Figure 10 illustrates such an example. An extensive bibliography can be found in Vydra *et al.* (1976). Berger and Fleet (1977) have automated a.s.v. using the wall-jet cell and the co-deposition technique of Florence (1970). Because of the high efficiency of the thin-layer cell the times required for deposition are very short. For similar pre-electrolysis times (ca 1 minute) with differential pulse stripping, the signal obtained was some 30 times larger than for the comparable manual procedure. Using a simple digital timing system to override the manual controls on the Princeton 174A polarograph, between 15 and 30 samples per hour can be analysed.

The technique can also be applied to stripping of organometallics as free radicals, e.g. triphenyltin compounds which form radicals which can be stabilized by adsorption on mercury (Booth and Fleet, 1970a, b). Applications to other organic molecules are discussed in Chapter 1.

Cathodic stripping. This is a technique where the pre-electrolysis is carried out at anodic potentials and the stripping step consists of a cathodic voltage scan. This method has been applied to the determination of halides as anions (Kemula and Kublic 1963), organics which form insoluble mercury derivatives such as the dithiocarbamate pesticides (Brand and Fleet, 1967, 1968) and more recently to a range of organosulphur drugs (Davidson and Smyth, 1977). Metal ions such as Mn^{2+} and Pb^{2+} can also be determined

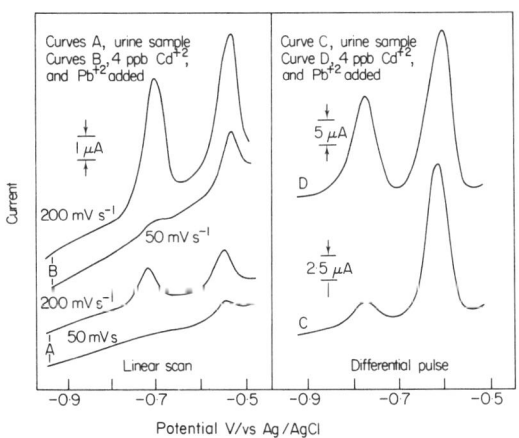

FIG. 10. Linear scan and differential pulse anodic stripping of cadmium and lead in urine.

via cathodic stripping of the respective oxides on either a carbon paste or rotating platinum electrode (Hrabankova *et al.*, 1965; Monnien and Zinke, 1968).

E. COULOMETRIC TECHNIQUES

Coulometric techniques nowadays play an important role as analytical techniques in their own right; their use in providing supporting mechanistic evidence for polarographic and voltammetric studies covers only a small fraction of their total range of applications.

(i) *Controlled current coulometry*

In this technique a constant cell current is used to generate a reagent in a well stirred solution which will stoichiometrically titrate the species to be determined. Some suitable means of monitoring the titration end point such as amperometry, potentiometry or colorimetry must be employed. The basic experimental assembly for the technique consists of a galvanostat, constant current source and a timer. The galvanostat is simply a high-gain, high current operational amplifier with the two-electrode coulometry cell placed in its feedback loop. The constant current derived from a precision constant current source causes a current of equal magnitude but opposite polarity to flow through the cell to the summing point of the amplifier. The timer scaler is usually a simple mains frequency operated device using a clock or electromechanical scaler. The overall precision of galvanostat/timer units approaches the 0.001% level.

A detailed discussion of the controlled current technique is found in Milner and Phillips (1967) and Janata and Mark (1969). The principal advantages of the technique over the conventional titrimetric procedures lie in its high precision, ability to titrate microgram samples and to electrogenerate unstable or uncommon titrants such as Cr(II), Cu(I), Ag(II), Cl_2, Br_2, BrCl, Ti(III) etc., and the relative ease of automation of the technique.

(ii) *Controlled potential coulometry*

In this technique the potential of the working electrode is controlled at a suitable potential on the plateau of the limiting current of the sample species of interest. The cell current decays exponentially with time to a constant background level. The instrumental requirements for this technique comprise a potentiostat for controlling the potential and an integrator for measuring the current.

The first practical potentiostat was designed by Hickling (1942) and pioneer studies in the development of the principle as an analytical technique were carried out by Lingane (1945). However it was not until the development of the amplifier based potentiostat–current integrator by Booman (1957) that the method achieved any real status. Developments and improvements in the technique resulted from Harrar and Shain's work (1966) on potential gradients in controlled potential electrolysis cells and with the introduction of commercial potentiostat/coulometry systems in the 1970s (Siegerman et al., 1975).

The instrumentation required for controlled potential coulometry is schematically shown in Fig. 11. Current integration may be carried out in several ways; via analogue integration based on operational amplifiers, with voltage to frequency converters followed by a counter or, as in some

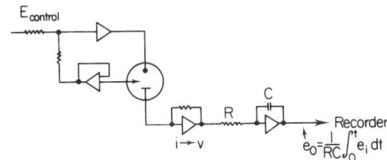

FIG. 11. Schematic circuit of potentiostat-integrator for controlled potential coulometry.

early studies, with a chemical coulometer. Most commercial systems are based on V–F converters. Detailed reviews of the technique including the requirements for cell design and potentiostat control have been given by Harrar (1975) and by Bard (1971).

Coulometric methods for mechanistic studies fall into two groups. Firstly, there are the absolute methods where the coulometric measurement is carried out at the microelectrode of interest, for example, at the dropping mercury electrode (microcoulometry) (Gilbert et al., 1951; Bogan et al., 1951; Manousek 1965*). In the second group of macro coulometric procedures the measurement is carried out at a large area electrode, such as the mercury pool (macro coulometry) and the evidence in terms of observed n values and identification of products is used as additional supporting evidence for the proposed mechanism of the electrode process of interest. In the latter approach the mechanisms of complex reactions at a large area electrode are often quite different from analogous results obtained at a microelectrode due to the differing concentrations of reactive intermediates at the electrode surface using the two methods.

---

* Private communication.

F. APPLICATION OF COMPUTERS IN ELECTROCHEMICAL INSTRUMENTATION

(i) *Minicomputer based systems*

The most significant trend in recent years, as in other areas of analytical instrumentation, has been the development of digital and computerized polarographs.

In the digital electrochemical system developed by Clem and Goldsworthy (1971) all the electrochemical data after the analogue to digital conversion is stored in a pulse height analyser. Since the results from sequential scans are held in different memory locations subtractive polarography is carried out simply by transferring data between memory locations and therefore allowing the same capillary to be used for both sample and blank scans. In the linear sweep mode the sensitivity of the technique is increased from 10^{-7}M to 5×10^{-9}M. The system can also be used for controlled current and controlled potential coulometry. The same authors (Clem and Goldsworthy 1971) have also described an interesting design of digital potentiostat which employs digital current feedback.

On-line computer systems. Off-line treatment of electrochemical data for complex calculations of thermodynamics, kinetics, etc. has long been used but of more direct interest is the use of on-line computer control of electrochemical systems (Perone and Jones, 1973; Mattson *et al.*, 1972, 1973).

The most extensive work in this area has been carried out by Perone and co-workers. In their initial studies on fast sweep derivative polarography (Perone *et al.*, 1968) the computer was programmed to locate and measure the peak heights on the current voltage curve, as well as to indicate the noise level and possible presence of broad peaks. By using this system it was possible to obtain reliable analytical data over a concentration range of 3–5 orders of magnitude. In an extension of this work (Perone *et al.*, 1971) the computer was programmed to carry out data acquisition and processing of fast sweep voltammograms but in addition the operator could oversee data processing functions by graphical communication with the computer through an oscilloscopic display terminal. This made it possible for the operator to select appropriate data regions to be analysed, e.g. to select the regions of a fast sweep polarogram to be used to calculate the extrapolated base line or the part of a peak to be used for calculating its contribution to a subsequent peak. The system was evaluated for the data processing of a multi-component fast-sweep polarogram.

Problems often arise in the polarographic analysis of mixtures because the reduction signal from one component may distort or conceal the reduction signal of another. In an interesting study, Perone *et al.* (1969) devised an interrupted voltage-sweep technique in order to minimize the interference caused by the reduction current of a more positively reduced species on a more negatively reduced one. By interrupting the sweep after each reduction step the more positively reduced species is effectively removed from the diffusion layer. The on-line computer, acting in real time, senses the reduction peak, interrupts the voltage sweep, computes from the peak height the time required to allow sufficient depletion of the electroactive species in the diffusion layer and then restarts the voltage sweep. Quantitative determinations of up to 1000:1 mixtures with widely separated peaks were reported to be successful but the computer-optimized approach failed for peaks separated by less than 150 mV. As the dedicated use of a minicomputer is not economically feasible a hardware device based on the earlier study has also been developed (Jones and Perone, 1970).

For the resolution of mixtures with peak separations of less than 150 mV Gutnecht and Perone (1970) have described a numerical deconvolution procedure using an on-line computer. An empirical equation was first developed to describe the shape of the stationary-electrode polarograms and a standard set of constants obtained for each ion by measurements on pure solutions. When a mixture was analysed these standard constants were used to regenerate the polarograms which were then fitted to the signal obtained from the unknown mixture. This approach enabled quantitative resolution of overlapping peaks of similar magnitude with peak separations of less than 40 mV. Peak overlap however sometimes causes difficulties in the qualitative resolution of multiple peak polarograms. A pattern recognition procedure (Sybrandt and Perone, 1971) has been applied to this problem. After studying the effects of concentration ratios, degree of peak overlap and peak potential variation these authors concluded that ideally overlapping polarograms with a 2 mV peak separation and a 20 fold ratio of concentration could be resolved but in practice the precision of the experimental data was the limiting factor.

Computer optimization of experimental parameters. The use of a digital computer to optimize experimental parameters for electroanalytical measurements has been described by Thomas *et al.* (1976). In normal experimental situations the approach to optimization has been based on variation of individual parameters and monitoring the observed signal change. Using the technique of linear sweep anodic stripping voltammetry as a model the approach employed was to first specify the desired performance criteria and allow the computer to calculate the value of the various experimental parameters required to meet these criteria. The parameters chosen for optimization were the signal-to-noise ratio and the signal-to-background (capacitive) ratio. These parameters were defined as follows. The signal S is defined by

$$S = l_p = KC_0^* t_p V^{\frac{1}{2}} \tag{11}$$

where C_0^* is the bulk concentration of electroactive species of interest, t_p is the electrolysis (plating) time, V is the sweep rate and K a hydrodynamic constant. The noise level, N, is defined as $4.0 \times$ standard deviation of the imperturbed signal, σ. The background signal, B, due primarily to charging of the double layer but also including terms due to reduction or oxidation of residual impurities is defined as

$$B = \phi V + \beta V^{\frac{1}{2}} + \delta \tag{12}$$

where the experimentally determined constants, ϕ, β and δ can be determined by measuring B at varying scan rates.

The signal-to-noise and signal-to-background ratios are defined by the terms m and b respectively, where

$$m = \frac{S}{N} = \frac{i_p}{4\sigma} \tag{13}$$

and

$$b = \frac{S}{B - \delta} = \frac{i_p}{B - \delta} \tag{14}$$

It was further shown that the optimum scan rate, V_2 and plating time, t_{opt}, could be obtained from

$$V_2 = \left(\frac{-\beta + \left[\beta^2 + 4\phi \left(\frac{S_1 m_2}{b_2 m_1} \right) \right]^{\frac{1}{2}}}{2\phi} \right)^2 \tag{15}$$

and

$$t_{opt} = t_p \cdot \frac{m_2}{m_1} \left(\frac{V_1}{V_2} \right)^{\frac{1}{2}} \tag{16}$$

where m_2 and b_2 are the user-specified performance criteria, and m_1, S_1, t_p, and V_1 are experimentally measured or definable quantities.

Experimentally the approach is first to determine the values of m_1, ϕ, β and δ for a given solution by carrying out a series of runs at various scan rates. Then, using equations, the computer calculates the optimum values and proceeds to perform the optimized experiment. The entire operation is carried out under computer control with only minimal user interaction.

The basic principles for the use of on-line computer control of electrochemical systems is now well established and detailed experimental information on the use of such systems is readily available. Thomas et al. (1976) have discussed the computer hardware/software and interface requirements for anodic stripping and Kryger et al. (1975a, b) and Jagner and Kryger (1975) have also described a computerized electroanalysis system for anodic stripping and its application to the analysis of trace metals in sea water; computer controlled multiple electrode monitoring has also been described by Zipper et al. (1974).

(ii) *Microprocessor based systems*

Within a very few years it will be common practice to find in-board microprocessors or microcontrollers in most analytical instruments. The truly astonishing rate of progress in the development of these devices with its concurrent lowering of costs will mean that these devices will be commonplace not only in multiparameter sophisticated instruments such as atomic absorption spectrophotometers, gas chromatographs, etc., but also in the low-cost dedicated single element analysers such as blood, lead or glucose monitors as well as in automatic titrators.

The first example of the new generation of electroanalytical instruments has already appeared. In 1976 Princeton Applied Research (PAR, 1976a) introduced the model 374 microprocessor based polarograph. This system, based on a National Semiconductor microprocessor provides two significant advances in polarograph design. Firstly the microprocessor controller unit is used to control the electrochemical experiment, to select the operational technique, apply the input signal and acquire and interpret data. At the same time it has solved one of the main problem areas of classical polarography by redesigning and re-engineering the dropping mercury electrode assembly. The d.m.e. and its associated mercury reservoir has long been a source of problems, being susceptible to mechanical vibrations and also acting as an excellent antennae for stray electrical and r.f. noise. The electrode assembly on the Model 374 consists of a gas pressure operated d.m.e. with a short (0.1 second) drop time, which can also function as a hanging mercury drop electrode. The capillary has a small reservoir, which can be sealed off from the main mercury compartment and the mercury drops are extruded by a motor driven piston.

The data processing system on the 374 consists of 2000 words of random access memory (RMA) of which 1500 are used to store two 750 point polarographic curves, the remaining 500 being used for control and data processing functions. The operation of the instrument is an interactive sequence, with the user selecting the technique and the various experimental parameters. After each data entry the controller checks for validity before accepting the data. The controller can also monitor the status of the electrode/cell assembly and identify faults such as absence of mercury, dry reference electrode, etc. A sliding point data averaging routine is used, with each point on the polarogram being checked for validity before entering into memory file and the applied potential incremented to the value required for the next data point. The final polarograph readout is automatically attenuated and peaks identified in terms of half-wave potential, peak current in microamps and p.p.m. concentrations reported. With this approach improvements in sensitivity over conventional analogue instruments are due partly to the use of signal averaging and also due to the more favourable discrimination against background charging current. The method employed to eliminate the d.c. background current is to sample the background current prior to the current sample measurement, invert this value and use it to zero the current voltage converter. This allows the full dynamic range of the i–E converter to be used for each measurement. In the differential pulse mode of operation, for example, metal ions such as Cd(II) and Pb(II) can be determined down to the 10 p.p.b. level, an improvement of 10–15 × over conventional analogue instruments. As this very elegant instrument has shown, improvement in sensitivity and accuracy are only two of the features which can be achieved by using dedicated computer control. The use of the computer to monitor the experimental system and "intelligently" modify the course of the experiment to provide optimum measurement conditions is a far more valuable and exciting prospect.

G. CONCLUDING REMARKS

During the last ten years the widespread acceptance of polarographic and voltammetric techniques as analytical tools have largely resulted from improvements in instrumentation. With the introduction of the microprocessor controlled multi-mode polarograph this development might be considered to have reached its plateau and most of the most promising recent advances have been in the applications area. However, novel techniques still continue to appear. Barker (1966; Barker et al., 1973), who has undoubtedly made by far the most valuable contribution to the development of modern polarographic techniques, has recently introduced a multi mode polarograph. In this system four different polarographic techniques are provided— square wave polarography, radio-frequency polarography, linear-scan voltammetry and a new technique known as "amplitude modulation polarography". The latter technique measures the amplitude modulation of the high frequency component of the voltage developed across the polarographic cell when a low frequency voltage and high frequency current are applied simultaneously to the working electrode. Selection of these techniques was made on the grounds that they are all selective to varying degrees of reversibility.

REFERENCES

Adams, R. N. (1969). "Electrochemistry at Solid Electrodes". Marcel Dekker, New York.
Alder, J. F., Fleet, B. and Kane, P. O. (1971). *J. Electroanal. Chem.* **30**, 427.
Bard, A. J. (1971). *Pure Appl. Chem.* **25**, 379.
Bard, A. J. and Flego, U. S. (1975). *Anal. Chem.* **47**, 2321.
Barendrecht, E. (1967). *In* "Electroanalytical Chemistry (A. J. Bard, ed.), Vol. 2, p. 53. Marcel Dekker, New York.
Barker, G. C. (1966). *In* "Polarography 1964" (G. J. Hills, ed.). Macmillan, London.
Barker, G. C. and Gardener, A. W. (1960). *Z. Anal. Chem.* **173**, 79.
Barker, G. C., Gardener, A. W. and Williams, M. J. (1973). *J. Electroanal. Chem.* **42**, 21.
Beilby, A. L. (1964). *Anal. Chem.* **36**, 22.
Berger, T. A. and Fleet, B. (1977). *Anal. Chem.* (in press).
Bieder, A. and Brunel, P. (1971). *Ann. Pharm. Franc.* **29**, 461.
Blaedel, W. J. and Boyer, S. L. (1971). *Anal. Chem.* **43**, 1538.
Blaedel, W. J. and Jenkins, R. A. (1974). *Anal. Chem.* **46**, 1952.
Bogan, S., Meites, L., Peters, E. and Sturtevant, J. M. (1951). *J. Amer. Chem. Soc.* **73**, 1584.
Bond, A. M. (1972). *J. Electroanal. Chem.* **36**, 235.
Bond, A. M. and Canterford, J. H. (1971a). *Anal. Chem.* **43**, 228.
Bond, A. M. and Canterford, J. H. (1971b). *Anal. Chem.* **43**, 393.
Bond, A. M. and Canterford, J. H. (1972). *Anal. Chem.* **44**, 741.
Booman, G. L. (1957). *Anal. Chem.* **29**, 213.
Booth, M. D. and Fleet, B. (1970a). *Anal. Chem.* **42**, 825.
Booth, M. D. and Fleet, B. (1970b). *Talanta*, **17**, 491.
Brand, M. J. D. and Fleet, B. (1967). *J. Polarogr. Soc.* **13**, 77.
Brand, M. J. D. and Fleet, B. (1968). *Analyst*, **93**, 498.
Breyer, B. and Bauer, H. H. (1963). "Alternating Current Polarography and Tensammetry". Interscience, New York.
Breyer, B. and Gutmann, F. (1946). *Trans. Faraday Soc.* **42**, 650.
Brezina, M. and Volke, J. (1975). *In* "Progress in Medicinal Chemistry" (Ellis and West, eds.), Vol. 12. North-Holland, Amsterdam.
Brooks, M. A. and Hackman, M. R. (1975). *Anal. Chem.* **47**, 2059.
Brooks, M. A., de Silva, J. A. F. and Hackman, M. R. (1973). *Amer Lab.* **10**, 126.
Christie, J. H. and Osteryoung, R. A. (1974). *Anal. Chem.* **49**, 301.
Christie, J. H., Jackson, L. L. and Osteryoung, R. A. (1976). *Anal. Chem.* **48**, 242.
Clem, R. G. and Goldsworthy, W. W. (1971). *Anal. Chem.* **43**, 1718.
Clem, R. G. and Sciamanna, A. F. (1975). *Anal. Chem.* **47**, 276.
Clem, R. G., Litton, G. and Ornelas, L. D. (1973). *Anal. Chem.* **45**, 1306.
Copeland, T. R., Christie, J. H., Osteryoung, R. A. and Skogerboe, R. K. (1973). *Anal. Chem.* **45**, 995.
Davidson, I. E. and Franklin Smyth, W. (1977). *Anal. Chem.* **49**, 1195.
Delahay, P. (1954). "New Instrumental Methods in Electrochemistry". Interscience, New York.
EDT Research (1976). LCA-10 Detector. (Address: 65 Ivy Crescent, London W.4).
ESA (1974). Environmental Sciences Associates Inc., 175 Bedford Street, Burlington, Massachusetts 01803 (see also *Interface*, **12**, Nos 4, 5, 1974).
Fleet, B. and Fouzder, N. B. (1975a). *J. Electroanal. Chem.* **63**, 59.
Fleet, B. and Fouzder, N. B. (1975b). *J. Electroanal. Chem.* **63**, 69.
Fleet, B. and Fouzder, N. B. (1975c). *J. Electroanal. Chem.* **63**, 79.
Fleet, B. and das Gupta, S. (1976). *Nature*, **263**, 122.

Fleet, B. and Jee, R. D. (1969). *Talanta*, 16, 1561.
Fleet, B. and Jee, R. D. (1976). *In* "Selected Annual Reviews of the Analytical Sciences" (L. S. Bark, ed.), Vol. 4, p. 1. The Chemical Society, London.
Fleet, B. and Little, C. J. (1974). *J. Chromatog. Sci.* 12, 747.
Fleet, B., Ho, A. Y. W. and Tenygl, J. (1972a). *Talanta*, 19, 317.
Feet, B., Ho, A. Y. W. and Tenygl, J. (1972b). *Anal. Chem.* 44, 2157.
Fleet, B., Gunasingham, H., de Damia, G., Berger, T. A., das Gupta, S. and Little, C. J. (1977). "Proceedings of Int. Conf. on Physical Chemistry and Hydrodynamics". Hemisphere, Washington, D.C.
Fleischman, M., Korinkek, K. and Pletcher, D. (1971). *J. Electroanal. Chem.* 31, 39.
Florence, T. M. (1970). *J. Electroanal. Chem.* 27, 273.
Florence, T. M. and Farrar, Y. J. (1973). *J. Electroanal. Chem.* 41, 127.
Gilbert, G. A. and Rideal, E. K. (1951). *Trans. Farad. Soc.* 47, 396.
Gruendler, P. and Choschzich, H. (1972). *Z. Chem. Lpz.* 12, 274.
Gutnecht, W. F. and Perone, S. P. (1970). *Anal. Chem.* 42, 906.
Harrar, J. E. (1975). *In* "Electroanalytical Chemistry" (A. J. Bard, ed.), Vol. 8, p. 1. Marcel Dekker, New York.
Harrar, J. E. and Shain, I. (1966). *Anal. Chem.* 38, 1148.
Hasebe, K. and Osteryoung, J. G. (1975). *Anal. Chem.* 47, 2412.
Heyrovsky, J. and Kuta, J. (1965). "Principles of Polarography". Publishing House of Czechoslovak Academy of Sciences, Prague.
Heyrovsky, J. and Zuman, P. (1968). "Practical Polarography". Academic Press, London.
Hickling, A. (1942). *Trans. Faraday Soc.* 38, 27.
Hume, D. N. and Carter, J. N. (1972). *Chimia Analityezna*, 17, 747.
Hrabankova, E., Dolezal, J. and Masin, V. (1965). *J. Electroanal. Chem.* 22, 195.
Interface (1974). *Electrochemistry Newsletter*, 12, Nos 4, 5. Chemistry Department, Purdue University, Lafayette, Indiana.
Iversen, P. and Lund, H. (1969). *Anal. Chem.* 41, 1322.
Jagner, D. and Kryger, L. (1975). *Anal. Chim. Acta.* 80, 255.
Janata, J. and Mark, H. B., Jr (1967). *Anal. Chem.* 39, 1896.
Janata, J. and Mark H. B. Jr (1969). *In* "Electroanalytical Chemistry" (A. J. Bard, ed.), Vol. 3, p. 1. Marcel Dekker, New York.
Jee, R. D. (1971). Ph.D. Thesis, University of London.
Jones, D. G. and Perone, S. P. (1970), *Anal. Chem.* 42, 1151.
Kemula, W. and Kublik, Z. (1963). *Adv. Anal. Chem. Instr.* 2, 123. Interscience, New York.
Kissinger, P. T. (1976). *Anal. Chem.* 48, 17R.
Kryger, L. and Jagner, D. (1975). *Anal. Chim. Acta.* 78, 251.
Kryger, L., Jagner, D. and Skov, H. J. (1975) *Anal. Chim. Acta* 78, 241.
Lane, R. F. and Hubbard, A. T. (1976). *Anal. Chem.* 48, 1287.
Levich, V. G. (1962). "Physicochemical Hydrodynamics". Prentice Hall, New Jersey.
Lindquist, J. (1973). *Anal. Chem.* 45, 1006.
Lingane, J. J. (1945). *J. Amer. Chem. Soc.* 67, 1976.
Maironovskii, S. G. (1965). *Talanta* 12, 1299.
Mairanovskii, S. G. and Gultyai, V. P. (1965). *Elektrokhimya*, 1, 460.
Mann, C. K. (1969). *In* "Electroanalytical Chemistry" (A. J. Bard, ed.), Vol. 3, p. 57. Marcel Dekker, New York.
Mason, W. D. and Sandmann, B. (1976). *J. Pharm. Sci.* 65, 599.
Matson, W. R., Roe, D. K. and Carritt, D. E. (1965). *Anal. Chem.* 37, 1598.
Mattson, J. S., Mark, H. B. Jr and Macdonald, H. C. (Eds) (1973). "Computers in Chemical Instrumentation", Vol. 2: Electrochemistry—Calculations, Simulation and Instrumentation. Marcel Dekker, New York.

Mattson, J. S., Mark, H. B. Jr and Macdonald, H. C. (eds) (1972). "Computers in Chemical Instrumentation", Vol. 1: Basic Concepts in Computer Chemistry. Marcel Dekker, New York.
McAllister, D. H. and Dryhurst, G. (1972). *Anal. Chim. Acta.* **58**, 273.
Michielli, R. F. and Downing, G. R. (1974). *J. Agr. Food. Chem.* **22**, 449.
Milner, G. W. C. and Phillips, G. (1967). "Coulometry in Analytical Chemistry". Pergamon, New York.
Monien, H. and Zinke, K. (1968). *Z. Anal. Chem.* **240**, 32.
Neeb, R. (1965). *Z. Anal. Chem.* **208**, 168.
Newman, J. (1973). *In* "Electroanalytical Chemistry" (A. J. Bard, ed.), Vol. 6, p. 187. Marcel Dekker, New York.
Novak, J. V. A. (1962). *In* "Progress in Polarography", Vol. 2, p. 569. Interscience, New York.
Oesterling, T. O. and Olson, C. L. (1967a) *Anal. Chem.* **39**, 1543.
Oesterling, T. O. and Olson, C. L. (1967b), *Anal. Chem.* **39**, 1546.
Osteryoung, J. G. and Hasebe, K. (1976). *Rev. Polarog. Japan* **22**, 1.
*PAR (1972). Modern Analytical Polarography Workshop Manual.
PAR (1974a). Technical Note 108: "Why Deaeration and How."
PAR (1974b). Technical Note 109a: "Stripping Voltammetry Some Helpful Techniques".
PAR (1975). Lock-In Amplifier Primer.
PAR (1976a). Model 374 Polarograph, Technical Data.
PAR (1976b). Technical Note 110: "Applications Bibliography".
Parry, E. P. and Osteryoung, R. A. (1965). *Anal. Chem.* **37**, 1634.
Perone, S. P., Harrar, J. E., Stephens, F. B. and Anderson, R. E. (1968). *Anal. Chem.* **40**, 899.
Perone, S. P., Jones, D. O. and Gutnecht, W. F. (1969). *Anal. Chem.* **41**, 1154.
Perone, S. P., Frazer, J. W. and Kray, A. (1971). *Anal. Chem.* **43**, 1485.
Perone, S. P. and Jones, D. O. (1973). "Digital Computers in Scientific Instrumentation". McGraw-Hill, New York.
Rooney, R. C. (1962). "Principles and Applications of Cathode-ray Polarography". Southern Analytical *et al.*, Surrey, England.
Ross, S. D., Finkelstein, M. and Rudd, E. J. (1975). "Anodic Oxidation". Academic Press, New York.
Ryan, T. H. Koryta, J., Hofmanova-Matejkova, A. and Brezina, M. (1974). *Anal. Letters*, **7** (5), 335.
Siegerman, H. D. and O'Dom, G. (1972). *Amer. Lab.* June.
Siegerman, H. D. Chang, J. and Thompson, J. (1975). *Amer. Lab.* February.
Smith, D. E. (1966). *In* "Electroanalytical Chemistry" (A. J. Bard, ed.), Vol. 1, p. 1. Marcel Dekker, New York.
Smith, D. E. (1972). *In* "Electrochemistry, Calculations, Simulation and Instrumentation" J. S. Mattson, H. B. Mark Jr and H. C. Macdonald, eds), Vol. 2, p. 369. Marcel Dekker, New York.
Smoler, I. (1954). *Coll. Czech. Chem. Commun.* **19**, 238.
Stock, J. T. (1976). *Anal. Chem.* **48**, 1R.
Stradyn, Y. P., Reikhmanis, G. O. and Gavar, R. A. (1965). *Elektrokhimiya* **1**, 935.
Strafelda, F. and Dolezal, J. (1967). *Coll. Czech. Chem. Commun.* **32**, 2707.
Stulik, K. and Stulikova, M. (1973). *Anal. Letters*, **6**, 441.
Stulikova, M. (1973). *J. Electroanal. Chem.* **48**, 33.
Sybrandt, L. B. and Perone, S. P. (1971). *Anal. Chem.* **43**, 382.
Telupilova, O. and Masinova, V. (1953). *Czechoslov Farmacie*, **2**, 226.
Tenygl, J. (1978). *Talanta* (in press).

* PAR = Princeton Applied Research Corporation, P.O. Box 2565, Princeton, New Jersey, U.S.A.

Thomas, Q. V., Kryger, L. and Perone, S. P. (1976). *Anal. Chem.* **48**, 762.
Vydra, F., Stulik, K. and Julakova, E. (1976). "Electrochemical Stripping Analysis". Ellis Horwood, Chichester, U.K.
Westmorland, P. G., Day, R. A. and Underwood, A. L. (1972). *Anal. Chem.* **44**, 737.
Woodson, A. L. and Smith, D. E. (1970). *Anal. Chem.* **42**, 242.
Yoshimori, T., Arakowa, M. and Takeuchi, T. (1965). *Talanta*, **12**, 147.
Zipper, J., Fleet, B. and Perone, S. P. (1974). *Anal. Chem.* **46**, 2111.
Zittel, H. E. and Miller, F. J. (1965). *Anal. Chem.* **37**, 300.
Zuman, P. (1964). "Organic Polarography", Pergamon Press, Oxford.
Zuman, P. and Perrin, C. L. (1969). "Organic Polarography". Interscience, New York.

APPLICATIONS

In Pharmacy and Pharmacology

Chapter 3

THE POLAROGRAPHIC DETERMINATION OF PSYCHOTROPIC, HYPNOTIC AND SEDATIVE DRUGS

MARVIN A. BROOKS

*Department of Biochemistry and Drug Metabolism,
Hoffmann-La Roche Inc., Nutley, N.J., U.S.A.*

I. INTRODUCTION

The majority of voltammetric methods published have dealt more with psychotropic (which include psychotomimetic, psychosedatives and antidepressants), hypnotic and sedative drugs than with any other group of pharmaceuticals. This chapter will be principally concerned with a review of the voltammetric studies reported in the literature for these classes of drugs and to show by example how voltammetry has been used to solve a wide variety of pharmaceutical problems.

II. PSYCHOTOMIMETIC DRUGS

Atropine (I), an anticholinergic agent, has been examined by d.c. and a.c. polarography using a supporting electrolyte of acetonitrile containing 0.1M tetrabutylammonium perchlorate (Woodson and Smith, 1970). Brooks *et al.* (1974) have determined atropine using differential pulse polarography at the dropping mercury electrode (d.m.e.) after nitration at room temperature with 10% KNO_3 in concentrated sulphuric acid. As little as 200 ng atropine ml^{-1} of 1N NaOH supporting electrolyte can be determined.

Oxyphencyclimine hydrochloride (II) is also an anticholinergic which has atropine-like properties with longer duration of action. The compound is chemically similar to atropine in that it contains a tertiary-amine common to most anticholinergics. Mikolajek (1967) reported the polarographic determination of oxyphencyclimine in 0.1N KCl at the d.m.e. in the concentration range of 12 to 60 μg ml^{-1} of solution, using a cathodic wave, presumably corresponding to the reduction of an azomethine group.

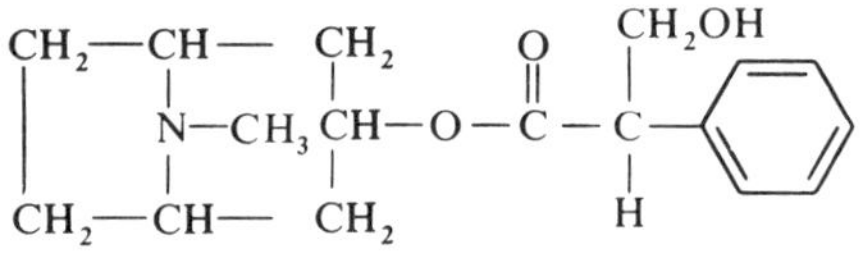

(I) *Atropine*

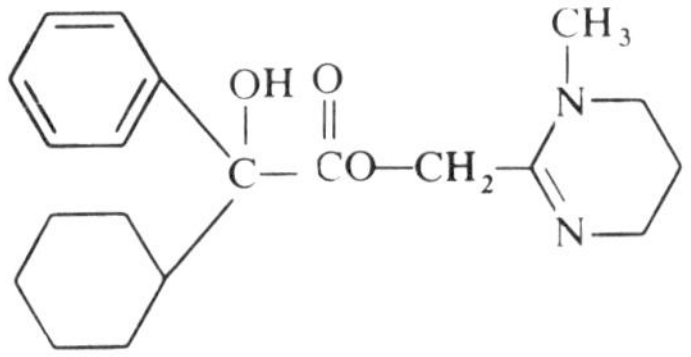

(II) *Oxyphencyclimine*

III. PSYCHOSEDATIVE DRUGS

Antipsychotic drugs and antianxiety drugs are two subdivisions of the class of psychosedative drugs. The antipsychotic drugs consist principally of drugs used to treat schizophrenia, and sometimes classified as major tranquilizers or neuroleptics. The antianxiety drugs are also described as anxiolytics or minor tranquilizers.

A. ANTIPSYCHOTICS

(i) *Phenothiazines*

Chlorpromazine (III) is the most widely used phenothiazine drug. The compound itself is not reducible but has been assayed as its sulphoxide (Porter, 1964; Oelschlager and Bunge, 1974), nitroso (Tur'yan et al., 1970) and nitro derivatives (Dumortier and Patriarche, 1973). The nitration procedure was also applied to N-substituted derivatives of chlorpromazine including promethazine, diethazine and prochlorperazine. The derivatives, which were prepared in mixtures of acetic and nitric acids at 0°C, underwent structural elucidation using u.v., i.r. and n.m.r. spectroscopy and mass spectrometry and showed the presence of two nitro groups in the phenothiazine molecule. These two groups were polarographically reduced in two

3. DETERMINATION OF PSYCHOTROPIC, HYPNOTIC AND SEDATIVE DRUGS

steps to give the corresponding amino compound, consuming 12 electrons (Fig. 1).

The sulphoxide derivative formed by bromine oxidation was used to assay the drug in parenteral, tablet and syrup dosage forms in the range 1–8 µg ml^{-1} (Porter, 1964). Compared to spectrophotometric methods, the procedure is both more sensitive and specific.

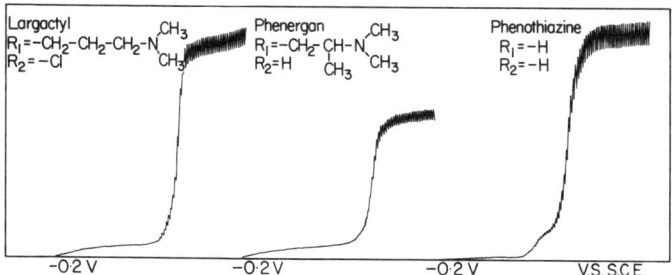

FIG. 1. D.c. polarograms of di-nitro derivatives of chlorpromazine (largactyl), promethazine (phenegran), and phenothiazine in 1% KOH. Reprinted from Dumortier and Patriarche (1973) by courtesy of Springer-Verlag.

The drug and its sulphoxide metabolite have also been analysed in mixtures by determining the sulphoxide prior to oxidation with bromine and both compounds after oxidation (Porter and Beresford, 1966). The concentration of chlorpromazine was then calculated by difference. Porter (1967) has described a procedure employing twin cell substractive linear sweep voltammetry to determine chlorpromazine sulphoxide, the oxidative decomposition product of chlorpromazine. Concentrations of 0.5–2.0 µg of sulphoxide ml^{-1} were determined in the presence of 100 µg of chlorpromazine ml^{-1}. The method was found to be extremely rapid and highly specific, and could be applied also to the determination of the sulphoxides of fluphenazine (IV) and promethazine (V). Recently, Oelschlager and Bunge (1974) have employed oxidation with 2.5% nitric acid to yield a sulphoxide derivative suitable for measurement of chlorpromazine in 25 mg sugar coated tablets. The assay was compared to a u.v. method and was found to be more accurate and of equal precision. These authors studied the mechanism of oxidation of chlorpromazine in nitric acid and examined the d.c. and a.c. polarographic properties of the derivative in Britton–Robinson buffers in the pH range of 1.8–7.0. Constant potential coulometry was used by the authors to demonstrate that the polarographic activity in acidic solution was the result of a two-electron reduction of the sulphoxide to the parent compound.

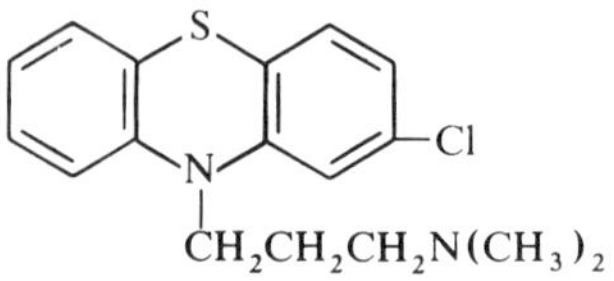

(III) Chlorpromazine

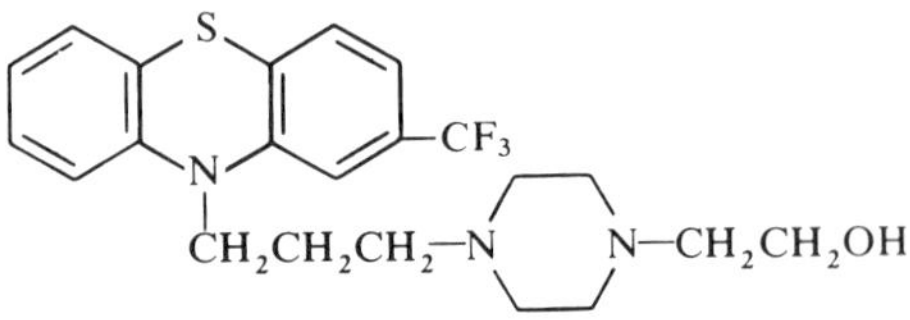

(IV) Fluphenazine

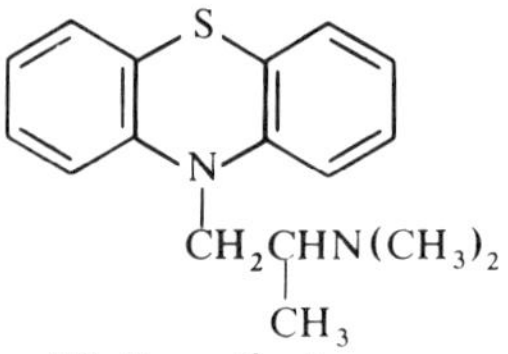

(V) Promethazine

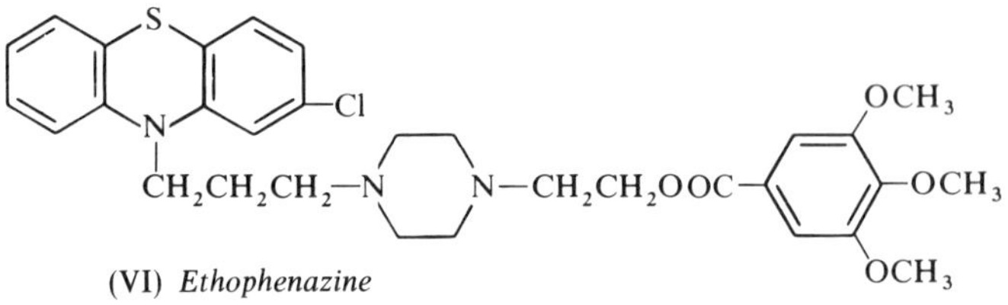

(VI) Ethophenazine

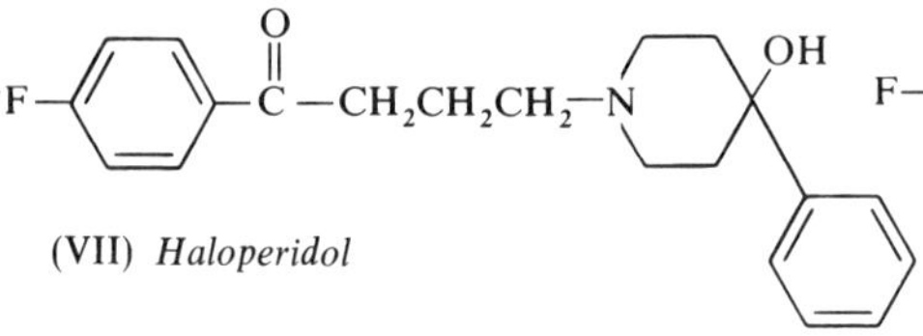

(VII) Haloperidol

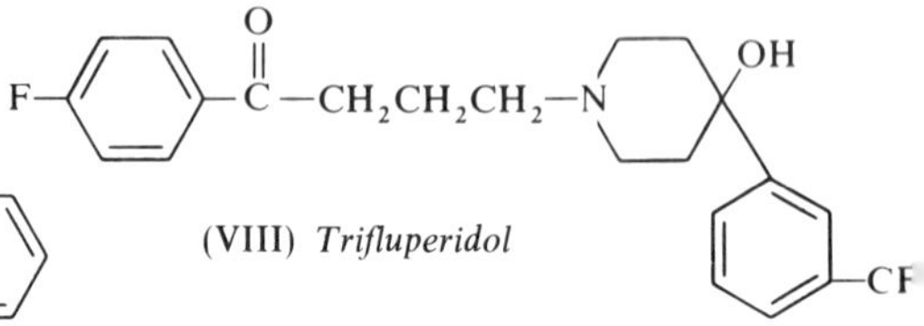

(VIII) Trifluperidol

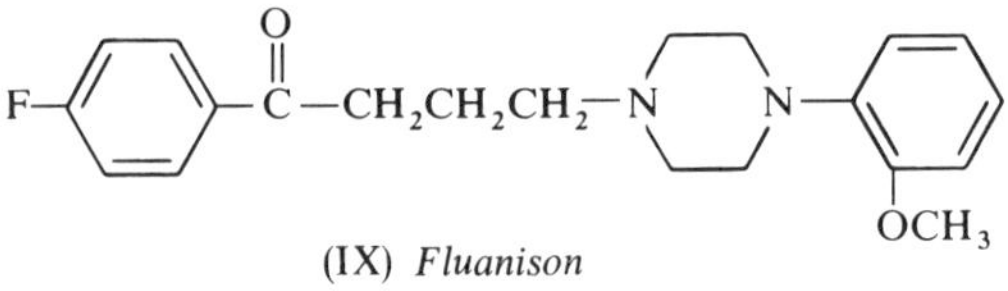

(IX) Fluanison

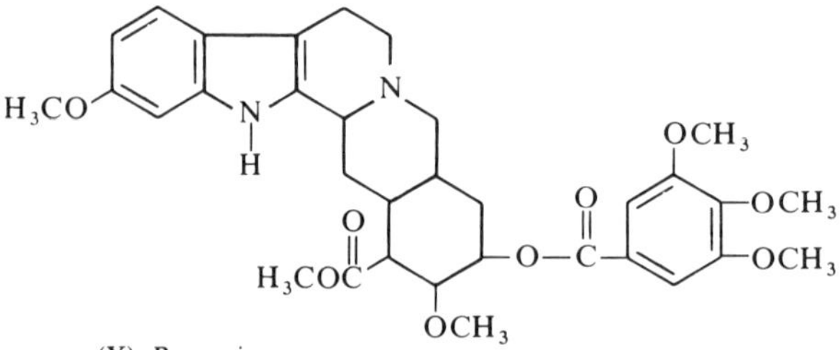

(X) Reserpine

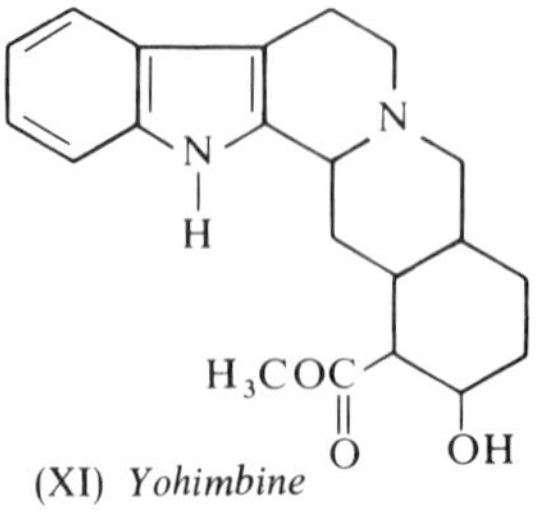

(XI) Yohimbine

The sulphoxide derivative of the compound was also used for the quantitative differentiation of the parent drug and its sulphoxide metabolite in urine from subjects who received an oral dose of 100 mg of chlorpromazine hydrochloride (Porter and Beresford, 1966). The methodology involved removal of interfering naturally occurring constituents in the urine with the use of Amberlite resin IRA-400-chloride column, followed by linear sweep voltammetry. Beckett et al. (1974) have reported a method to measure the N-oxide, N-oxide-sulphoxide and sulphoxide of chlorpromazine in aqueous mixtures and in biological samples using ion-pair extraction, selective solvent extraction, selective reduction and linear sweep voltammetry. The authors reported that the three compounds showed well defined peaks in acidic media (pH 1–6) in the concentration range of 10^{-5} to 5×10^{-8}M.

Kabasakalian and McGlotten (1959) studied the voltammetric behaviour of 14-phenothiazine tranquilizers at the gold electrode and found regular, reproducible, well-defined anodic waves. The effects of pH, concentration, and temperature on the potential and height of the wave were reported. The effects of substituents in the 10- and 2-positions on half-wave potential in aqueous and alcoholic 0.1N sulphuric acid were also reported. Merkle and Discher (1964a, b) observed that electrolytic oxidation of seven phenothiazines at a platinum electrode proceeded in two discreet one-electron steps in 9N sulphuric acid. The first step was the loss of an electron from the compound to produce a cationic free radical which has an intense red colour. The second step is the further oxidation by a one-electron mechanism to give the colourless sulphoxide.

Pungor et al. (1971a, b) examined the voltammetric behaviour of chlorpromazine and methophenazine (VI) at the silicone-rubber based graphite electrode in aqueous and nonaqueous 0.1M KCl supporting electrolytes using an Ag/AgCl reference electrode. Values for the $E_{\frac{1}{2}}$ and the analysis of these drugs in formulations were reported.

(ii) *Butyrophenones*

Substituted butyrophenones have antipsychotic properties very similar to substituted phenothiazines, yet they possess a completely different chemical structure. Haloperidol (VII) is a widely used antipsychotic and was the first of the butyrophenone tranquilizers. Volke et al. (1971) have determined haloperidol and two related *p*-fluoro-substituted butyrophenone derivatives, trifluperidol (VIII) and fluanisol (IX) by direct current and oscillographic polarography. The polarographic activity is due to the reduction of the carbonyl group, common to all three of these compounds. Concentrations from 10^{-3}M to 10^{-5}M of each of the compounds were determined in 0.25N KOH in 50% ethanol, and in pH 5.3 acetate buffer in 50% DMF (Fig. 2). Mikolajek et al. (1974) have recently described a rapid and simple

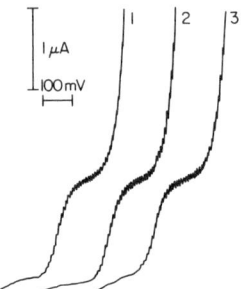

FIG. 2. D.c. polarographic scan from -1.1 V to -1.4 V vs s.c.e. for 5×10^{-4} (1) Trifluperidol (2) Fluanison and (3) Haloperidol in pH 5.3 buffer in 50% DMF. Reprinted from Volke et al. (1971) by courtesy of VEB Verlag Volk and Gesundheit.

toxicological procedure for the determination of trifluperidol, haloperidol and haloperidid in blood. The assay involved selective extraction of the compounds from alkalinized blood into cyclohexane, the residue of which was dissolved in Britton–Robinson buffer (pH 4.6) and methanol (2:3 v/v) for polarographic analysis. The assay has a sensitivity of $10\ \mu g\ ml^{-1}$ with a relative standard deviation of $\pm 4\%$, and was used to measure blood levels in the rat following an $80\ mg\ kg^{-1}$ dose.

(iii) *Rauwolfia alkaloids*

Reserpine (X), a member of the Rauwolfia alkaloid class of antipsychotics, is a drug isolated from the plant *Rauwolfia serpentina* or snakeroot. Vorel et al. (1961) performed oscillopolarography on this compound and its nitro

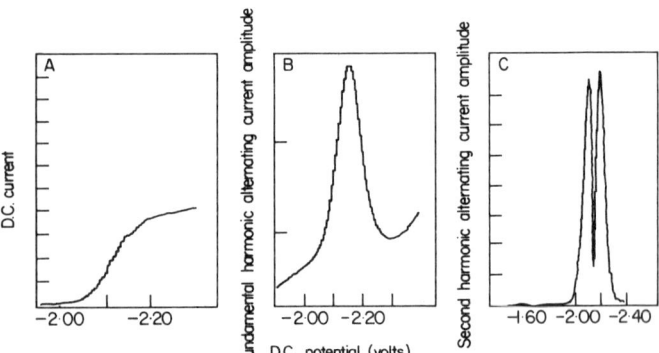

FIG. 3. Polarograms of 1.73×10^{-4}M reserpine in acetonitrile—0.10 M TBAP. (A) D.c. polarogram, (B) fundamental harmonic a.c. polarogram and (C) second harmonic a.c. polarogram. Reprinted from Woodson and Smith (1970) by courtesy of the American Chemical Society.

derivative and suggested a semiquantitative determination. Woodson and Smith (1970) observed direct current and alternating current polarographic responses of reserpine and deserpidine in acetonitrile containing 0.1M tetrabutylammonium perchlorate (Fig. 3). D.c. and a.c. polarography demonstrated that the electrode reactions were diffusion controlled. Quantitative determination of each drug was possible at a lower level of 5×10^{-5}M using direct current polarography, and was limited by background current due to electroactive impurities in the solvent.

Yohimbine (XI), another antipsychotic Rauwolfia alkaloid, has been used to produce experimental anxiety. Yoshino et al. (1960) have reacted yohimbine hydrochloride with tungstosilicic acid and tungstophosphoric acid in potassium hydroxide and hydrochloric acid to form complexes which were used to determine the compound amperometrically. Dusinsky (1962) has reported on the alkaline hydrolysis of the indole alkaloids, reserpine and yohimbine, by oscillopolarography.

B. ANTIANXIETY AGENTS

Antianxiety drugs are differentiated from sedative-hypnotics because they tend to produce calming effects without the adverse side effects, such as drowsiness and sleep, which are common to the barbiturates.

(i) *Miscellaneous*

Meprobamate (XII) and benactyzine (XIII) are two commonly administered antianxiety drugs. Vorel et al. (1961) performed linear sweep voltammetry on the two compounds and their nitro derivatives. It was noted that benactyzine underwent hydrolysis under the conditions of assay, and the polarographic activity was lost several minutes after the inception of the analysis. Hynie et al. (1965) also performed oscillographic polarography on benactyzine in aqueous and nonaqueous supporting electrolytes.

(ii) *1,4-Benzodiazepines*

1,4-Benzodiazepines are a relatively new group of antianxiety drugs of which chlordiazepoxide (XIV) diazepam (XV) and oxazepam (XVI) are some of the principal derivatives in current use. Because of their intrinsic polarographic activity, perhaps more polarography has been reported with this series of compounds than any other.

Oelschlager (1963), Senkowski et al. (1964), Oelschlager et al. (1967a) and Jacobsen and Jacobsen (1971) have examined the polarographic reduction of chlordiazepoxide in a variety of supporting electrolytes using several different polarographic techniques. In acid solutions, three well-defined polarographic reduction waves attributed to the reduction of the N-4-oxide.

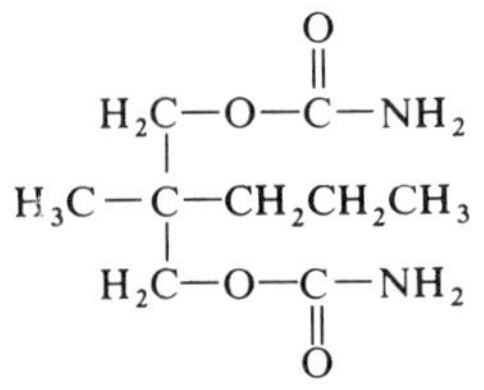

(XII) *Meprobamate*

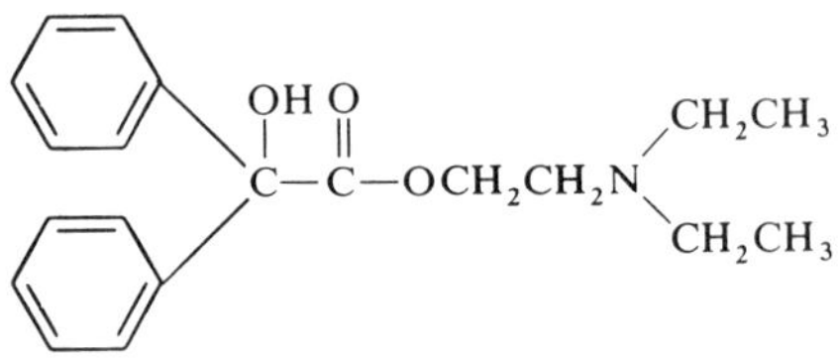

(XIII) *Benactyzine*

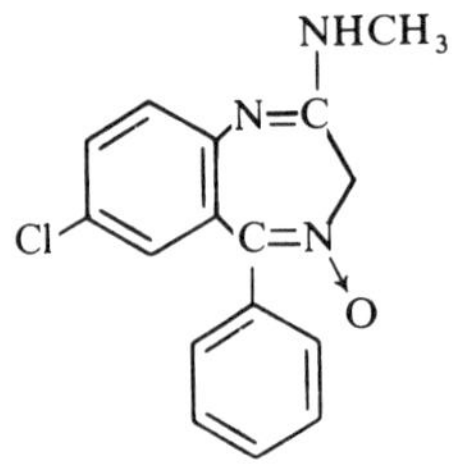

(XIV) *Chlordiazepoxide*

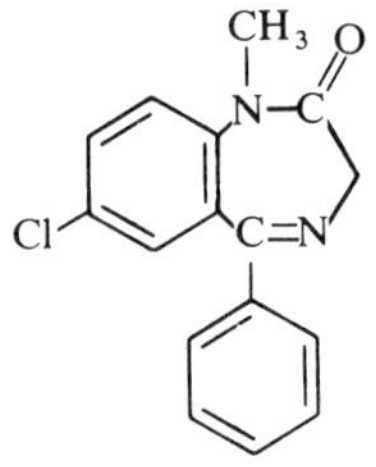

(XV) *Diazepam*

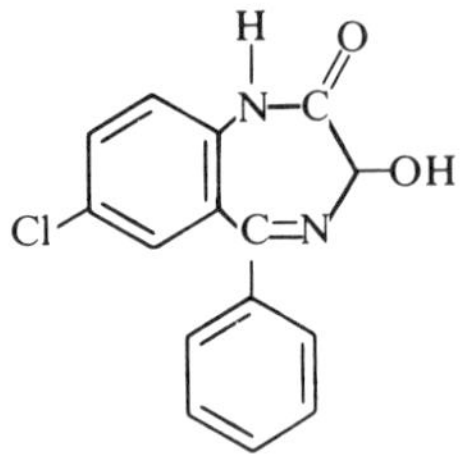

(XVI) *Oxazepam*

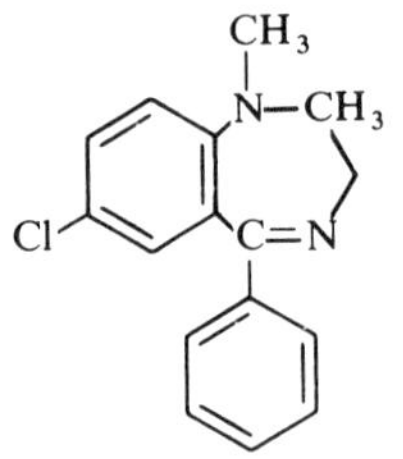

(XVII) *Medazepam*

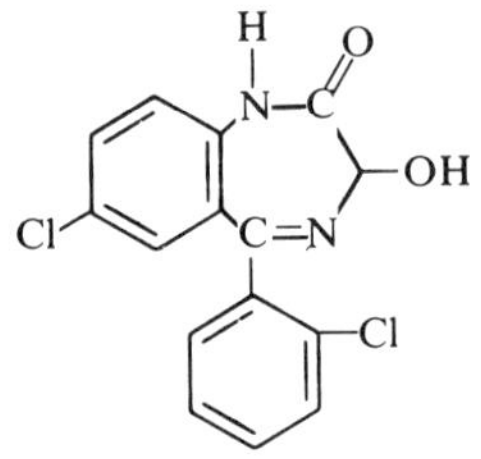

(XVIII) *Lorazepam*

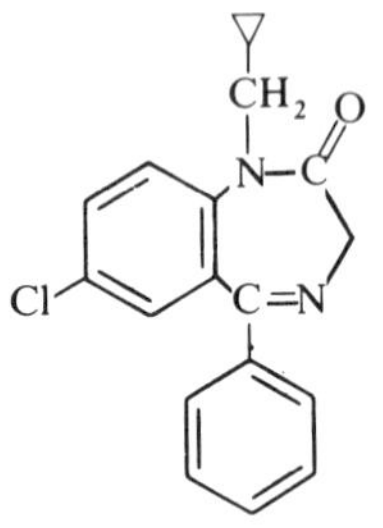

(XIX) *Prazepam*

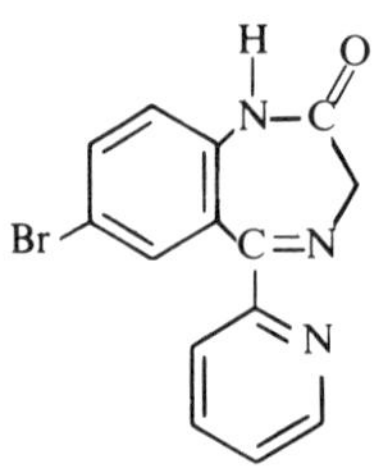

(XX) *Bromazepam*

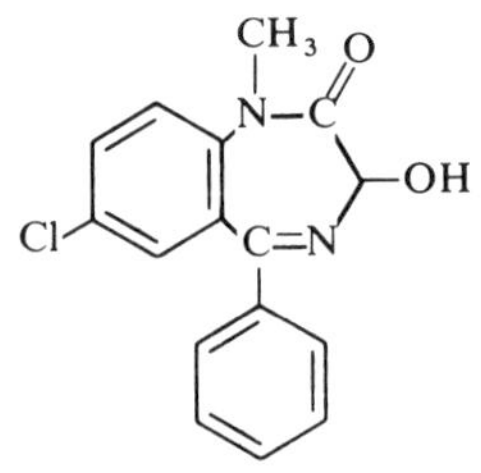

(XXI) *Temazepam*

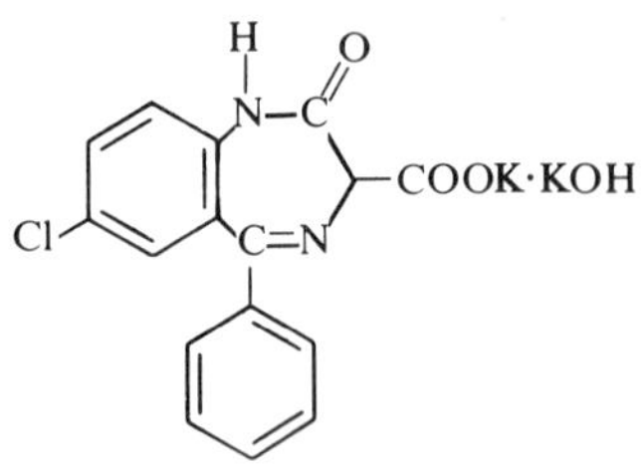

(XXII) *Chlorazepate (Dipotassium)*

the 4,5-azomethine and the 1,2-azomethine functional groups were reported (Fig. 4). Saber and Fahmy (1972) have studied the polarographic reduction mechanism of chlordiazepoxide and its acid hydrolysis product, 2-amino-5-chlorobenzophenone, in aqueous and nonaqueous supporting electrolytes.

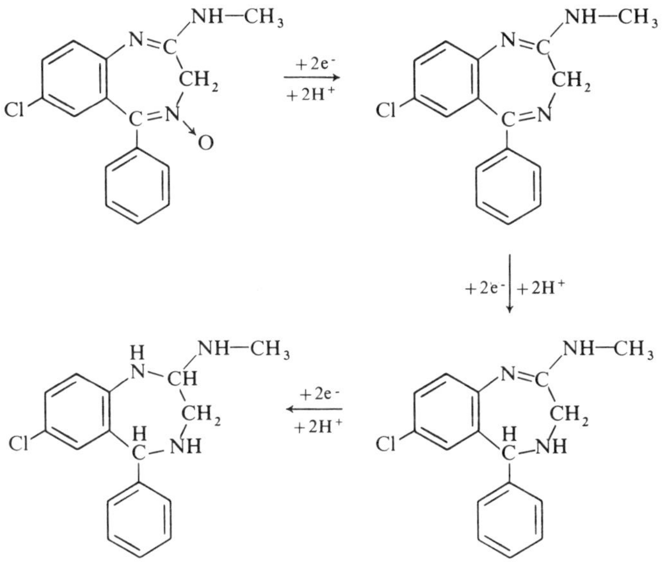

FIG. 4. Polarographic reduction mechanism of chlordiazepoxide in acidic supporting electrolyte at the d.m.e. Reprinted from Oelschlager et al. (1967a) by courtesy of Verlag Chemie GmbH.

Caille et al. (1970) have assayed chlordiazepoxide in dosage forms by spectrofluorometric and polarographic methods and found both methods yielded comparable results. Jacobsen and Jacobsen (1971) have reported an assay for chlordiazepoxide in dosage forms which has the distinct advantage over the u.v. method in that extraction of the drug into organic solvent and removal of insoluble material was not required. The assay was used for tablets also containing propantheline bromide which showed no polarographic activity but showed u.v. absorbance similar to chlordiazepoxide.

The polarographic determinations of chlordiazepoxide in biological fluids employ either dilution of the sample with the supporting electrolyte and measurement (Jacobsen and Jacobsen, 1971) or selective extraction from an alkalinized sample followed by measurement (Cimbura and Gupta, 1965; Hackman et al., 1974). The procedure of Jacobsen and Jacobsen (1971) had a limit of detection of 13 µg chlordiazepoxide ml^{-1} plasma due to masking of the polarographic wave by protein adsorbed on the surface of the mercury drop. The method is only suitable as a rapid toxicological assay of the drug in serum due to its lack of sensitivity. The assay of Cimbura and Gupta (1965) was applied to the analysis of chlordiazepoxide in plasma, urine or gastric contents and is also only suitable for toxicological analysis. The d.p.p. procedure of Hackman et al. (1974) is capable of determining the drug and

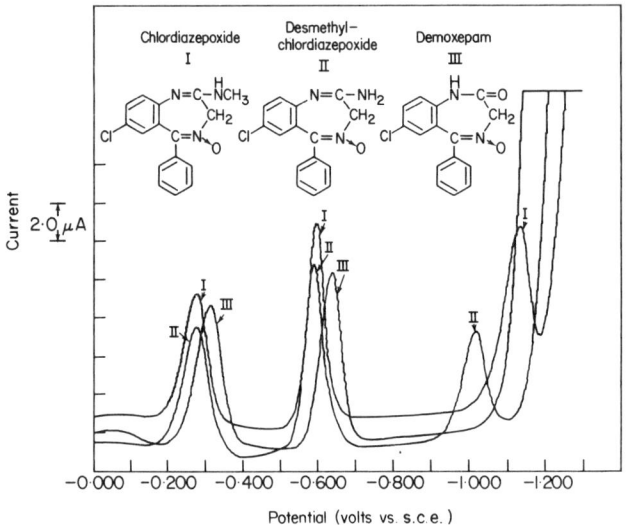

FIG. 5. Differential pulse polarography of chlordiazepoxide and its metabolites in 0.1 N H_2SO_4. Reprinted from Hackman et al. (1974) by courtesy of American Chemical Society.

its N-desmethyl and lactam metabolites following therapeutic administration. Figure 5 shows the d.p.p. behaviour of these substances in 0.1N H_2SO_4. The sensitivity of this latter assay is 0.05–0.1 μg of each compound ml^{-1} of serum using a 2 ml sample. The assay is rapid and selectivity is improved by the thin-layer chromatographic step. It was used to measure levels of parent

TABLE I. Plasma levels (μg/ml) of chlordiazepoxide and its metabolites following chronic administration[a] of a single daily 30 mg oral dose

Study interval		Chlordiazepoxide		N-desmethylchlordiazepoxide		Demoxepam	
Day	Hour	D.p.p.[b]	Fluor.[c]	D.p.p.	Fluor.	D.p.p.	Fluor.
22	1	1.65	1.42	0.41	0.33	0.43	0.42
22	2	1.63	1.46	0.41	0.29	0.46	0.42
26	1	1.45	1.36	0.40	0.32	0.50	0.48
26	2	1.28	1.18	0.33	0.26	0.53	0.49
30	1	1.57	1.48	0.35	0.33	0.49	0.48
30	2	1.58	1.35	0.37	0.32	0.50	0.55
34	1	1.43	1.50	0.31	0.26	0.43	0.52
34	2	1.33	1.35	0.41	0.27	0.46	0.47
36	1	1.34	1.41	0.36	0.28	0.40	0.48
36	2	1.23	1.23	0.37	0.27	0.41	0.62

[a] The subject received 10 mg oral doses three times a day for 21 days, and a single 30 mg oral dose once a day thereafter. [b] Differential pulse polarographic procedure. [c] Spectrofluorometric procedure.

3. DETERMINATION OF PSYCHOTROPIC, HYPNOTIC AND SEDATIVE DRUGS

drug and metabolites following a single 30 mg i.v. and oral dose and following chronic oral administration. A comparison of the data obtained using this assay and the standard spectrofluorometric assay (Table I) showed good correlation between the methods. In addition, the d.p.p. assay measured the intact compounds, having the advantage over the spectrofluorometric assay in that photochemical conversion of the compounds to fluorescent derivatives was eliminated.

Diazepam (XV) is another major antianxiety drug of the 1,4-benzodiazepine class of compounds which differs from chlordiazepoxide in that it contains a methyl group at the N-1 position, and it lacks both the basic side chain at the C-2 and the N-4-oxide functional group.

The polarographic reduction mechanism of diazepam in acidic solution involves the 2-electron reduction of the 4,5-azomethine bond (Oelschlager et al., 1964; 1966a; Senkowski et al., 1964). Studies were also reported on the polarographic reduction of 2-methylamino-5-chlorobenzophenone (MACB), the acid hydrolysis product of diazepam (Oelschlager et al., 1964; Cimbura and Gupta, 1965). Polarographic activity was attributed to the 2-electron reduction of the carbonyl group of MACB.

Assay of the drug in dosage forms has been reported by Caille et al. (1970) and Jacobsen and Jacobsen (1972). The latter authors described a procedure in which 2 or 5 mg tablets were dissolved in 0.2M sulphuric acid and analysed directly. Further experiments showed that excipients in the tablets (starch, cellulose, talc, etc.) did not affect the determination of the drug. Thus the assay is more rapid than the u.v. assay which requires the removal of the excipients. Lund and Opheim (1977) have devised a method for the automated polarographic analysis of diazepam and chlordiazepoxide in tablets.

Toxicological analysis for diazepam in blood, urine or gastric content (Cimbura and Gupta, 1965) and in blood (Fidelus et al., 1972) have been reported. The determination of therapeutic levels of the drug in plasma by linear sweep voltammetry (Berry, 1971) and in serum by d.p.p. (Jacobsen et al., 1973) have also been reported. These assays have sensitivities of 0.02–0.03 μg diazepam ml^{-1} sample; however, they are considered "total" assays since they also measure levels of the N-desmethyl metabolite which will be co-extracted using the described procedures. Brooks et al. (1973a) have developed a d.p.p. method incorporating a thin layer chromatographic separation for the determination of both diazepam and N-desmethyl-diazepam extracted from blood. The method differed from the other reported assays in that diethyl ether replaced petroleum ether or benzene as the extraction solvent. The more polar solvent was found necessary to quantitatively extract both the parent drug and its metabolite in order to measure both compounds with high sensitivity. Recent work (Brooks and de Silva, 1975) has demonstrated that diazepam and its desmethyl metabolite can be measured with a sensitivity of 0.05–0.1 µg ml^{-1} using a 2 ml plasma sample. Thus the assay is capable of determining the levels of diazepam arising after 10 mg single dose administration, and of determining both the parent drug and the metabolite with the necessary selectivity for therapeutic or toxicological examination.

The urinary metabolites of diazepam in man have been evaluated by Dugal et al. (1973) using rapid scan polarography and by Brooks and de Silva (1975) using d.p.p. The former procedures require conversion of the parent drug and metabolites to their respective benzophenones prior to polarography. The d.p.p. procedure employed enzymatic deconjugation of the oxazepam, N-desmethyldiazepam and 3-hydroxydiazepam metabolites, selective extraction, and thin layer chromatography prior to polarographic analysis. The assay measured all three compounds with a sensitivity of approximately 0.25–0.5 µg ml^{-1} urine and monitored the urinary levels following therapeutic doses of diazepam.

Oxazepam (XVI) is a metabolite of diazepam and differs from it by a hydroxy group replacing the hydrogen on the C-3 atom and a hydrogen replacing the methyl group at N-1. Oelschlager et al. (1969b, 1970), Volke et al. (1970) and Goldsmith et al. (1973) have studied the polarographic reduction mechanism for this compound. These authors reported that the reduction of the compound in acidic medium (pH < 6) showed a four-electron step corresponding to simultaneous reduction of the 4,5-azomethine bond and reductive splitting of the carbon-hydroxyl bond in the diazepine ring. In strongly alkaline pH (> 12), a two-electron wave due to only the reduction of the azomethine bond was noted.

TABLE II. *Determination of Medazepam in 10 mg capsules*

Number	Polarography (mg)	(%)	UV-Method (mg)	(%)
1	10.13	101.3	10.21	102.1
2	10.02	100.2	10.07	100.7
3	10.27	102.7	9.96	99.6
4	10.22	102.2	9.98	99.8
5	10.11	101.1	9.93	99.3
6	10.37	103.7	10.26	102.6
7	10.02	100.2	10.03	100.3
8	9.96	99.6	10.08	100.8
9	9.89	98.9	9.91	99.1
10	10.07	100.7	10.02	100.2
11	10.03	100.3	9.83	98.3

Assays for the determination of the drug in dosage forms have been reported by Oelschlager et al. (1969b), Fazzari and Riggleman (1969), Caille et al. (1970) and Goldsmith et al. (1973). These assays employed extraction of the drug into water or methanol, dilution with supporting electrolyte and polarographic assay. The procedure of Goldsmith et al. (1973) was also used to monitor contaminations of oxazepam in the form of quinazolinecarboxyaldehyde (2–10%) which showed polarographic activity in pH 4.7 acetate buffer ($E_{\frac{1}{2}} = -0.72$ V vs s.c.e.).

Medazepam (XVII) is another 1,4-benzodiazepine which possesses antianxiety activity, and which is currently marketed in Europe. Oelschlager and Oehr (1970) determined medazepam at the dropping mercury electrode by d.c. polarography and oscillopolarography. The change in $E_{\frac{1}{2}}$ with pH was studied, including an anomalous behaviour in buffers with pH values greater than 8. The polarographic activity is based on the reduction of the 4,5-azomethine bond in the molecule. Determination of medazepam in samples of capsules was carried out with a precision of $\pm 2.0\%$. This data was compared to a spectrophotometric method (Table II).

Lorazepam (XVIII), a new 3-hydroxy-1,4-benzodiazepin-2-one, analogous to oxazepam, is currently being investigated for its antianxiety properties. Goldsmith et al. (1973) and Oelschlager and Senguen (1974c, 1975) have examined the polarographic reduction mechanism of lorazepam and have found it to be identical to that of oxazepam. Cullen et al. (1973) have presented a method for the determination of this compound using a sensitive and versatile automated procedure capable of performing voltage-scanning polarographic analysis. The technique interfaced a polarograph with a unique continuous flow system and a polarographic cell which completely removed interfering dissolved oxygen. The system permitted high resolution at a rapid assay rate. It was operated in d.c., sampled d.c., pulse or differential pulse polarographic modes and a broad selection of electrode configurations could be used. Fifteen tablet or capsule formulations of lorazepam per hour were analysed at the 1.0 mg level with a relative standard deviation of $\pm 1.4\%$. Polarography of the sample in the d.c. mode was performed in a methanolic sodium acetate–acetic acid supporting electrolyte. Assays have also been reported by Goldsmith et al. (1973) and Oelschlager and Senguen (1974c) for the measurement of the compound in dosage forms.

de Silva et al. (1974a) have recently described a d.p.p. assay to measure levels of lorazepam glucuronide in the urine following a 4 mg dose of 7-chloro-1,3-dihydro-5-(2'-chloro-phenyl)-2H-1,4-benzodiazepin-2-one. The assay required deconjugation of lorazepam at pH 5.3, followed by extraction of the drug into diethyl ether after adjustment of the pH to 9.0. The residue of the ether extract was dissolved in 0.1N HCl containing 5% methanol, and analysed by d.p.p. The assay was rapid, and had a sensitivity of 0.05 µg ml^{-1}

using a 5 ml sample. The assay possessed greater sensitivity and selectivity than the standard Britton–Marshall spectrophotometric assay.

Prazepam (XIX) which differs from diazepam by a N-1 cyclopropyl group in place of the N-1 methyl group was polarographically examined using Britton–Robinson buffers containing 10% DMF, in the pH range from 2 to 10 by Oelschlager and Senguen (1973). The polarographically active functional group, the 4,5-azomethine, was used to determine the compound in 10 and 25 mg tablets. Further mechanistic studies by these authors (1974b) have shown that the 4,5-azomethine bond is reduced to yield 4,5-dihydrodiazepam at both the d.m.e. and Hg pool electrodes. In addition the catalytic hydrogenation of prazepam occurred less specifically than the electrochemical reduction, in that the cyclopropane ring was cleaved as well as eliminated during the hydrogenation.

Bromazepam (XX) 7-bromo-1,3-dihydro-5(2-pyridyl)-2H-1,4-benzodiazepin-2-one, is an antianxiety drug which is presently under study. Differential pulse polarography of the compound at the d.m.e. in pH 5.4 phosphate buffer gives three peaks at -0.565 V, -1.280 V and -1.350 V vs s.c.e., the first due to the reduction of the 4,5-azomethine and the latter two were believed to be due to reduction processes occurring in the pyridinyl group (Brooks et al., 1974).

de Silva et al. (1974b) have determined the urinary excretion products of bromazepam in man using d.p.p. following the administration of a single 12 mg oral dose. The assay involved selective extraction of bromazepam and the 2-amino-5-bromobenzoyl-pyridine metabolite from the deconjugated metabolites 3-hydroxybromazepam and 2-amino-3-hydroxy-5-bromo-benzoyl pyridine into separate diethyl ether fractions. The residues of the respective diethyl ether extracts were dissolved in 1M phosphate buffer (pH 5.4) and analysed by d.p.p. which yielded two distinct well-resolved reduction peaks for the 4,5-azomethine functional group of the benzodiazepine-2-one and for the carbonyl functional group of the benzoyl pyridine in each fraction. The overall recoveries of the compounds in the conjugated and uncongugated fractions were $80 \pm 5\%$ and $45 \pm 5\%$, respectively, with a sensitivity for the four compounds of 50–100 ng per five ml of urine.

In a recent study, Brooks and Hackman (1975) have determined bromazepam in blood with a sensitivity of 10–20 ng ml^{-1} using a 0.5 ml micro-cell for d.p.p. analysis. The assay employed selective extraction of bromazepam from alkalinized blood into benzene:methylene chloride (90:10). The residue of this extract was dissolved in phosphate buffer (pH 7.0) for analysis. The d.p.p. and g.l.c. (electron capture) assays measure blood levels in man following chronic single daily oral administrations of 3 mg of bromazepam. The two assays yielded comparable results (see Table III). The chromatographic and d.p.p. methods have comparable sensitivity, although the

TABLE III. *Blood levels of bromazepam following chronic administration of a single daily 3 mg oral dose*

Day	Concentration, ng/ml	
	D.p.p.[a]	E.c.–g.l.c.[b]
1	26	27
4	48	53
9	69	92
10	71	77
12	123	125
15	119	105

[a] Differential pulse polarographic assay. [b] Electron capture–gas liquid chromatographic assay.

chromatographic method is more selective. Another recent study described the determination of bromazepam in blood by linear sweep voltammetry with a sensitivity of 0.05 µg ml^{-1} (Senguen and Oelschlager, 1975).

Smyth et al. (1977) have made a spectral and polarographic study of the acid-base and complexing behaviour of bromazepam. An indirect analytical method, based on the complexation of bromazepam with Cu(II) ions, was investigated by d.p.a.s.v. (Fig. 6).

Temazepam (XXI) the 3-hydroxy analogue of diazepam has also been examined by polarography (Oelschlager and Oehr, 1974). The compound is relatively unstable in strong acidic or alkaline buffers and consequently the optimum pH range for assay is 5–7. The polarographic reduction of the compound consumed four electrons in acidic buffers. The only product

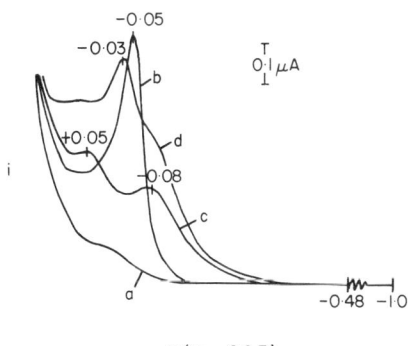

FIG. 6. Effect of bromazepam on stripping current of Cu^{2+} ions: (a) blank BR buffer pH 9.0; (b) 1 × 10^{-5}M Cu^{2+}; (c) 1 × 10^{-5}M Cu^{2+} + 1 × 10^{-5}M bromazepam; (d) 2 × 10^{-5}M Cu^{2+} + 1 × 10^{-5} bromazepam. Reprinted from Smyth et al. (1977) by courtesy of Elsevier Press.

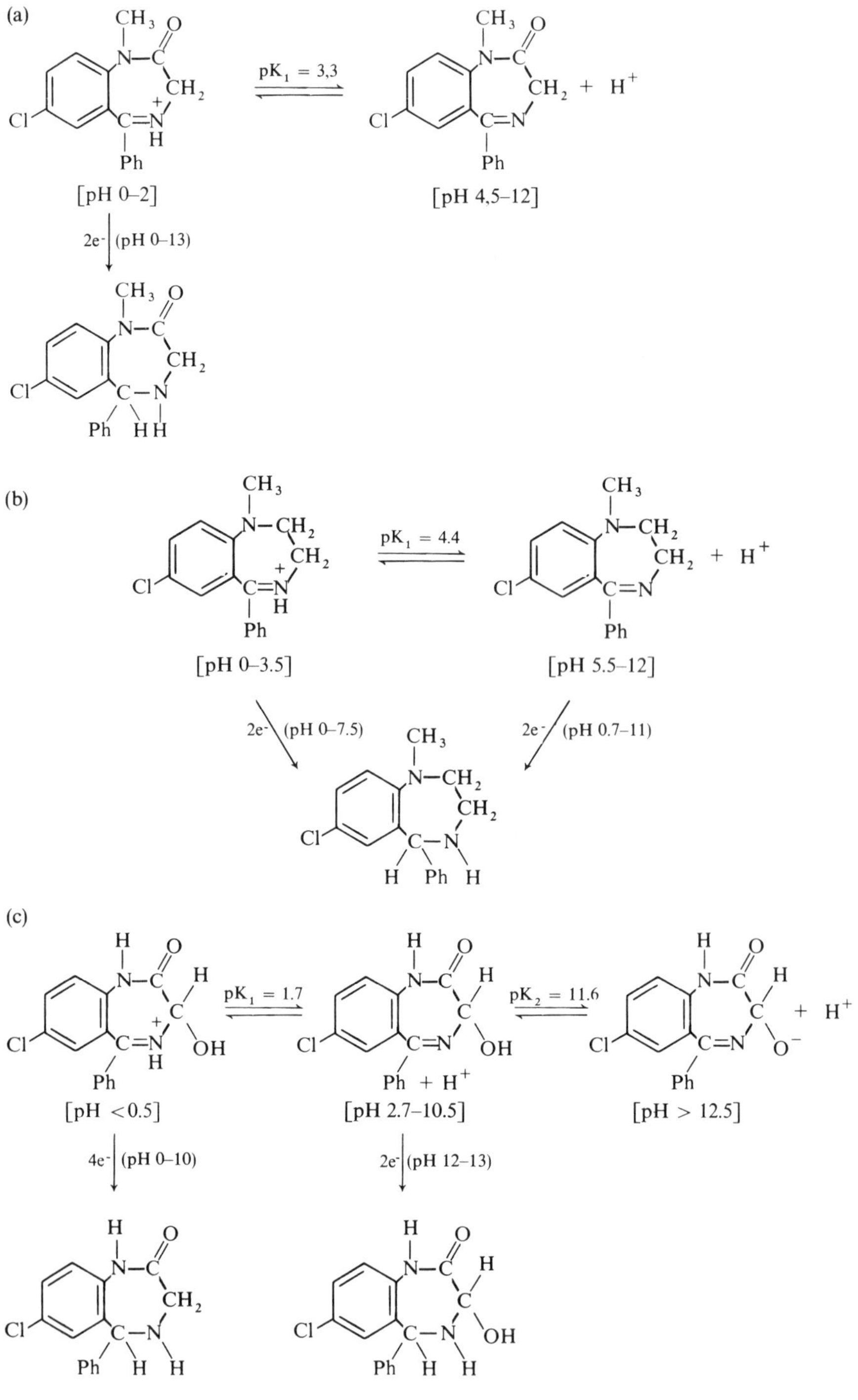

FIG. 7. Polarographic reduction mechanism and acid-base equilibrium for (a) diazepam, (b) medazepam, (c) oxazepam. Reprinted from Clifford and Franklin Smyth (1973) by courtesy of Springer-Verlag.

found after reduction at the mercury pool electrode was 7-chloro-1-methyl-5-phenyl-1,3,4,5-tetrahydro-2H-1,4-benzodiazepin-2-one.

Potassium clorazepate (XXII), was recently introduced as an antianxiety agent. The compound is the dipotassium carboxylic acid analogue of N-desmethyldiazepam. Oelschlager and Senguen (1974a) have described an assay for the compound in acetate buffer (pH 4.7) containing 10% DMF which can be applied to the assay of 5 and 10 mg dosage forms. The compound was found to be unstable in this medium and yielded N-desmethyldiazepam which was measured polarographically. Abruzzo et al. (1976) reported a d.p.p. assay to measure the stability of the compound in aqueous media in the pH range of 1.5–12 and in blood. This work demonstrated that the rate of degradation decreased with increase in pH with relatively instantaneous breakdown below pH 4 and relative stability at pH 7.4. Smyth and Leo (1975) described a differential extraction procedure to determine clorazepate in the presence of other benzodiazepines.

Several recent papers have been concerned with the polarographic properties of the 1,4-benzodiazepines as a class. Clifford and Smyth (1973) have examined the acid-base equilibria and polarographic reduction mechanisms of chlordiazepoxide, diazepam (Fig. 7a), medazepam (Fig. 7b), oxazepam (Fig. 7c), and lorazepam. The compounds typically show the presence of at least three species, the protonated, neutral and anionic forms. The protonated form is reducible over a wide pH range and is suggested for analytical determinations in dosage forms or after selective extraction from body fluids. The general range of utility is $10^{-4}M$–$10^{-7}M$ for the six compounds. These acid-base equilibria studies were later extended to 7-amino and 7-acetamido nitrazepam, desmethyldiazepam and the lactam metabolite of chlordiazepoxide (Barrett et al., 1974) in aqueous solutions, pH 1–14. Based on these data the authors were able to develop a solvent extraction procedure to selectively extract diazepam in the presence of desmethyldiazepam. Cleghorn (1972) has reviewed the polarographic behaviour of the 1,4-benzodiazepines, including diazepam, oxazepam, medazepam, chlordiazepoxide and nitrazepam.

Bogatskii et al. (1971) investigated the polarographic reduction of 1,3,5,7-substituted 1,3-dihydro-2H-1,4-benzodiazepin-2-ones and benzodiazepin-2-thiones at the d.m.e. in dimethylformamide–water solution at pH values of 4.80 and 9.40. They showed that a correlation existed between the $E_{\frac{1}{2}}$ values for the reduction of 7-substituted 1,3-dihydro-5-phenyl-2H-1,4-benzodiazepin-2-ones and the Hammett constants $[\sigma_m]$ of substituents in the 7-position. A correlation was suggested between the ability of these compounds to undergo electrochemical reduction and their psychopharmacological activities. Benzodiazepines such as diazepam, N-desmethyldiazepam, oxazepam, and nitrazepam were included in this study.

Brooks et al. (1975) examined the relationship between E_p and the Hammett function (σ_m) in three series of 1,4-benzodiazepines related to chlordiazepoxide, medazepam and N-desmethyldiazepam in pH 3.0 and 7.0 phosphate buffers as supporting electrolytes. Good correlations were established between σ_m and the potential for the reduction of the 4,5-azomethine functional group for the three series of compounds in both supporting electrolytes. An example of this type of data for chlordiazepoxide is

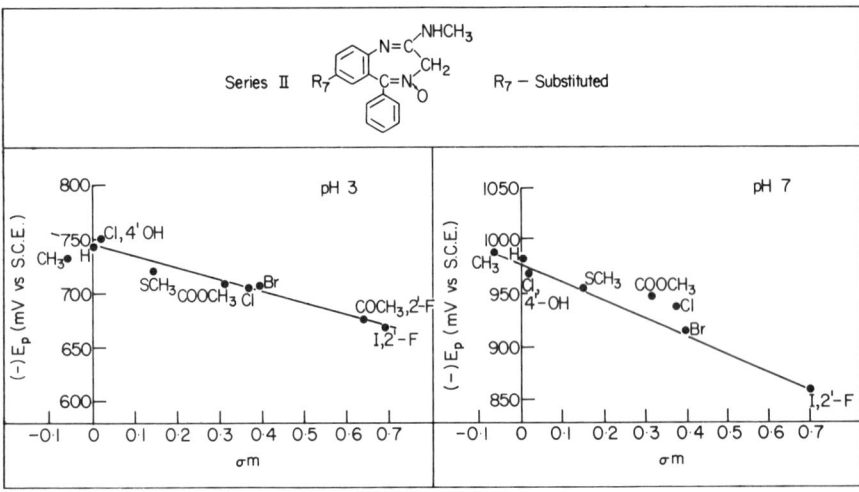

FIG. 8. Hammett plot for R-7 substitution in chlordiazepoxide at pH 3.0 and pH 7.0. Reprinted from Brooks et al. (1975) by courtesy of Elsevier Press.

shown in Fig. 8. No correlation could be established for the reduction of the 1,2-azomethine and N-4-oxide in the series of compounds structurally related to chlordiazepoxide. Polarographic data was also presented for two groups of 1,4-benzodiazepines structurally related to medazepam and N-desmethyldiazepam with heterocyclic substituents in the 5 position.

IV. ANTIDEPRESSANT DRUGS

Antidepressant drugs are generally divided into three categories, tricyclics, monoamine oxidase inhibitors (MAOI), and sympathomimetic stimulants. Dextroamphetamine and methamphetamine are typical examples of the sympathomimetic stimulants. No polarographic methods have appeared in the literature for these compounds.

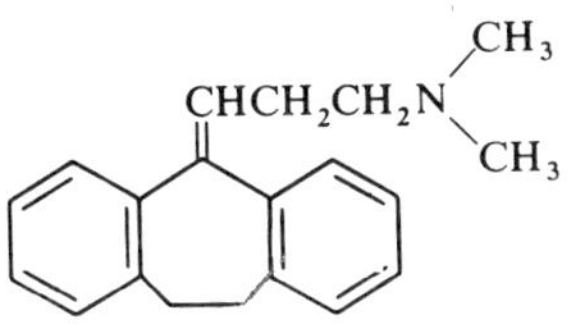

(XXIII) *Amitriptyline*

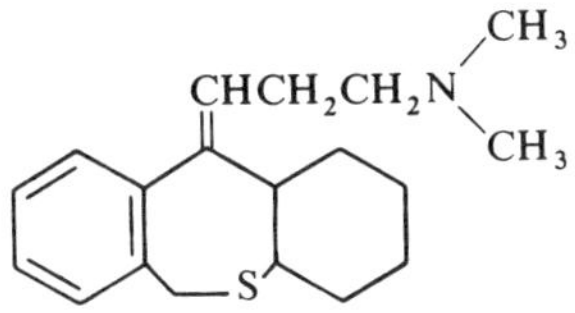

(XXIV) *Prothiaden*

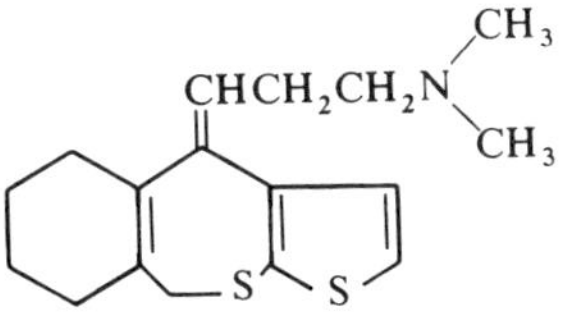

(XXV) *Dithiaden*

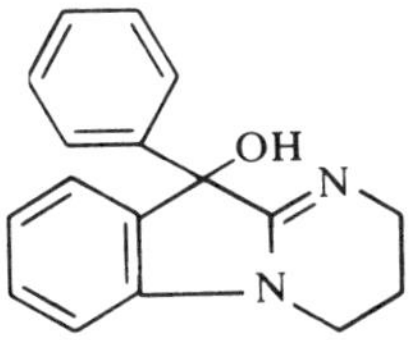

(XXVI) *Wy 23409*

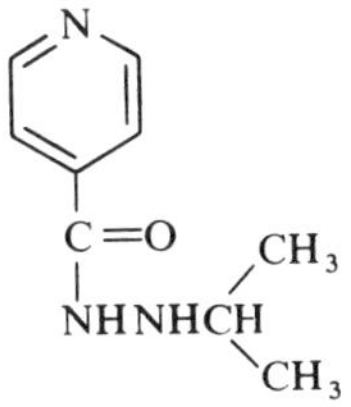

(XXVII) *Iproniazid*

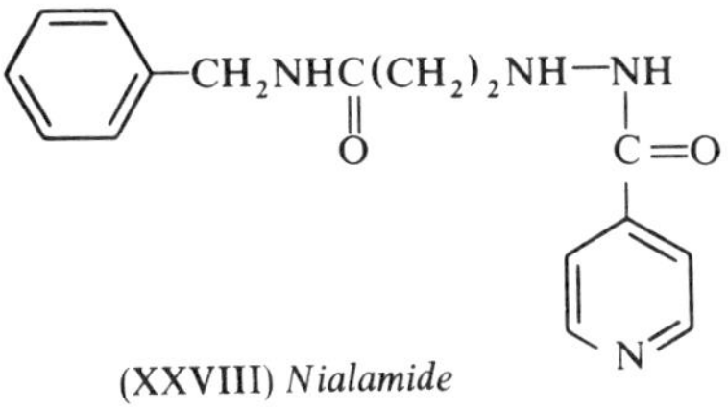

(XXVIII) *Nialamide*

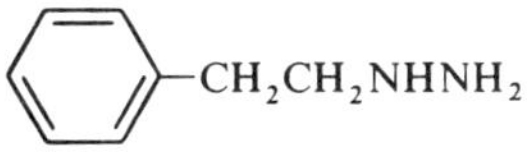

(XXIX) *Pehnelzine*

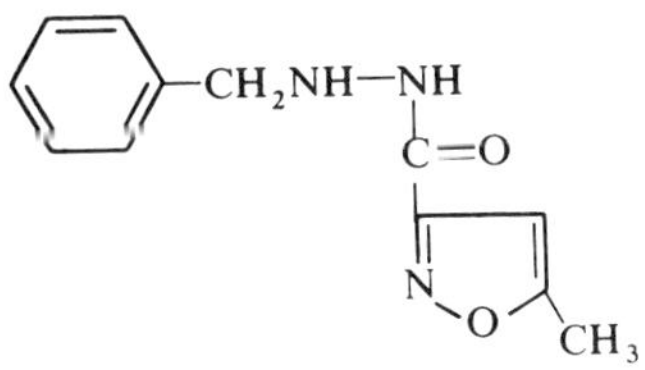

(XXX) *Isocarboxazide*

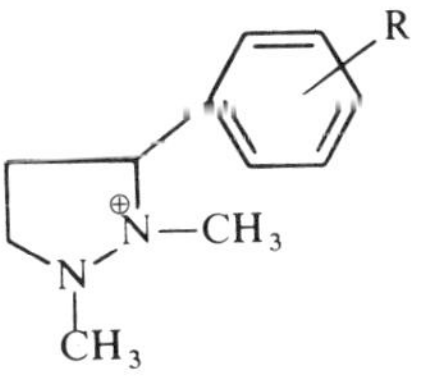

(XXXI) *1,2-Dimethyl-3-arylpyrazolinium salt*

A. TRICYCLICS

Volke et al. (1975) have reported on the anodic oxidation of three dibenzocycloheptane derivatives at the platinum and gold rotating disc electrodes. The compounds studied were amitriptyline (XXIII) prothiaden, (XXIV) and dithiaden (XXV) in a nonaqueous supporting electrolyte of 0.1M tetrabutylammonium perchlorate in acetonitrile. A linear calibration was obtained over a concentration range of 2.5 to 15×10^{-4}M; the currents of anodic oxidation were diffusion controlled at a rotation speed of 1300 r.p.m. and scan rate of 500 mv min^{-1}. The data suggested that the common mechanism is an attack on the amino group and may be either a fission of the double bond between the external chain and the seven-membered ring or an oxidation of the nitrogen atom to the N-oxide state. Both processes would involve two electrons. For prothiaden and dithiaden this primary process was probably followed by one electron radical cation formation.

Chan (1974) has reported a highly specific assay method for the determination of 10-(m-chlorophenyl)-2,3,4,10-tetrahydropyrimido[1,2a]indol-10-ol hydrochloride (Wy 23409) (XXVI) an experimental antidepressant, in the presence of its immediate precursor in its chemical synthesis or in the presence of thermal degradation products. The polarographic assay was preferred over thin-layer chromatographic methods which were too lengthy for routine analysis, and g.l.c. methods which suggested decomposition upon analysis.

B. MONOAMINE OXIDASE INHIBITORS

The MAOI drugs are divided into two groups: the hydrazids and nonhydrazids.

Iproniazid (XXVII) was the first MAOI drug introduced, but has been withdrawn from the market because of adverse toxicity. Brandys (1964) proposed a polarographic method for iproniazid in ground tablets. The tablet was first extracted with water to which Britton–Robinson buffer solution pH 6.59 and 0.5% aqueous gelatin (3 drops) were added. In twelve polarographic determinations 22.5 µg of iproniazid per ml were determined with a standard deviation of 0.62 µg.

Nialamide (XXVIII), a hydrazid MAOI, has also been determined by Brandys (1964) in ground tablets using the method described above. In twelve determinations, 35 µg of nialamide per ml were determined with a standard deviation of ± 0.74 µg.

Schlitt et al. (1967) have determined the hydrazid MAOI drugs, nialamide and phenelzine (XXIX). Nialamide gave cathodic phenomena at the d.m.e. in pH 5.9 phosphate buffers in two steps at -0.87 V and -1.01 V vs s.c.e.

A linear response between concentrations of 1×10^{-5}M to 1.5×10^{-4}M and wave height was noted. Phenelzine was determined at a level of 4×10^{-4} M in phosphate buffer (pH 5.9) as an acetone derivative using a 4% acetone solution containing 0.01% gelatin. Isocarboxazid (XXX), another hydrazid of the MAOI class of drugs, was studied by El-Darawy et al. (1975) in aqueous buffers and in nonaqueous methanolic solutions. In both media, reduction occurred in the isoxazole ring resulting in a single diffusion-controlled irreversible wave corresponding to the transfer of four electrons per molecule.

A polarographic study of seven 1,2-dimethyl-3-arylpyrazolinium salts (XXXI) which demonstrate monoamine oxidase inhibiting activity similar to nialamide has been reported (Omar and El-Rabbat, 1974). In acidic and neutral media, the electrode process involves an irreversible two-electron transfer and the formation of a pyrazolidine. The polar effects of substitution at the 3-phenyl position on the half-wave potential demonstrated good correlation with the Hammett constants. In addition a good correlation was obtained between antidepressant activity of these compounds and the values of $E_{\frac{1}{2}}$ which suggested a redox limiting factor that influences the *in vivo* effectiveness in mice.

V. HYPNOTIC OR SEDATIVE DRUGS

As a general rule hypnotics are used to induce sleep, whereas sedatives usually are administered to bring about a milder degree of central nervous system depression.

A. MISCELLANEOUS

Chloral hydrate (XXXII) was first introduced in the mid-19th century, when only alcohol, opium and cannabis were available for hypnosis and sedation. It is used primarily in the treatment of insomnia. Elving and Bennett (1954) reported that the reduction mechanism of chloral hydrate was due to a composite of diffusion and kinetic processes. Their data indicated that the compound was reduced to dichloroacetaldehyde hydrate, which subsequently dehydrated and reduced to chloroacetaldehyde, which then reduced to acetaldehyde. This was followed by reduction to ethyl alcohol or 2,3-dihydroxybutane, or both. The overall process yielded one major wave corresponding to the reduction to acetaldehyde ($E_{\frac{1}{2}} = -1.4$ V vs s.c.e.). A secondary wave corresponding to the reduction of acetaldehyde was observed at -1.7 V vs s.c.e. in ammonia buffers.

Barlot and Albisson (1956) have analysed the compound as its oxime derivative after reaction with hydroxylamine. The derivative yielded two reduction waves in a pH 2 supporting electrolyte due to the one-electron reductive splitting of a chlorine atom, followed by the two-electron reduction of the hydroxylamine to the amine.

Chlorbutanol (XXXIII) a hypnotic chemically related to chloral hydrate is not generally administered at the present. Birner (1961) reported a polarographic method in which the compound was determined using benzethonium chloride as a supporting electrolyte. The height of the wave was proportional to a concentration of between 5 and 500 µg ml^{-1}. The method was applied to the determination of the drug in horse serum. In brief, the method required steam distillation of the compound from serum in the presence of tungsto-

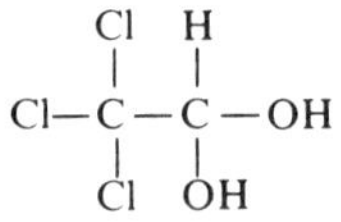

(XXXII) Chloral Hydrate 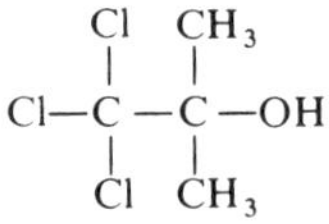   (XXXIII) Chlorobutanol 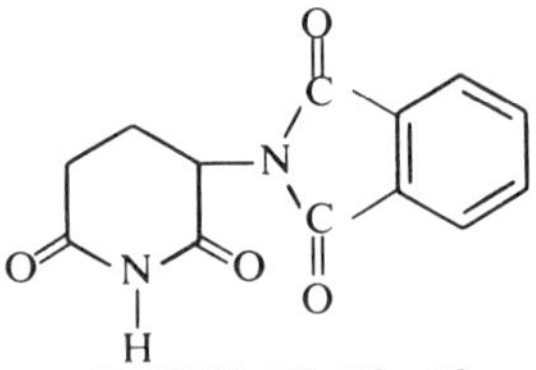   (XXXIV) Thalidomide

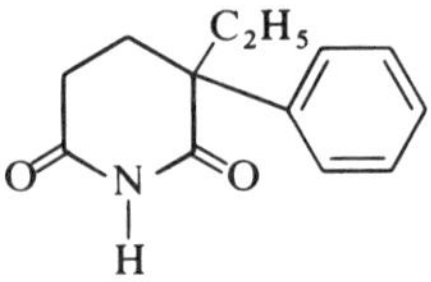

(XXXV) Glutethimide 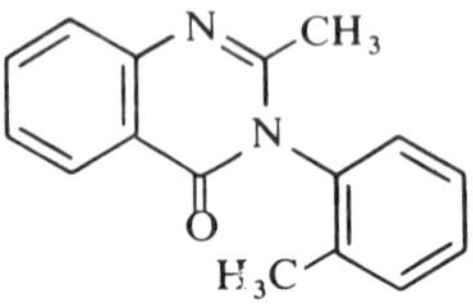   (XXXVI) Methaqualone

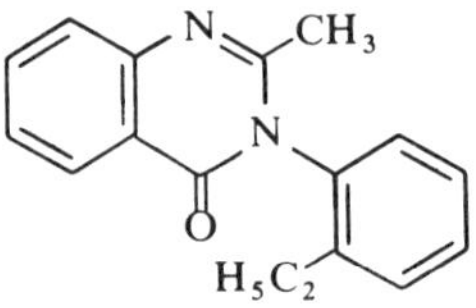

(XXXVII) Ethinazone 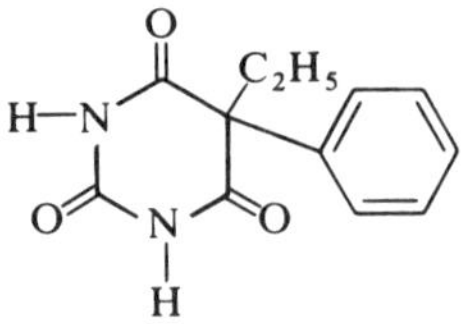   (XXXVIII) Phenobarbital

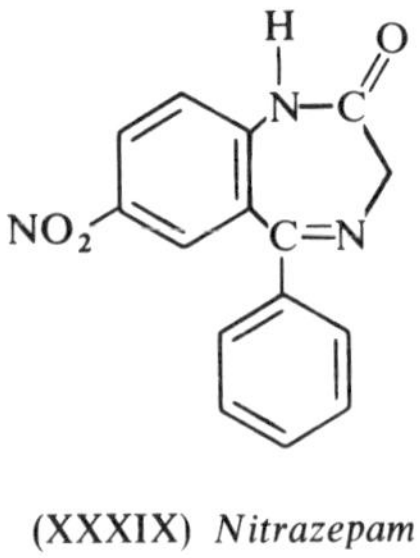

(XXXIX) Nitrazepam 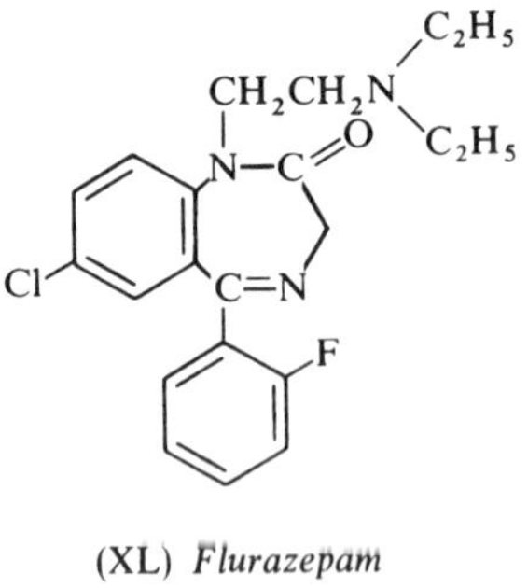   (XL) Flurazepam

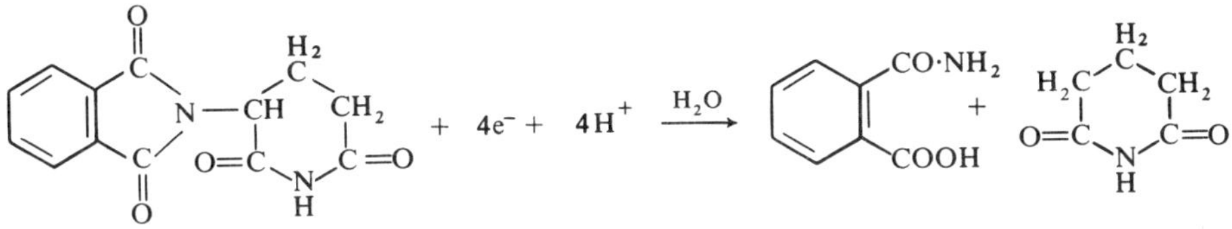

FIG. 9. Proposed reduction mechanism of thalidomide. From Hetman (1964) by courtesy of Elsevier Press.

silicic acid. A portion of the distillate is mixed with 0.05 M benzethonium chloride and Na_2SO_3, and then analysed polarographically. The standard deviation for samples containing 4 mg of added chlorbutanol was 3%, with recoveries greater than 98%.

Thalidomide (XXXIV), a hypnotic which was removed from clinical use due to its teratogenicity, was determined by Hetman (1964) by conventional d.c. polarography. A solution containing tetrahydrofurfuryl alcohol: methanol (2:3), barbital and LiCl which was adjusted to pH 3.5 by addition of H_3PO_4 was used as the supporting electrolyte. The sensitivity limit of the assay was $2 \, \mu g \, ml^{-1}$ employing a 10 ml cell. The proposed four-electron reduction mechanism is shown in Fig. 9.

Glutethimide (XXXV) was determined by Danek and Strozik (1968) following nitration in conc HNO_3:conc H_2SO_4 (1:1) by d.c. polarography at the d.m.e. The method was applied to determination of hot ethanolic extract of ground tablets. The extract was evaporated, the residue of which was nitrated, and diluted with 20% aqueous NaOH containing 0.5% aqueous gelatin (0.2 ml) and water.

Methaqualone (XXXVI) a member of the sedative and hypnotic class of compounds was determined employing d.c. polarography (Pflegel and Wagner, 1967a) in Britton–Robinson buffer (pH 2.6) and showed an analytical wave for the two-electron reduction of the azomethine bond. Methaqualone and ethinazone (XXXVII) an ethyl derivative of methaqualone were also determined in tablets (Pflegel and Wagner, 1967b).

B. BARBITURATES

Phenobarbital (XXXVIII) was determined by Lordi et al. (1960) by a.c. polarography in dosage forms. A solution of phenobarbital (0.2 to 0.4 mg), extracted from a solid or liquid dosage form, into M KNO_3 in borate buffer (pH 7.9) was analysed by a.c. polarography using an applied alternating current of 0.03 V r.m.s. at a rate of $0.00062 \, V \, s^{-1}$. The concentration of phenobarbital was calculated with an accuracy of $\pm 2\%$ by reference to a standard curve. The a.c. polarographic behaviour appeared to be due to the reaction of phenobarbital with mercurous ion formed by a polarographic electrode reaction; the resultant mercurous phenobarbital was adsorbed at the mercury–solution interface giving rise to the anodic wave.

Cohen (1965) studied amobarbital, barbital, barbituric acid, diallylbarbituric acid, pentobarbital and phenobarbital as a function of depolarizer concentration, pH, a.c. amplitude and a.c. frequency. In addition, the influence of drop time, buffer composition, surfactant and polymer concentration on the barbituric acid wave was reported. An investigation of direct current polarography of these compounds was also made. Application to pharmaceutical preparations appears to be useful only in cases where direct analysis without prior separation was possible.

Woodson and Smith (1970) studied phenobarbital by direct current and alternating current polarography in acetonitrile containing 0.1 M tetrabutylammonium perchlorate. The drug was determined with a sensitivity of approximately 5×10^{-5}M for both methods.

Zuman and Brezina (1962) have reviewed voltammetric methods for barbiturates and thiobarbiturates, employing their anodic behaviour and list a large number of references to some earlier works.

Kato and Dryhurst (1975) have examined the electrochemical oxidation of barbituric acid, 1-methylbarbituric acid and 1,3-dimethylbarbituric acid at the pyrolytic graphite electrode at pH 1 in the presence of chloride ion. The three compounds showed a single well-defined oxidation peak at approximately 1.0 V vs s.c.e. with a calculated n value of approximately 4 based on coulometry. The details of the processes responsible for the voltammetric peaks were not reported. Preparative scale synthesis under the same conditions yielded four products, the three major products, accounting for more than 80–90% of the oxidized barbituric acid, were the appropriate N-methylated 5′,5′-dichlorohydurilic acids, 5,5-dichlorobarbituric acids and alloxans.

Brooks et al. (1973b) developed a differential pulse polarographic method for the determination of phenobarbital in blood to measure therapeutic and overdose levels. Phenobarbital was selectively extracted from blood buffered to pH 7.0 into an ethyl acetate:butanol (80:20) mixture, subjected to a "clean-up" procedure, nitrated and polarography was then carried out in pH 7.0 phosphate buffer. Quantitation was based on the measurement of the reduction peak for the nitro derivative. 1 µg ml^{-1} of blood can be determined with an overall recovery of 72.3 ± 6.5%. A modified polarographic method for the determination of both phenobarbital and diphenylhydantoin, an anticonvulsant drug, employing a thin-layer chromatographic separation of the initial plasma extract was also reported.

A recent publication (Wiegrebe and Wohrhahn, 1975) also employed the formation of the nitro derivative for the polarographic determination of phenobarbital, primidone, diphenylhydantoin, methylphenobarbital and mephenytoin. The assay differed from that previously reported (Brooks et al., 1973b) in that the compounds were separated by t.l.c. after nitration. The assay could be used to measure the five drugs following therapeutic administration as well as in the toxicological situation.

C. 1,4-BENZODIAZEPINES

Nitrazepam (XXXIX) an hypnotic and sedative, is similar in structure to chlordiazepoxide, diazepam and oxazepam. The compound is identical to N-desmethyldiazepam with the exception that the 7-chloro group has been replaced by a 7-nitro group in the molecule. Senkowski et al. (1964) were the first to report the polarography of this compound. The authors noted two analytical waves in 20% methanol in 0.1N HCl for the reduction of the 7-nitro group and the reduction of the 4,5-azomethine bond respectively. Further studies (Oelschlager et al., 1966b) demonstrated that the first polaro-

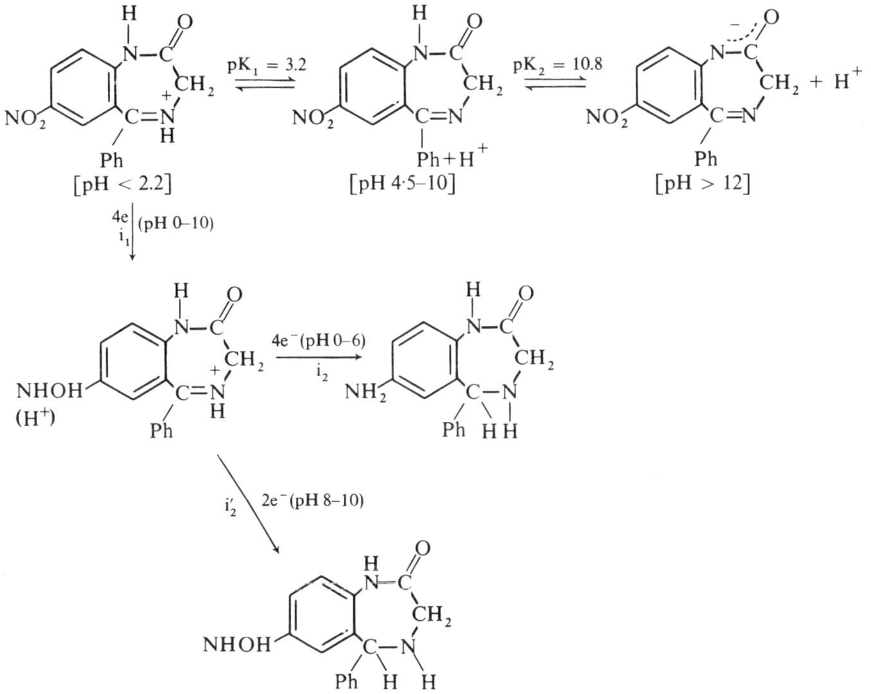

FIG. 10. Polarographic reduction mechanism and acid-base equilibrium of nitrazepam. Reprinted from Clifford and Franklin Smyth (1973) by courtesy of Springer-Verlag.

graphic wave was due to the four-electron reduction of the 7-nitro substituent to the hydroxylamine and the second due to the simultaneous four-electron reduction of the 4,5-azomethine and the intermediate hydroxylamine to the amine. Under alkaline conditions, four electrons were involved in the reductions of the 7-nitro group and only two electrons for the reduction of the 4,5-azomethine. Oelschlager et al. (1969a), Halvorsen and Jacobsen (1972) and Clifford and Smyth (1973) also studied in detail the reduction mechanism of nitrazepam in acid, neutral and basic media and proposed several polarographic reduction intermediates. The reduction mechanism is shown in Fig. 10 and a cyclic voltammogram in Fig. 11.

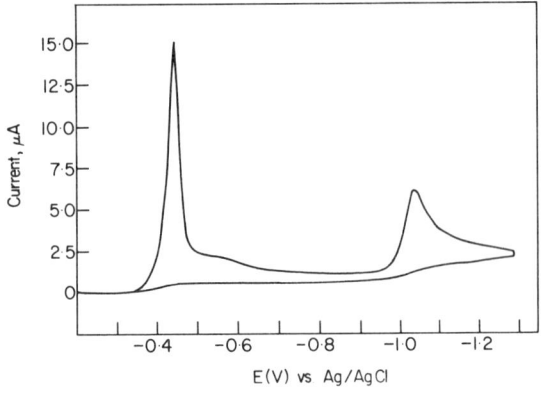

FIG. 11. Cyclic voltammogram of 10^{-4}M nitrazepam in phosphate buffer pH 6.9. Scan rate 0.1 V sec^{-1}. Reprinted from Halvorsen and Jacobsen (1972) by courtesy of Elsevier Press.

Oelschlager et al. (1967b) also described a method for the determination of nitrazepam, and its 7-amino and 7-acetamido metabolites, by incorporating a thin-layer chromatographic separation into the analysis. All three compounds in the range of 30 to 90 µg were separated, eluted from the thin-layer adsorbent, and determined by d.c. polarography with a precision of ±4%. The polarography of the two metabolites showed the typical reduction of the 4,5-azomethine bond. These metabolites were examined in Britton–Robinson buffers in the concentration range of 5×10^{-5} to 10^{-3}M in supporting electrolytes of pH 2–10.

Oelschlager et al. (1966b) described a d.c. polarographic method [(accuracy ±3%) in pH 4.7 Britton–Robinson buffer containing 10% dimethylformamide] to determine the active substance contained in a 5 mg tablet. Halvorsen and Jacobsen (1972) have reported a toxicological assay for the direct measurements of 0.5–80 µg nitrazepam ml^{-1} of serum.

TABLE IV. Urinary excretion data on the major metabolites of flurazepam in humans who received single oral 90 mg doses of dalmane

Subject	Excretion Period Hours	Form Excreted	Compound Measured[a]	Concentration (µg/ml)	Milligram Equivalent as Flurazepam		Percent of Dose Excreted	Direct Assay for total Benzodiazepines (%)
1	0–24	free	[I-A]	0.17	0.35		0.49	
			[I-B]	1.14	2.75		3.63	
		bound	[II]	9.30	22.5		29.70	
						Total	33.82	42.2
2	0–24	free	[I-A]	1.83	2.26		2.98	
			[I-B]	6.57	8.78		11.6	
		bound	[II]	18.1	24.3		32.10	
						Total	46.68	50.2
3	0–24	free	[I-A]	2.14	1.73		2.28	
			[I-B]	11.4	9.95		13.1	
		bound	[II]	25.8	22.6		29.80	
						Total	45.18	43.8

[a] I–A = mono-desethylflurazepam; I–B = di-desethylflurazepam; II = N-1-hydroxyethylflurazepam.

Flurazepam (XL), is an hypnotic which has recently been introduced for the treatment of insomnia. A urinary assay developed by de Silva et al. (1974c) can determine the parent drug and its major metabolites. The assay involved the determination of the major unconjugated metabolites, N-mono-desalkyl- and N-di-desalkylflurazepam, and the conjugated metabolite, N-1-hydroxyethylflurazepam, and employed a combination of selective extraction, thin-layer chromatographic separation, and differential pulse polarography. The supporting electrolyte was pH 4 phosphate buffer, and the compounds all exhibited polarographic E_p values of -0.80 V $\pm$ 0.03 V vs s.c.e. The assay was applied to the measurement of urinary levels of the metabolites of flurazepam following a single 90 mg dose (Table IV). An assay in which 0.1 ml of urine was diluted with 1M phosphate buffer (pH 7.0) and measured ($E_p = -0.955$ V vs s.c.e.) directly was also presented. This assay with a sensitivity limit of 5 µg ml^{-1} was designed to be used as a rapid means of measuring "total" benzodiazepines to confirm ingestion of flurazepam. The authors also demonstrated how polarographic methods could be used to study chemical equilibria. The rate of hydrolysis of flurazepam was monitored by simultaneously measuring the height of the reduction peaks for the 4,5-azomethine group and for the carbonyl group of its benzophenone derivative at -0.595 V and -0.715 V vs s.c.e. respectively, and indicated a reaction half-life of 44.5 minutes. This reaction was found to be completely reversible by adjustment of the pH to 7.0.

Two more recent reports have dealt with the measurement of flurazepam and its metabolites in plasma (Clifford et al., 1974) and in blood (Brooks and Hackman, 1975) by differential pulse polarography. The assays have similar limits of sensitivity (10–20 mg ml^{-1}) and could be used to measure levels of N-desalkylflurazepam following chronic administration or an overdose of flurazepam. Levels of the parent drug and N-1- hydroxylethyl metabolite are below the limits of detection.

Groves (1976) has made a detailed study of the acid-base and polarographic behaviour of flurazepam and its metabolites and his comprehensive work should be consulted for further reference.

VI. CONCLUSION

Voltammetric analysis is a rapid and sensitive technique for the measurement of drugs in pure solutions, bulk materials, dosage forms and biological fluids. The technique should be considered as an alternative to chromatographic and photometric methods routinely employed to solve pharmaceutical problems. The voltammetric studies discussed in this chapter for psychotropic, hypnotic and sedative compounds demonstrate the excellent utility of the method. Further research using voltammetry to solve pharmaceutical problems should deal with increasing the selectivity and specificity of voltammetric measurements. This will enable the drug of interest to be measured in the presence of breakdown products or metabolites which are usually structurally related and show similar electrochemical activity. The design of reliable voltammetric detectors for liquid chromatographic analysis to improve the selectivity of the methods is a major area of present electroanalytical research.

REFERENCES

Abruzzo, C., Brooks, M. A., Cotler, S. and Kaplan, S. A. (1976). *J. Pharm. Biopharm.* **4**, 29.
Barlot, J. and Albisson, C. (1956). *Chim. Anal. (Paris),* **38**, 313.
Barrett, J., Smyth, W. F. and Hart, J. P. (1974). *J. Pharm. Pharmacol.* **26**, 9.
Beckett, A. H., Essien, E. E. and Franklin Smyth, W. (1974). *J. Pharm. Pharmacol.* **26**, 399.
Berry, D. J. (1971). *Clin. Chim. Acta,* **32**, 235.
Birner, J. (1961). *Anal. Chem.* **33**, 1955.
Bogatskii, A. V., Andronati, S. A., Gul'tyai, V. P., Vikhlyaev, Y. I., Galatin, A. F., Zhilina, Z. I. and Klygul, T. A. (1971). *J. Gen. Chem. U.S.S.R.* **41**, 1364.
Brandys, J. (1964). *Dissnes Pharm. Warsz.* **16**, 355. *Anal. Abstr.* (1966). **13**, 361.
Brooks, M. A., Bel Bruno, J. J., de Silva, J. A. F. and Hackman, M. R. (1975). *Anal. Chim. Acta,* **74**, 367.
Brooks, M. A. and de Silva, J. A. F. (1975) *Talanta,* **22**, 844.
Brooks, M. A., de Silva, J. A. F. and Hackman, M. R. (1973a). *Amer. Lab.* **5**(9), 23.
Brooks, M. A., de Silva, J. A. F. and Hackman, M. R. (1973b). *Anal. Chim. Acta* **64**, 165.
Brooks, M. A., de Silva, J. A. F. and Hackman, M. R. (1974). Hoffmann-La Roche Inc., unpublished data.
Brooks, M. A. and Hackman, M. R. (1975) *Anal. Chem.* **47**, 2059.
Caille, G., Brown, J. and Mockle, J. A. (1970). *Can. J. Pharm. Sci.* **5**, 78.
Chan, H. K. (1974). *J. Pharm. Pharmacol.* **26** Suppl., 37.
Cimbura, G. and Gupta, R. C. (1965). *J. Forens. Sci.* **10**, 228.
Cleghorn, H. P. (1972). *J. Sci. Res. Council. Jam.* **3**, 150.
Clifford, J. M. and Franklin Smyth, W. (1973). *Z. Anal. Chem.* **264**, 149.
Clifford, J. M., Smyth, M. R. and Franklin Smyth, W. (1974). *Z. Anal. Chem.* **272**, 198.
Cohen, E. M. (1965). "Alternating current polarography of barbituric acid and some 5,5-disubstituted derivatives", Ph.D. Thesis, Rutgers University. University Microfilms, Ann Arbor, Michigan.
Cullen, L. F., Brindle, M. P. and Papariello, G. J. (1973). *J. Pharm. Sci.* **62**, 1708.
Danek, A. and Strozik, H. (1968). *Dissnes Pharm. Warsz.* **18**, 519.
de Silva, J. A. F., Dukersky, I. and Brooks, M. A. (1974a) *J. Pharm. Sci.* **63**, 1943.
de Silva, J. A. F., Bekersky, I., Brooks, M. A., Weinfeld, R. E., Glover, W. and Puglisi, C. V. (1974b). *J. Pharm. Sci.* **63**, 1440.
de Silva, J. A. F., Puglisi, C. V., Brooks, M. A. and Hackman, M. R. (1974c). *J. Chromatogr.* **99**, 461.
Dugal, R., Caille, G. and Cooper, S. F. (1973). *Un. Med. Can.* **102**, 2491.
Dumortier, A. G., and Patriarche, G. J. (1973). *Z. Anal. Chem.* **264**, 153.
Dusinsky, G. (1962). Abhandlungen der Deutschen Akademie der Wissenschaften zu Berlin, Janaer Symposium, 176 pp.

El-Darawy, Z. I., El-Makkawi, H. K. and Saber, T. M. H. (1975). *Pharmazie*, **30**, 94.
Elving, P. J. and Bennett, C. G. (1954). *J. Electrochem. Soc.* **101**, 520.
Fazzari, F. R. and Riggleman, O. H. (1969). *J. Pharm. Sci.* **58**, 1530.
Fidelus, J., Zietek, M., Mikolajek, A. M. and Grochowska, Z. (1972). *Microchim. Acta*, 84.
Goldsmith, J. A., Jenkins, H. A., Grant, J. and Smyth, W. F. (1973). *Anal. Chim. Acta*, **66**, 424.
Groves, J. A. (1976). Ph.D. thesis, London University.
Hackman, M. R., Brooks, M. A., de Silva, J. A. F. and Ma, T. S. (1974). *Anal. Chem.* **46**, 1075.
Halvorsen, S. and Jacobsen, E. (1972). *Anal. Chim. Acta*, **57**, 127.
Hetman, J. S. (1964). *Anal. Chim. Acta*, **30**, 313.
Hynie, I., Prokes, J. and Kacl, K. (1965). *Cslka. Farm.* **14**, 466.
Hynie, I., Prokes, J. and Kacl, K. (1967). *Anal. Abstr.* **14**, 273.
Jacobsen, E. and Jacobsen, T. V. (1971). *Anal. Chim. Acta*, **55**, 293.
Jacobsen, E. and Jacobsen, T. V. (1972). *Anal. Chim. Acta*, **60**, 472.
Jacobsen, E., Jacobsen, T. V. and Rojahn, T. (1973). *Anal. Chim. Acta*, **64**, 473.
Kabasakalian, P. and McGlotten, J. (1959). *Anal. Chem.* **31**, 431.
Kato, S. and Dryhurst, G. (1975). *J. Electroanal. Chem.* **62**, 415.
Lordi, N. G., Cohen, E. M. and Taylor, B. L. (1960). *J. Amer. Pharm. Assoc. Sci. Ed.* **49**, 371.
Lund, W. and Opheim, L. N. (1977). *Anal. Chim. Acta* **88**, 275.
Merkle, F. H. and Discher, C. A. (1964a). *Anal. Chem.* **36**, 1639.
Merkle, F. H. and Discher, C. A. (1964b). *J. Pharm. Sci.* **53**, 620.
Mikolajek, A. (1967). *Acta Pol. Pharm.* **24**, 337.
Mikolajek, A. (1968). *Anal. Abstr.* **15**, 5595.
Mikolajek, A., Krzyanowska, A., and Fidelus, J. (1974). *Z. Anal. Chem.* **272**, 39.
Oelschlager, H. (1963). *Arch. Pharm. Berl.* **296**, 396.
Oelschlager, H. and Bunge (1974). *Arch. Pharm. Berl.* **307**, 410.
Oelschlager, H. and Oehr, H. P. (1970). *Pharm. Acta Helv.* **45**, 708.
Oelschlager, H. and Oehr, H. P. (1974). *Pharm. Acta Helv.* **49**, 179.
Oelschlager, H. and Senguen, F. I. (1973). *Arch. Pharm. (Weinheim)*, **306**, 737.
Oelschlager, H. and Senguen, F. I. (1974a). *Arch. Pharm. (Weinheim)*, **307**, 401.
Oelschlager, H. and Senguen, F. I. (1974b). *Arch. Pharm. (Weinheim)*, **307**, 909.
Oelschlager, H. and Senguen, F. I. (1974c). *Pharmazie*, **29**, 770.
Oelschlager, H. and Senguen, F. I. (1975). *Chem. Ber.* **108**, 3303.
Oelschlager, H., Volke, J. and Kurek, E. (1964). *Arch. Pharm. Berl.* **297**, 431.
Oelschlager, H., Volke, J. and Hoffmann, H. (1966a). *Coll. Czech. Chem. Commun.* **31**, 1264.
Oelschlager, H., Volke, J., Lim, G. T. and Frank, V. (1966b). *Arzneim.-Forsch.* **16**, 82.
Oelschlager, H., Volke, J., Hoffmann, H. and Kurek, E. (1967a). *Arch. Pharm. Berl.* **300**, 250.
Oelschlager, H., Volke, J. and Lim, G. T. (1967b). *Arzneim.-Forsch.* **17**, 637.
Oelschlager, H., Volke, J. and Lim, G. T. (1969a). *Arch. Pharm. Berl.* **302**, 241.
Oelschlager, H., Volke, J., Lim, G. T. and Sprang, R. (1969b). *Arch. Pharm. Berl.* **302**, 946.
Oelschlager, H., Volke, J., Lim, G. T. and Bremer, V. (1970). *Arch. Pharm. Berl.* **303**, 364-370.
Omar, N. M. and El-Rabbat, N. A. (1974). *Can. J. Pharm. Sci.* **9**, 57.
Pflegel, P. and Wagner, G. (1967a). *Pharmazie*, **22**, 60.
Pflegel, P. and Wagner, G. (1967b). *Pharmazie*, **22**, 643.
Porter, G. S. (1964). *J. Pharm. Pharmacol.* **16** Suppl., 24T.
Porter, G. S. (1967). *J. Pharm. Pharmacol.* **19**, 176.
Porter, G. S. and Beresford, J. (1966). *J. Pharm. Pharmacol.* **18**, 223.
Pungor, E., Feher, Z. and Nagy, G. (1971a). *Magy. Kem. Foly.* **77**, 298.
Pungor, E., Feher, Z. and Nagy, G. (1971b). *Acta Chim. Hung.* **70**, 207.
Pungor, E., Feher, Z. and Nagy, G. (1972a). *Anal. Abstr.* **22**, 3501.
Pungor, E., Feher, Z. and Nagy, G. (1972b). *Anal. Abstr.* **23**, 747.

Saber, T. M. H. and Fahmy, A. M. (1972). *Egypt. J. Chem.* **15**, 123.
Schlitt, L., Rink, M. and Von Stackelberg, M. (1967). *J. Electroanal. Chem.* **13**, 10.
Senguen, F. I., and Oelschlager, H. (1975). *Arch. Pharm. Berl.*, **308**, 720.
Senkowski, B. Z., Levin, M. S., Urbigkit, J. R. and Wollish, E. G. (1964). *Anal. Chem.* **36**, 1991.
Smyth, W. F. and Leo, B. (1975). *Anal. Chim. Acta*, **76**, 287.
Smyth, M. R., Beng, T. S. and Franklin Smyth, W. (1977). *Anal. Chim. Acta.* **92**, 129.
Tur'yan, Y. I., Merkyukova, T. V. and Bogdanova, O. V. (1970). *Zh. Analit. Khim.* **25**, 384.
Tur'yan, Y. I., Merkyukova, T. V. and Bogdanova, O. V. (1971). *Anal. Abstr.* **21**, 1455.
Volke, J., El-Laithy, M. M. and Volkova, V. (1975). *J. Electroanal. Chem.* **60**, 239.
Volke, J., Oelschlager, H. and Lim, G. T. (1970). *J. Electroanal. Chem.* **25**, 307.
Volke, J., Wasilewska, L. and Ryvolova-Kejharova, A. (1971). *Pharmazie*, **26**, 399.
Vorel, V., Prokes, J. and Dolezal, V. (1961). *Soudni Lekarstui*, **5**, 49.
Vorel, V., Prokes, J. and Dolezal, V. (1962). *Anal. Abstr.* **9**, 3419.
Wiegrebe, W. and Wehrhahn, L. (1975). *Arzneim.-Forsch*, **25**, 517.
Woodson, A. L. and Smith, D. E. (1970). *Anal. Chem.* **42**, 242.
Yoshino, T., Seno, A. and Sugihara, M. (1960a). *J. Pharm. Soc. Japan* **80**, 1484.
Yoshino, T., Seno, A. and Sugihara, M. (1960b). *Anal. Abstr.* **9**, 4454.
Zuman, P. and Brezina, M. (1962). *In* "Progress in Polarography" (P. Zuman and I. M. Kolthoff, eds), Vol. II, pp. 687–701. Interscience, New York.

Chapter 4

QUANTITATIVE ANALYSIS OF MAJOR ANTIBIOTICS CONTAINING A NITRO GROUP

J. T. BROWNE,

Beecham Pharmaceuticals, Research Division, Brockham Park, Betchworth, Dorking, Surrey, England

I. POLAROGRAPHIC BEHAVIOUR OF THE NITRO GROUP

Although the precise polarographic behaviour of the nitro group is influenced by the nature of its environment, its reduction follows a generalized pattern. In acidic solution reduction to the hydroxylamine only, to the hydroxyalmine and then to the amine in two steps, or directly to the amine in one step can occur. In cases where two waves occur, the largest wave occurs at a small negative potential and has been shown to involve four electrons. This wave is generally constant in height over a wide pH range and is of good definition. In acidic solutions (pH < 4) a smaller two electron reduction wave occurs at more negative potentials and disappears in neutral and alkaline solutions.

The largest wave, which occurs at the least negative potential, is due to the reduction of the nitro group to a hydroxylamine and is irreversible electrochemically. The reaction involved in this reduction process is shown below:

$$R-NO_2 + 4H^+ + 4e^- \longrightarrow R-NHOH + H_2O$$

Although this reduction process is normally observed as a single four-electron reduction step it is really a combination of two separate reduction steps with a short lived intermediate. The first stage is a slow two-electron reduction of the nitro group to a nitroso group. Nitroso compounds are generally reduced at more positive potentials than nitro compounds, thus this nitroso compound is not isolated but reduced immediately to the hydroxylamine in a fast two-electron process. The main wave is therefore a combination of the following reactions:

$$R-NO_2 + 2H^+ + 2e^- \longrightarrow R-NO + H_2O \quad \text{Slow}$$
$$R-NO + 2H^+ + 2e^- \longrightarrow R-NHOH \quad \text{Fast}$$

With the exception of a few compounds, these two waves are combined to give a single four-electron polarographic reduction wave.

The polarographic behaviour of this nitro group wave is not constant for all nitro compounds but is governed by a number of factors including molecular structure, pH, composition of supporting electrolyte and drop time at the dropping mercury electrode (d.m.e.). The reduction potential is independent of concentration in small concentration ranges.

This main nitro group reduction wave is useful in the polarographic analysis of nitro compounds for the following reasons:

(1) It occurs at a small negative potential where a limited number of other polarographic reductions occur. Thus the determination of nitro compounds enjoys a measure of selectivity in the field of polarographic analysis. The selectivity of this technique for determination of such drugs in body fluids is also enhanced by the strong adsorption of some aromatic nitro compounds at the mercury surface in comparison to naturally occurring substances, e.g. proteins.

(2) The wave/peak height is relatively large because of the four-electron reduction step, thus making the polarographic determination of nitro-containing drugs sensitive. In particular differential pulse polarography (d.p.p.) has been reported to determine concentrations of aromatic nitro compounds down to 1×10^{-8}M in pure solutions. The peaks obtained are usually well-defined with half peak widths of the order of 60–70 mV.

(3). The wave/peak height is unaffected by minor changes in pH.

The smaller secondary wave which occurs at more negative potentials during the reduction of nitro compounds involves a two-electron reduction step. This generally corresponds to the reducible protonated hydroxylamine. The un-ionized hydroxylamine is generally polarographically inactive whereas the cation is reduced in a two-electron step to the corresponding amine.

$$R \overset{+}{N}H_2OH + 2H^+ + 2e^- \longrightarrow R \overset{+}{N}H_3 + H_2O$$

As the second wave is smaller and more pH sensitive than the main nitro reduction wave, the latter is more useful in the quantitative polarographic determination of nitro compounds.

Although considerable variations occur between compounds, the general metabolic route for aromatic nitro compounds, into which category most nitro-containing drugs fall, involves similar reduction processes to those occurring in polarographic reduction.

In summary, the most useful polarographic wave for the analysis of nitro-containing drugs is the four-electron reduction wave which generally occurs in the range $E_{\frac{1}{2}} - 0.1 \rightarrow -0.4$ V (vs s.c.e.) where many other compounds are polarographically inactive. The exact reduction potential for this wave is governed by the pH of the supporting electrolyte which, for analytical purposes, is commonly a buffer solution in the range pH 3.5–5.5.

The remainder of this chapter will be concerned with a review of the analytical methods used for the assay of some nitro-containing drugs in both formulations and biological fluids. A critical comparison of the techniques is made in each case.

II. APPLICATIONS

A. CHLORAMPHENICOL

Chloramphenicol, (I)d-threo-1-p-nitrophenyl-2-dichloracetamido-1,3-propanediol, one of the few naturally occurring nitro compounds, is a broad spectrum antibiotic. The compound is widely used, particularly in the treatment of typhoid fever and influenzal meningitis and is thought to act by a specific inhibition of amino acid incorporation into ribosomes. Since the compound was first isolated nearly thirty years ago a variety of analytical techniques, including polarography, have been applied to its determination.

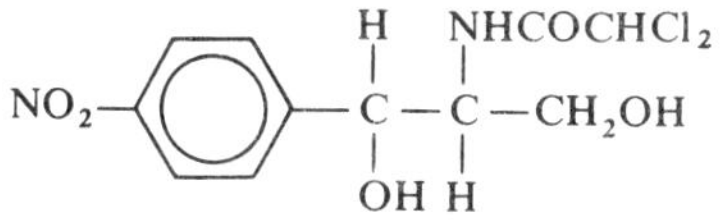

$$NO_2-\underset{}{\bigcirc}-\underset{\underset{OH}{|}}{\overset{\overset{H}{|}}{C}}-\underset{\underset{H}{|}}{\overset{\overset{NHCOCHCl_2}{|}}{C}}-CH_2OH$$

(I) Chloramphenicol

(i) *Polarographic analysis*

The polarographic behaviour of chloramphenicol has been studied by a number of workers and used as the basis for a quantitative analytical technique. The most useful property of this compound from the viewpoint of polarographic analysis is the four-electron nitro group reduction which manifests itself polarographically at a small negative potential. The compound gives rise to a smaller reduction wave at more negative potentials which is susceptible to changes in the composition of the supporting electrolyte and is therefore of little analytical interest. Chloramphenicol can be determined from the height of the nitro reduction wave by reference to a standard chloramphenicol preparation either by the standard addition method or with a calibration curve. This diffusion controlled wave is well characterized and highly reproducible, the current being proportional to concentration over a wide range.

The original polarographic method for the analysis of chloramphenicol was developed by Hess (1950) who used a buffered solution at pH 4 as supporting electrolyte. However, the large amounts of thymol added for maxima suppression caused the waves to become deformed and measurement was irreproducible. Later workers have recommended gelatine (Russu and Cruceanu 1965a) and methylene blue (Summa, 1965) as preferred maximum suppressors.

Faith and Hancak (1967) found that chloramphenicol could be determined in 37.5% acetic acid, using 1% gelatine for maximum suppression, with an error of less than $\pm 2\%$ by measurement of the height of the nitro group wave which occurred at $E_{\frac{1}{2}} = -0.33$ V. A more common supporting electrolyte, acetate buffer at pH 4.7 with 0.003% decylamine as maximum suppressor, was used by Fossdal and Jacobsen (1971) who determined chloramphenicol over a concentration range of 0.3–600 μg ml^{-1}. They also found it possible to determine chloramphenicol in milk down to 0.3 p.p.m. by mixing equal volumes of pH 6.4 acetate buffer with the milk samples before polarographic analysis under nitrogen. Chloramphenicol yielded a poorly formed second wave at pH 4.7 which was of little analytical utility and corresponded to reduction of the hydroxylamine group to the amine. Milk proteins were found to be surface active and caused some distortion of the polarographic wave. Proteins also caused the diffusion current to be depressed and the wave to be shifted to more negative potentials; Zuman (1975) reported a similar level of sensitivity (1 μg ml^{-1}) for the analysis of chloramphenicol in acetate buffer at pH 4. The superior sensitivity of differential pulse polarography over the standard d.c. technique was demonstrated by Siegerman (1975) (see Fig. 1) who achieved a significant improvement in detection limit for chloramphenicol to 0.1 μg ml^{-1} using the differential pulse technique and an acetate buffer solution (pH 4).

In an unpublished study, Siegerman and Kral were able to detect chloramphenicol in this acetate buffer using a commercial microprocessor controlled polarographic analyser which employed background subtraction techniques to remove current contributions from the supporting electrolyte. Van Bennekom (private communication), using a modified differential pulse polarograph, have determined chloramphenicol at levels of the order of ng ml^{-1} in 0.1M acetate buffer.

The selectivity of polarographic analysis has been exploited by Pflegel (1973) and Pflegel and Shoukrallah (1974) for the determination of chloramphenicol and its palmitate in a solution without prior separation. The

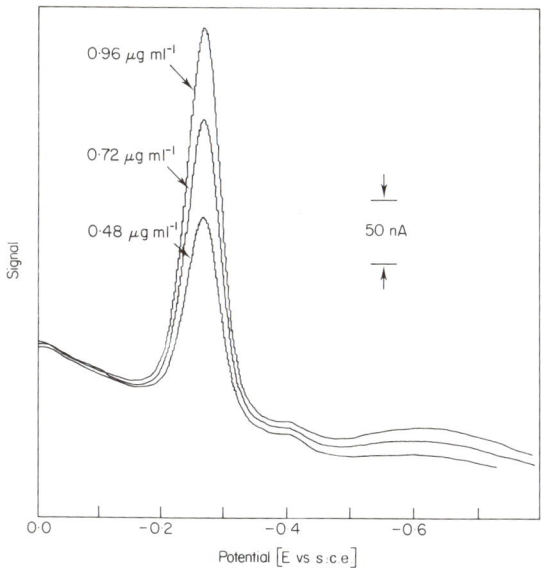

FIG. 1. Differential pulse polarograms of chloramphenicol in 0.1M acetate buffer pH 4. (Instrumental conditions: drop time 1 s; pulse amplitude 25 mV; scan rate 2 mV s^{-1}).

palmitate moiety was found to promote a marked adsorption of chloramphenicol palmitate on the dropping mercury electrode in the pH range 4–6, whereas chloramphenicol was not adsorbed. As a result of this adsorption chloramphenicol palmitate was more easily reduced than chloramphenicol itself with a difference in half wave potential of 140 mV. The optimum conditions for the determination of chloramphenicol in the presence of the palmitate were found to be a 1:1 mixture of 0.1M acetate buffer (pH 4.7) and ethanol whereas for simultaneous determination of the two compounds a Britton–Robinson buffer (pH 6.0) ethanol mixture was preferred.

Drug precursors and analogues are likely to occur as impurities in the drug itself and therefore must be analysed and controlled on a routine batch to batch basis. Mixtures of the chloramphenicol synthetic intermediates, 1-*p*-nitro-2-acetamido-3-hydroxypropiophenone (II), 1-*p*-nitrophenyl-2-acetamido-1,3-propanediol (III) and 1-*p*-nitrophenyl-2-amino-1,3-propanediol (IV), may be determined without complete separation in the reaction mixture. Dumanovic *et al.* (1971) found it possible to polarographically determine II and III, or II and IV simultaneously in mixtures. However, Summa (1965) found that it was not possible to differentiate polarographically between chloramphenicol and its precursor, the biologically inactive free amine 1-*p*-nitrophenyl-2-amino-1,3-propanediol (IV).

$O_2N-\underset{}{\bigcirc}-\underset{\underset{O}{\|}}{C}-\underset{\underset{NH-COCH_3}{|}}{CH}-CH_2OH$

(II) 1-*p*-nitro-2-acetamido-3-hydroxypropiophenone

$O_2N-\underset{}{\bigcirc}-CH-\underset{\underset{OH}{|}}{\underset{\underset{NH-COCH_3}{|}}{CH}}-CH_2OH$

(III) 1-*p*-nitrophenyl-2-acetamido-1,3-propanediol

$O_2N-\underset{}{\bigcirc}-CH-\underset{\underset{OH}{|}}{\underset{\underset{NH_2}{|}}{CH}}-CH_2OH$

(IV) 1-*p*-nitrophenyl-2-amino-1,3-propanediol

Other nitro compounds involved in the synthesis of chloramphenicol, N-(β-hydroxy-4-nitrophenethyl)-acetamide ($E_{\frac{1}{2}} = -0.66$ V) and N-(4-nitrobenzoylmethyl)-acetamide ($E_{\frac{1}{2}} = -0.45$ V), may also be determined polarographically in a carbonate buffer solution pH 10.5 where they give distinguishable waves proportional to concentration in the range 0.1–10 mM (Gromova et al., 1969.

Chloramphenicol can be determined directly in simple pharmaceutical formulations such as capsules, suppositories, injectable products, eye drops, ear drops and nose drops where the only sample preparation required is dissolution and dilution in alcohol, (Summa, 1965; Russu and Cruceanu, 1965a). For lipid-containing preparations, e.g. ointments and creams, a preliminary extraction of the drug with organic solvents is required before the standard polarographic determination in buffered solution can be applied (Russu and Cruceanu, 1965b). The standard U.S.P. XVI extraction procedure is suitable for application to ointment formulations whereas a modified extraction procedure is required for creams which form an oil in water emulsion (Summa, 1965). This extraction procedure is also suitable for isolation of chloramphenicol from blood and urine (Levine and Fischbach, 1951). Syrup formulations commonly contain a number of excipients which are polarographically active and may interfere in the analysis, e.g. antioxidants, thickening, sweetening, colouring and flavouring agents. Extraction of chloramphenicol cinnamate from a syrup formulation containing other antibacterial agents may be achieved by boiling for a short period with absolute ethanol before dilution and analysis (Corsi and Mecarelli, 1971).

For the majority of drugs containing a nitro group, the primary metabolic pathway involves reduction of the nitro group to an amine. As the nitro group is polarographically active and amenable to quantification while the amino group is inactive, it is often possible to determine the drug in the presence of its metabolites without prior separation. Although chloramphenicol is largely metabolized in man, with only 5–10% of the drug being excreted unchanged, it appears that little reduction of the nitro group occurs in the metabolic process. The two major metabolites excreted are reported to be the 3-glucuronide and the deacylated chloramphenicol (Fig. 2).

Although a procedure for the extraction of chloramphenicol from blood and urine prior to polarographic determination had been developed by Levine and Fischbach (1951) an extraction step was found to be unnecessary by Brezina and Zuman (1958) who mixed blood samples containing chloramphenicol with sodium hydroxide containing capryl alcohol to prevent

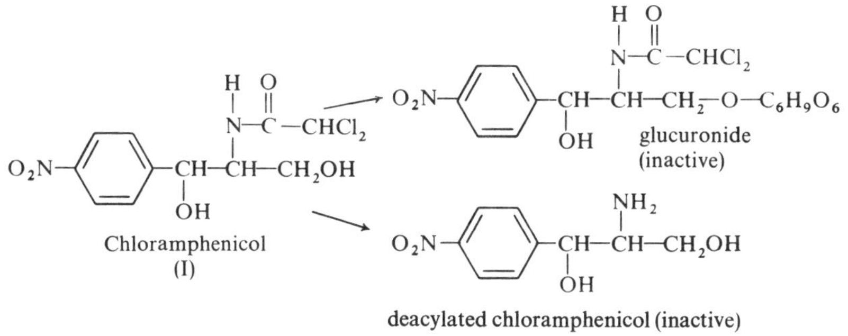

FIG. 2. Major metabolic pathway of chloramphenicol.

foaming, prior to polarographic analysis. For the determination of chloramphenicol in urine the same authors found that the only sample preparation required was simple dilution in a buffer solution.

Polarographic analysis has been used to the limits of its sensitivity and selectivity by different workers for the analysis of chloramphenicol. Sensitivity limits have been reduced to 1 ng ml^{-1} for the raw material by the use of a pH 4 acetate buffer and modified differential pulse polarography while the selectivity of the technique has been exploited for the simultaneous determination of the chloramphenicol and related compounds in the same solution without prior separation.

(ii) *Comparison with other analytical techniques*

Of all the analytical methods which can be applied to the determination of antibiotics, microbiological assay probably gives the most reliable measurement of true drug potency. Although microbiological methods are generally slow and imprecise, variations on the basic technique have reduced the time required for chloramphenicol analysis down to a minimum of two hours. Automation of the microbiological assay method for chloramphenicol is also an advantage, and has helped to make microbiological analysis of chloramphenicol competitive with the more rapid and precise instrumental techniques available. However, the technique still requires considerable care and is not popular as a routine control analytical method.

Chloramphenicol is unsuitable for analysis by direct titrimetric methods. A preliminary reaction or complexation stage is usually required prior to the final quantitative determination. This restriction largely negates the main advantages of titrimetric analysis which are speed and simplicity.

Direct spectrophotometry is one of the simplest and most rapid quantitative analytical methods available for the determination of chloramphenicol. Ultraviolet spectrophotometry is accepted as an official method for the determination of chloramphenicol potency (Federal Register, 1961). However, the direct u.v. spectrophotometric method is non-selective and insensitive to degradation as several decomposition products absorb at the same wavelength as the parent drug. Thus, in order to use this method for determination of the stability of chloramphenicol as a raw material or formulated products, it is necessary to employ a separation stage, normally chromatographic, prior to u.v. spectrophotometric measurement. Colorimetric analytical methods are generally more selective than direct measurement by u.v. spectrophotometry and therefore more widely applicable. In general the nitro group is reduced to an amine prior to diazotization and coupling to give a coloured azo dye for photometric determination.

Chloramphenicol may be determined by gas–liquid-chromatography with minimal sample preparation and a detection limit of 0.01 μg (Yamamoto et al., 1967). However, this procedure does not give complete separation of chloramphenicol from related compounds. For more complex samples it is therefore necessary to employ derivatization prior to chromatography. Although hexamethyldisilazane has been used successfully for this derivatization, N, O-bis(trimethylsilyl)-acetamide (BSA) is the preferred reagent due to its rapid action and the formation of a single phase system indicating the completion of the reaction (Margosis, 1974).

In contrast to these methods, the majority of which require a preliminary reaction of chloramphenicol prior to quantitative determination, chloramphenicol can be determined directly by polarography. It would appear that polarographic analysis has major advantages in terms of minimal sample preparation and high selectivity over the majority of alternative analytical methods which can be used in the determination of chloramphenicol.

B. NITROFURANS

The bactericidal properties of nitrofurans (Fig. 3) were demonstrated in 1944 and this group of compounds have been used widely as antibacterials ever since. There are a wide variety of derivatives on the basic nitrofuran

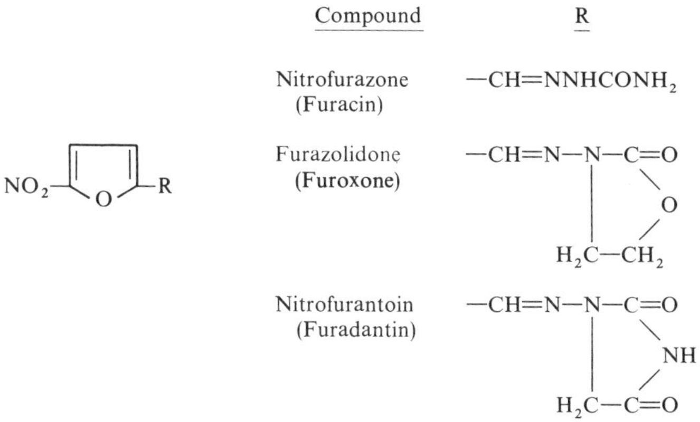

Compound	R
Nitrofurazone (Furacin)	—CH=NNHCONH$_2$
Furazolidone (Furoxone)	—CH=N—N—C=O, |, O, /, H$_2$C—CH$_2$
Nitrofurantoin (Furadantin)	—CH=N—N—C=O, |, NH, /, H$_2$C—C=O

FIG. 3. Nitrofuran structures.

structure. However, the nitro group in the 5 position is essential for activity and is therefore common to all drugs in the series.

(i) *Polarographic analysis*

The presence of the nitro group, which is susceptible to polarographic reduction, renders this class of compounds suitable for polarographic analysis. The effect of substituents in the furan ring on the potential of reduction of the nitro group has been studied by Stradins and Hillers (1964) who found that little qualitative information could be gained by an examination of the $E_\frac{1}{2}$ value, since they could not measure the reduction potential accurately enough.

The effects of pH and variation in supporting electrolyte may be used to advantage in selecting the optimum conditions for determination of a particular nitrofuran in a mixture of closely related compounds. However, for quantitative analysis it is important to ensure that samples and standards are prepared in identical media. Individual determination of furazolidone, nitrofurantoin, nitrofurazone, acinitrazole and metronidazole may be achieved by careful choice of the pH between 4.2 and 7.8 (Vignoli and Cristau, 1963). This method was found to be suitable for the determination of stability in solution and the drug content of tablet formulations. The Ilkovic equation was obeyed over the range of 10–40 µg of drug ml^{-1} and reproducible results were found with a relative standard deviation of ±3%.

Furazolidone has been determined polarographically in a buffer containing sodium salicylate and salicylic acid by Pasich and Lehmann (1964). They preferred the use of sodium carboxymethylcellulose as an internal standard rather than the standard addition technique used by Jones et al. (1965) in the polarographic determination of other nitrofurans. Extraction of 3,5 dinitrofurfurylidene-3-oxazolidone from a drug combination with acrinol may be achieved with methyl cellosolve at 40–50°C prior to polarographic determination at pH 7 with potassium chloride supporting electrolyte in the concentration range 10^{-4}–10^{-5}M (Suggii et al. 1965). Furidene, a thio compound, can be determined polarographically in ammonia buffer solution from the characteristic nitro group reduction wave (Milch and Hollos, 1965).

Nitrofurantoin may be extracted from oral suspensions and tablets with dimethylformamide prior to dilution with an equimolar solution of ammonium chloride and ammonium hydroxide for polarographic analysis. Summa (1962) found that this polarographic method was less susceptible to interference from formulation excipients than the standard U.S.P. XVI spectrophotometric method for determination of nitrofurantoin in oral suspensions. Simple dissolution in sodium salicylate and dilution with buffer is the only sample preparation required for the polarographic analysis of N-(5-nitro-2-furfurylideneamino)-imidazoline-2-one. The standard microbiological assay for these antibacterial agents requires a minimum of 21 hours whereas the polarographic method is a more sensitive assay and results could be obtained within one hour of receipt of samples. Preliminary drug extraction from urine is not required as no polarographically active substances interfere within the voltage range -0.1 to -0.6 V. The only sample preparation necessary is dilution with supporting electrolyte. In order to avoid differences in diffusion current which may occur with urine samples from different subjects, the standard addition technique is preferred for measurement of drug concentration (Jones et al., 1965). Other nitrofurans may also be analysed in urine and blood plasma by polarography, e.g. furazoline (Stradins et al., 1963).

The determination of nitrofurantoin in urine has been investigated by Mason and Sandman (1976) who used a rotating platinum electrode placed directly in urine samples taken from clinical patients. This method was found faster, more precise and accurate than the commonly used colorimetric assay. The levels encountered in the patients' urine varied from 3.5 to 71.0 mg ml^{-1} and the method employed simple and inexpensive d.c. instrumentation.

Burmicz et al. (1976) have made a detailed study of the acid-base behaviour and polarographic reduction mechanism of nitrofurantoin and suggested that hydroxylamine and amino metabolites could be measured simultaneously with the parent compound following certain corrections.

Nitrofurans are commonly included in animal feeds for medication and may be analysed polarographically after ether extraction from the feed.

Nitrofurazone and furazolidone may be determined either alone or in combination by an initial determination of total nitro compound, followed by a repeat analysis after selective destruction of furazolidone with alkali (Fricke et al., 1964).

(ii) *Comparison with other analytical methods*

Microbiological methods may be used to determine low levels of nitrofurans in a variety of samples, from pharmaceutical formulations to milk. At least one author (Monciu et al., 1971) claims that the microbiological method is much simpler than alternative chemical or physical analytical methods for the determination of nitrofurans.

Titrimetric analysis may also be used in the determination of nitrofurans and titanium trichloride would appear to be the most useful volumetric reagent (Cere et al., 1968).

Apart from the unselective nature of spectrophotometric analysis, the isomerism of nitrofurans which occurs on exposure to light makes their spectrophotometric determination inaccurate. Colorimetric analysis appears to be a very useful method for the determination of nitrofurans. The drug is normally extracted with nitromethane and reacted with hyamine to form a complex for colorimetric measurement. This method appears to be highly selective for the determination of nitrofurans in the presence of other antibacterial agents and is applicable to body fluids (Conklin and Hollifield, 1966).

Nitrofurans are generally unsuitable for direct g.l.c. analysis and require chemical modification prior to chromatography and determination. This makes the g.l.c. analytical precedures long and tedious.

Polarographic analysis appears to be very suitable for the determination of nitrofurans. The selectivity of the technique is useful for the determination of different nitrofurans in the same sample. This method may also be applied to nitrofurans in formulations and body fluids with a minimum of sample preparation.

C. NITROIMIDAZOLES

The most important nitroimidazoles with trichomonicidal activity are metronidazole and dimetridazole (Fig. 4). Nitroimidazoles are commonly used as antimicrobial agents in feed additives as a means of treating histomoniasis in poultry, swine dysentry and bovine venereal trichomoniasis.

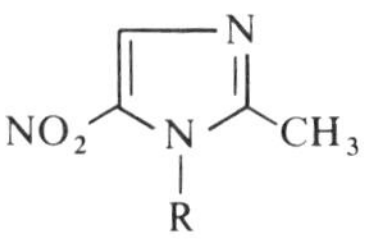

Compound	R
2-Nitroimidazole	H
Dimetridazole	CH_3
Metronidazole (Flagyl)	CH_2CH_2OH
Tinidazole	$CH_2CH_2SO_2CH_2CH_3$
Ornidazole	$CH_2CHOHCH_2Cl$

Nitroimidazole

FIG. 4. Structures of nitroimidazoles

(i) *Polarographic analysis*

As this group of compounds all contain a nitro group they are suitable for polarographic analysis by measurement of the nitro group reduction wave. Polarographic analysis appears to be one of the most widely used analytical procedures for the determination of nitroimidazoles in pharmaceutical or agricultural products. Metroinidazole may be determined polarographically in a reaction mixture containing 1-(2-hydroxyethyl)-2-methyl-4-nitroimidazole and 2-methyl-5-nitroimidazole. Although all compounds are analysed by reduction of the nitro group, it is possible to achieve individual quantitative determination of all three compounds, without prior separation, by careful control of pH. The half wave potentials of the three compounds are well separated only at high pH values where the compounds become unstable. It is therefore necessary to compromise between separation of half wave potentials and instability in the choice of experimental conditions. The optimum conditions require the use of 0.63N sodium hydroxide as solvent, the analysis being performed within 15 minutes of sample preparation (Dumanovic *et al.*, 1966).

Although high pH was required for the above separation the routine analysis of metronidazole is best carried out in acidic solutions where instability is not a problem. For raw material and tablet analysis a pH of 2.4 was preferred (Danek, 1961).

Polarography has been used to determine the position of the nitro group in N-substituted-2-methyl-nitroimidazoles by the measurement of half wave potentials in pH 4.1 Britton–Robinson buffers (Slamnik, 1976).

Tinidazole, a lesser known antitrichomonal agent, can also be determined in tablet formulations by d.c. polarography (Slamnik *et al.*, 1967).

Veterinary antibiotics are commonly administered in animal feeds and polarographic analysis is one of the most suitable procedures for determination of these drugs in the latter preparation. Dimetridazole can be extracted from feed samples and determined in saturated borax solution with a silver wire anode (Kane 1961). However, by this method extraneous materials are extracted from the feed which do not specifically interfere at the polarographic reduction value, but generally depress the diffusion current by 30 to 35%.

These materials may be removed by methanol extraction of the drug from the feed, with subsequent purification on an alumina chromatographic column prior to polarography at a dropping mercury electrode (Daftsios, 1964). Although the latter extraction method has a high recovery, average 98.5%, with a low standard deviation of 0.98 on three sample feeds tested, it is a more lengthy and tedious procedure than the former saturated borax method. Animal feeds commonly contain metal ions and the presence of copper enhances the peak height of the dimetridazole polarographic reduction wave. Cooper and Hoodless (1967) developed a polarographic method for the determination of dimetridazole in saturated borax solution, in which copper interference is prevented by complexation with cyanide. This procedure has been slightly adapted to involve sample extraction and treatment with sodium tetraborate and potassium cyanide before polarographic measurement by the Prophylactics in Animals Feeds Sub-Committee of the Society for Analytical Chemistry (1969).

Metronidazole and dimetridazole may be assayed directly in urine, serum, saliva, milk and eggs without prior extraction or separation from the biological material. However, this determination is not selective and measures metabolites with a nitro group together with the main drugs (Kane, 1961).

Metronidazole is metabolized in man to the extent of approximately 35% by oxidation of the 2-methyl group to the hydroxymethyl derivative. A lesser 1-acetic acid metabolite is formed to the extent of approximately 10% while approximately 35% of the drug is excreted unchanged (Fig. 5).

The nitro group in metronidazole and its two major metabolites behave

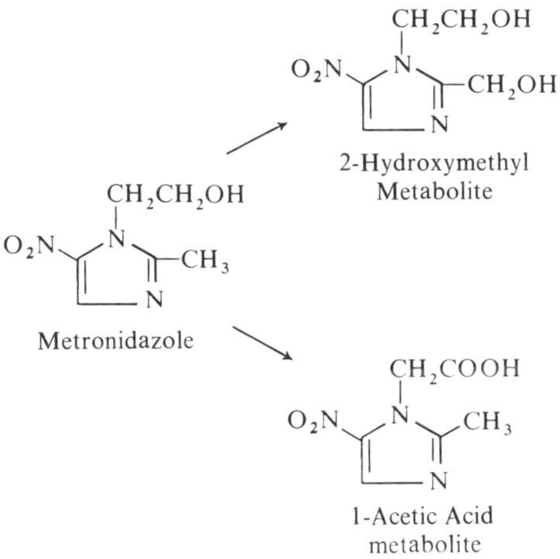

FIG. 5. Major metabolic pathway of metronidazole.

similarly under the conditions of polarographic assay, thus rendering these compounds indistinguishable polarographically. In order to carry out a specific determination of the drug substance, a preliminary extraction from the body fluid with ethyl acetate is required, with subsequent thin layer chromatographic separation. The sample is eluted from the silica gel with methanol prior to evaporation and dissolution in sodium hydroxide for polarographic analysis (de Silva et al., 1970). This extraction and clean-up procedure may be shortened to eliminate the t.l.c. separation stage by the use of protein precipitation. The compound is selectively extracted from protein free filtrate of plasma or urine buffered to pH 7. The residue of the ethyl acetate extract is dissolved in sodium hydroxide and analysed polarographically, with a sensitivity of 0.1 $\mu g\, ml^{-1}$ fluid (Brooks et al., 1976).

Dimetridazole is commonly used in veterinary medicine and its distribution in animal tissues and body fluids is therefore of interest. Polarography has been used successfully to determine dimetridazole in tissues, blood plasma, faeces and urine of guinea pigs with a detection limit of 2 $\mu g\, g^{-1}$ in solid matter and 0.1 $\mu g\, ml^{-1}$ in fluids (Allen and McLaughlin, 1972). However, Parnell (1973) found that the detection limit for dimetridazole in pig tissue could be extended to 0.1 $\mu g\, g^{-1}$ by the use of Tast polarography.

Lesser known nitroimidazoles such as tinidazole (Taylor et al., 1969) and impronidazole (Macdonald, private communication) have also been determined satisfactorily in animal tissues.

(ii) *Comparison with other analytical methods*

The nitroimidazole class of compounds may be determined by microbiological analysis. Serum samples may be analysed by the standard agar plate diffusion method after removal of interfering antibiotics by inactivation.

The standard non-aqueous titrimetric method may be applied to the determination of nitroimidazoles and the use of acetic anhydride as solvent with malachite green indicator to give a sharper end point than the British Pharmacopoeia method. However, as this method is non-selective, its applications are limited.

In order to overcome the non-selectivity of spectrophotometric analysis, it is necessary to employ a preliminary column chromatographic separation. This can be used to remove interfering agents and separate similar drugs.

In the colorimetric determination of nitroimidazoles a preliminary reaction step is required before determination. This reaction normally involves reduction to an amine.

Gas–liquid chromatography, which is one of the main methods for the determination of nitroimidazoles, can be applied to samples in animal feeds, pharmaceutical formulations and biological fluids. The method normally requires derivatization but can be used to determine ipronidazole and dimetridazole in the same sample without separation. The advantages of the g.l.c. method for determination of ipronidazole have led to its adoption as a standard method for determination of this compound in feeds by the Association of Official Analytical Chemists.

Polarography can be used to determine several nitroimidazoles in a range of samples without drug extraction, byt the careful control of pH. However, major drug metabolites must be chromatographically removed prior to biological fluid analysis. Animal feeds require extraction, chromatographic purification and complexation of interfering copper prior to polarographic analysis of nitroimidazoles.

REFERENCES

Allen, P. C. and McLoughlin, D. K. (1972). *J. Assoc. Off. Anal. Chem.* **55**, 1159.
Brezina and Zuman, P. (1958). "Polarography in Medicine Biochemistry and Pharmacy", p. 321. Interscience, Chichester and New York.
Brooks, M. A., D'Arconte, L. and de Silva, J. A. F. (1976). *J. Pharm. Sci.* **65**(1), 112.
Burmicz, J. S., Franklin Smyth, W. and Palmer, R. F. (1976). *Analyst*, **101**, 986.
Cere, L., Bomiouanni, G. and Mondino, A. (1968). *Annali. Chim.* **58**(11), 1268.
Conklin, J. D. and Hollifield, R. D. (1966). *Clin. Chem.* **12**(9), 632.
Cooper, P. J. and Hoodless, R. A. (1967). *Analyst*, **92**, 520.
Corsi, R. and Mecarelli, E. (1971). *Boll. Chim. Farm.* **110**(3), 147.
Daftsios, A. C. (1964). *J. Assoc. Off. Agric. Chem.*, **47**(2), 231.
Danek, A. (1961). *Dissert. Pharm. Krakow*, **13**(1), 107.
de Silva, J. A. F., Munno, H. and Stronjny, N. (1970). *J. Pharm. Sci.* **59**(2), 201.
Dumanovic, D., Volke, J. and Jouanovic, R. (1971). *J. Assoc. Off. Anal. Chem.* **54**(4), 882.
Dumanovic, D., Volke, J. and Vajgand, U. (1966). *J. Pharm. Pharmacol.* **18**, 507.
Faith, L. and Hancak, P. (1967). *Farm. Obzou*, **36**, 220.
Fossdal, K. and Jacobsen, E. (1971). *Anal. Chim. Acta*, **56**(1), 105.
Fricke, F. L., Keppel, G. E. and Hart, S. M. (1964). *J. Assoc. Off. Anal. Chem.* **47**(4), 787.
Gromova, E. V., Avrutskaya, I. A. and Fioshin, Ya. M. (1969). *Khim-farm. Zh.*, **3**(10), 51.
Hess, G. B. (1950). *Anal. Chem.* **22**, 5, 649.
Jones, B. M., Ratcliffe, R. J. and Stephens, S. G. (1965). *J. Pharm. Pharmacol.*, Suppl. 17, 52S.
Kane, P. O. (1961). *J. Polarog. Soc.* **8**, 58.
Levine, J. and Fischbach, H. (1951). *Antibiot. Chemotherapy.* **1**, 59.
Margosis, M. (1974). *J. Pharm. Sci.* **63**(3), 435.
Mason, W. D. and Sandman, B. (1976). *J. Pharm. Sci.* **65**(4), 599.
Milch, G. and Hollos, J. (1965). *Pharm. Zentralhalle*, **104**(8), 564.
Monoiu, D., Dobo, D., Constanta, V. and Economu, V. (1971). Gyogyszereszet, **15**(2), 10.
Parnell, M. J. (1973). *Pestic Sci.* **4**, 643.
Pasich, J. and Lehmann, M. (1964). *Farm. Polska*, **20**, 731.
Pflegel, P. (1973). *Pharmazie*, **28**(7), 483.
Pflegel, P. and Shoukrallah, I. (1974). *Pharmazie*, **29**(5), 316.
Russu, C. and Cruceanu, I. (1965a). *Farmaco. Ed. Prat.* **20**(1), 22.
Russu, C. and Cruceanu, I. (1965b). *Rev. Chim.* **16**, 347.
Siegerman, H. (1975). *Methods in Enzymology*, **43**, 373.

Slamnik, M. (1976). *J. Pharm. Sci.* **65**, 736.
Slamnik, M., Kajtez, F. and Sunjic, V. (1967). *J. Polarog. Sci.* **13**, 83.
Stradins, J. and Hillers, S. (1964). *In* "Nitro Compounds" (T. Urbanski, ed.). Pergamon Press, Oxford.
Stradins, J., Aizpuriete, I. and Hillers, S. (1963). *Latvijas, P.S.R. Zinatnu. Akad Vestis Kim Ser* **1**, 29.
Suggii, A., Kabasawa, Y. and Hasegewa, N. (1965). *Nippon Dingaka Yakugaka Henkya Hokoku* **7**, 1.
Summa, A. F. (1962). *J. Pharm. Sci.* **51**(5), 474.
Summa, A. F. (1965). *J. Pharm. Sci.* **54**(3), 442.
Taylor, J. A., Jr, Migliardi, J. K. and Schach Von Wittenan, M. (1969). *Antimicrob. Agents and Chem.* 267.
Vignoli, L. and Cristau, B. (1963). *Chim. Anal. (Paris)* **45**(10), 499.
Yamamoto, M., Iguchi, S. and Amoyama, T. (1967). *Chem. Pharm. Bull, Tokyo,* **15**(1), 123.
Zuman. P. (1975). *Proc. Soc. Anal. Chem.* **12**(7), 199.

Chapter 5

OXIDATIVE VOLTAMMETRIC ANALYSIS OF MOLECULES OF PHARMACEUTICAL IMPORTANCE

I. E. DAVIDSON

*Wyeth Laboratories, Huntercombe Lane South,
Taplow, Maidenhead, Berkshire, England*

I. INTRODUCTION

The most common electrochemical process used in organic analysis is the reduction of molecules at the dropping mercury electrode (d.m.e.). The application of oxidation processes is less common, principally due to the small anodic potential range given by mercury and the lack of reproducible results obtained at solid indicator electrode surfaces. However, bearing in mind that interferences occurring in certain reduction processes can be avoided by resorting to an analytical procedure based on an oxidation process and with the development of electrochemical detectors, particularly for use with h.p.l.c., the application of these oxidation processes to organic analysis is currently being reappraised.

In the following review of anodic oxidations, particularly of interest to workers in the pharmaceutical field, reference will be made not only to compounds used therapeutically in formulations, but also to those which are amenable to electrolytic oxidation and which are possible process materials or intermediates in the synthetic routes to pharmaceutical products, and are thus subject to analytical control. In addition, voltammetric processes involving oxidation used to determine drugs and metabolites in body fluids will be described as also will be the pharmaceutical applications of amperometric titrations based on anodic processes.

Compounds will be considered under headings based on drug indication and biological function, viz. vitamins, analgesics, central nervous system agents and tranquilizers/sedatives, antituberculins, naturally occurring compounds and miscellaneous compounds, and the chapter will be concluded with some general reviews on anodic oxidations.

II. VITAMINS

A. VITAMIN C (1-ASCORBIC ACID) (I)

This is perhaps the best-known example of the application of anodic oxidation to pharmaceutical analysis and much information has appeared in the literature on the polarography of this compound, and its related isomer erythorbic acid (isoascorbic or d-araboascorbic acid).

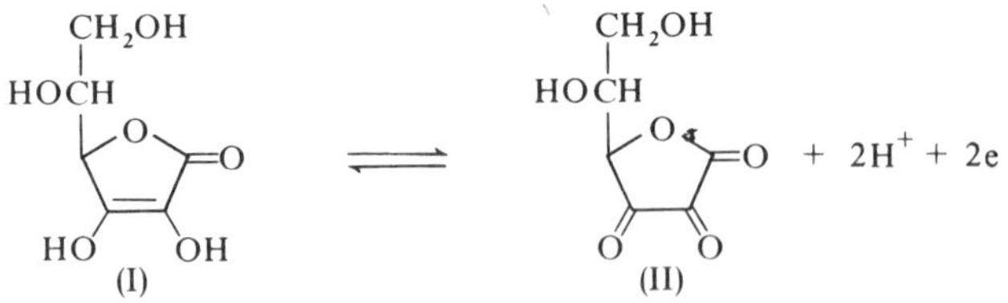

$$\text{(I)} \rightleftharpoons \text{(II)} + 2H^+ + 2e$$

The early work is well-documented in the monograph of Brezina and Zuman (1958) and details of conditions for the determination of the compound are given. Briefly the analysis is best performed in buffer solutions from pH 2.0–7.0 to prevent chemical oxidation by traces of oxygen. (Ascorbic acid is readily oxidized and is widely used in pharmacy to prevent oxidation of other pharmacologically active compounds.) A wide range of buffer systems has been employed and the above reference is best consulted for a full review. Complex formulations can be assayed for Vitamin C without separation of other vitamins and excipients. Further details of analytical methods for pharmaceutical preparations are provided by Heyrovsky and Zuman (1968), who describe the electrode process as one producing dehydroascorbic acid (II) by oxidation; the process is pH-dependent, half-wave potentials of + 0.18 V and −0.13 V occurring at pH 1.8 and 9.0 respectively in Britton–Robinson buffer. Perone and Kretlow (1966) studied the mechanistic aspects of the electroxidation of ascorbic acid using controlled potential techniques at the hanging mercury drop electrode, obtaining kinetic data on the electrode process.

An interesting amperometric titration for vitamin C was developed by Malik and Singh (1968) for analysing pharmaceuticals. Anodic waves in phosphate, borate, Britton–Robinson and acetate buffers were used and the best titration was obtained at + 0.2 V in Britton–Robinson buffer pH 5.2. Potassium aquopentacyanoferrate (III) was used as titrant.

5. OXIDATIVE VOLTAMMETRIC ANALYSIS OF MOLECULES 129

More recent work includes that of Schubert and Roland (1968) who determined the compound in oxalic acid by subjecting it to polarography at the dropping mercury electrode (d.m.e.) in an acetate buffer. Interference from sulphydryl compounds was found at pH 5.5.

Several workers have measured the vitamin C content of various foods by anodic oxidation. Kajita et al. (1973) measured 1-ascorbic acid in several foods. Interferents were removed by extraction of complexes of o-phenylene diamine with chloroform prior to determination. Kajita and Senda (1970) performed a simultaneous determination of 1-ascorbic acid and erythorbic acid in alkaline extracts of canned foods. Both gave two oxidation waves at pH 9 to 11. Grundova et al. (1973) determined both 1-ascorbic acid and 1-dehydroascorbic acid (II) in plant materials and preserved foods. Samples were diluted with 2% aqueous oxalic acid, filtered and chromatographically separated from anthocyanins on a polyamide column prior to polarographic determination. Owen and Franklin Smyth (1975) have used single sweep cathode ray polarography to analyse ascorbic acid in a wide range of dietary foods without prior separation. The differential pulse polarography of ascorbic acid was studied by Lindquist and Farroha (1975). Acetate buffer saturated with sodium oxalate was used, giving a pH of 5.5. Concentrations up to 25 µg ml^{-1} were studied and well-formed peaks obtained down to 0.2 µg ml^{-1}.

Work most recently performed describes the application of electrodes other than the d.m.e. to vitamin C analysis. Cescon and Montalti (1970) determined 1-ascorbic acid in aerated aqueous solutions at a vibrating platinum electrode. A study of the effect of X-rays on the compound was performed in solution and the radiolytic decomposition of vitamin C studied from the concentration before and after irradiation.

A recent paper by Lindquist (1975) describes the voltammetric determination of ascorbic acid, using a carbon paste electrode, by linear sweep and cyclic voltammetry. The half-peak potential was about 0.2 V greater than the half-wave potential at the dropping mercury electrode. The peak current was proportional to ascorbic acid concentration in the range 10^{-6} to 10^{-3}M and the relative standard deviation was better than $\pm 1\%$. Chloride and sulphur compounds, such as sulphides and thiols, did not interfere and sulphite could be determined at the same time as the ascorbic acid. Some substituted phenols interfered but could be detected by reversing the direction of polarization. Reductones interfered but tin (II) and manganese (II) did not and a method was developed for determining ascorbic acid in the presence of an excess of iron. The method was applied to ascorbic acid determination in fruits, vegetables and beverages. Peak height varied with scan rate and the half-peak potential varied with pH, giving a break corresponding to the pK_a value of 4.5.

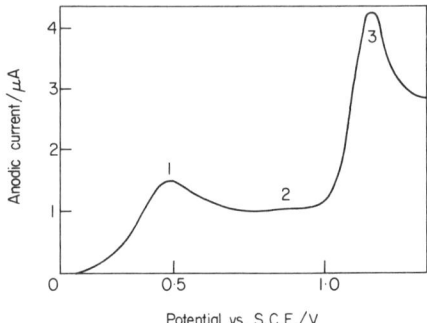

FIG. 1. Voltammogram of an extract of a multi-vitamin tablet (Nutrivimin) that contains vitamin C (100 mg) and iron (100 mg). 1: ascorbic acid; 2: pyridoxine; 3: iron complexed with 5-nitro-1,10-phenanthroline. (Lindquist, 1975).

An interesting application of this method was the determination of a multi-vitamin tablet (Nutrivimin) containing vitamin C, iron and pyridoxine. Figure I shows the voltammogram obtained, with the three peaks well separated.

A tubular carbon electrode was used by Mason *et al.* (1972) to determine vitamin C in pharmaceutical dosage forms. The method was highly specific and used to determine ascorbic acid in the presence of vitamins, A, D, B_1, B_6, B_{12}, biotin, niacinamide and calcium pantothenate without interference. No interference occurred from the excipients talc, dextrose, sucrose, gum tragacanth, acacia, gelatin, magnesium stearate, corn starch, sorbitan monoleate or polysorbate 40. Slight interference occurred from sodium lauryl sulphate (a 0.5% solution lowered the limiting current of 10^{-4}M ascorbic

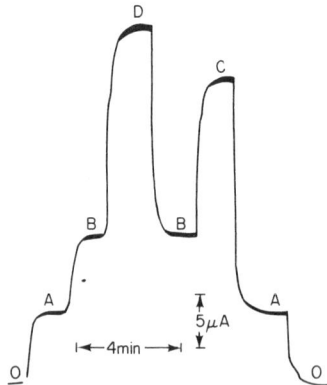

FIG. 2. Ascorbic acid current-time recorder trace. O: 0.05M acetic acid; A: 1×10^{-4}M B: 2×10^{-4}M; 4×10^{-4}M; D: 5×10^{-4}M. See text for details. (Mason, 1972).

acid by 20%). The current response was linear over the range 10^{-6} to 10^{-3}M, and the vitamin C content of tablets was determined with a relative standard deviation of 0.91%. Figure 2 shows the current time recorder trace obtained for different concentrations in 0.05 M acetic acid.

Other solid electrodes used in vitamin C analysis include a graphite electrode by Volova et al. (1972), while Brezina et al. (1972) studied the oxidation of ascorbic acid at platinum electrodes.

Several workers have been active in the analysis of vitamin C in biological fluids. Freude et al. (1953) showed that proteins, germ and yeast extracts, and chloride ion did not interfere in the polarographic analysis of the compound in body fluids.

A recent application of liquid chromatography with electrochemical detection has been described by Pachla and Kissinger (1976) for measuring ascorbic acid in foodstuffs, pharmaceuticals and body fluids. A thin layer amperometric detector with a carbon paste electrode was used. Improved sensitivity and selectivity was claimed over classical titration and colorimetric redox procedures.

A general review of vitamin C analysis by El-Sourady (1971) includes applications of the anodic polarography of the compound.

B. VITAMIN E (α-, β-, γ- AND δ-TOCOPHEROLS)

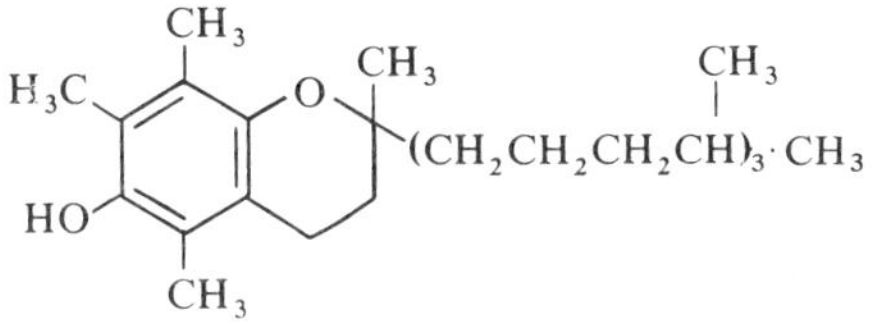

(III) Vitamin E (α-tocopherol)

Early studies performed by Smith et al. (1941, 1942) showed that α-tocopherol and three isomeric dimethylethyltocols are oxidized at the dropping mercury electrode in 75% ethanol with aniline–anilinium perchlorate or perchloric acid as buffers and supporting electrolytes. β- and γ-tocopherols are oxidized at the d.m.e. at more positive potentials than is α-tocopherol. It was shown that α-tocopherol could be determined polarographically in the presence of β- and γ-tocopherols.

The anodic oxidation of vitamin E was used by Niederstebruch and Hinsch (1967) for determining the tocopherols in oils and fats. Samples were saponified and the unsaponifiable fraction extracted with diethyl ether. The tocopherols were oxidized to tocopherylquinones with $Ce(SO_4)_2$ in ammoniacal ethanol. The precipitate was removed and the alcoholic solution analysed polarographically at pH 7.5 under nitrogen. The base electrolyte used was alcoholic 0.3M NH_4NO_3. Results reproducible to $\pm 3.5\%$ were obtained.

The carbon paste electrode has been recently used in the analysis of vitamin E by Atuma and Linquist (1973). Samples such as vegetable oils, foods and pharmaceuticals were analysed. No elaborate clean-up procedures such as chromatography were needed prior to analysis and reducing substances such as carotenoids, vitamin A, steroids, which interfere in other methods, were inactive in the potential range employed. Pharmaceutical samples (tablets) were ground, mixed, saponified and the unsaponifiable matter extracted and analysed using an electrolyte of 0.2M sulphuric acid in 75% ethanol. Peak potentials of +0.47 V, +0.55 V, +0.55 V and +0.62 V

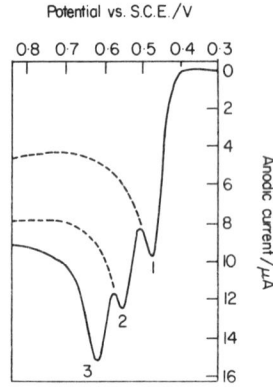

FIG. 3. A voltammogram of α-, β- and δ-tocopherol in a mixture. Electrolyte: 0.2M sulphuric acid solution in 75% ethanol. 1: α-Tocopherol; 2: β-tocopherol; 3: δ-tocopherol. (Atuma and Lindquist, 1973.)

were obtained for α-, β-, γ- and δ-tocopherol respectively using a linear-sweep voltammetric technique. Figure 3 shows the ability of the method to separate the major components in a mixture. Both α-tocopherol acetate and α-tocopherol have been determined simultaneously using a vitreous-carbon indicator electrode in a medium of acetonitrile containing 0.5M $LiClO_4$ (Shiozaki et al. 1971). α-Tocopherol acetate gave a diffusion-controlled one-electron oxidation peak at +1.33 V, and α-tocopherol a peak at +0.68 V under the same conditions.

Vitamin E has been measured by anodic polarography in various biological materials. Lucarini et al. (1970) performed a direct voltammetric determination of α-, β-, γ- and δ-tocopherols in animal tissues using a bubbling platinum microelectrode. A solution of dl, α-tocopherol (10^{-2}M) in ethanol/water (4:1 vol/vol) was standardized by amperometric titration with 0.01M

Ce(SO$_4$)$_2$. A calibration curve was constructed over the range 0.2–4.0 mM. An anodic scan between +0.30 and +0.80 V was used for analysis. Mixtures of the different forms exhibited multiple waves from which individual or total tocopherols could be measured. Cospito et al. (1969) performed a similar amperometric titration of mixed tocopherol concentrates with Ce(IV) sulphate. A bright platinum electrode was used, previously described by Cozzi et al. (1963). It was shown that even ten-fold amounts of vitamins K$_1$, K$_3$ and D$_2$ did not interfere, but vitamin A had to be absent. Schmandke (1965) and Schmandke and Gohlke (1965) described the polarographic determination of tocopherolactone (a metabolite of α-tocopherol) in body fluids, after thin layer chromatographic separation. The metabolite was extracted into diethyl ether, dried, dissolved in ethanol and acetate buffer (pH 5.33) and subjected to polarography from −0.4 → +0.2 V.

C. OTHER VITAMINS

Although vitamins C and E are the most studied compounds, several other vitamins have also been analysed using anodic polarography. Folic acid (IV) riboflavin (V) and vitamin C have been determined simultaneously by Kruze (1969). A borate–phosphate buffer solution was used. Half-wave potentials for 0.5% vitamin C, 0.01% folic acid and 0.2% riboflavin were +0.01 V, +0.7 V and +0.4 V respectively. The method was suitable for mono- or polyvitamin mixtures, with errors not exceeding 1%, 2% and 1.5% for the three compounds respectively.

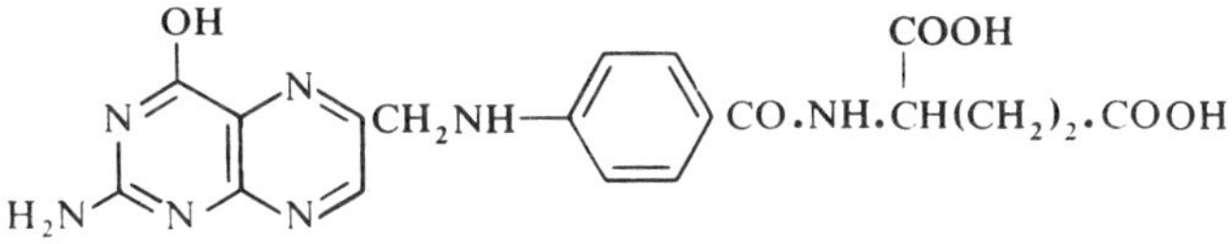

(IV) Folic acid

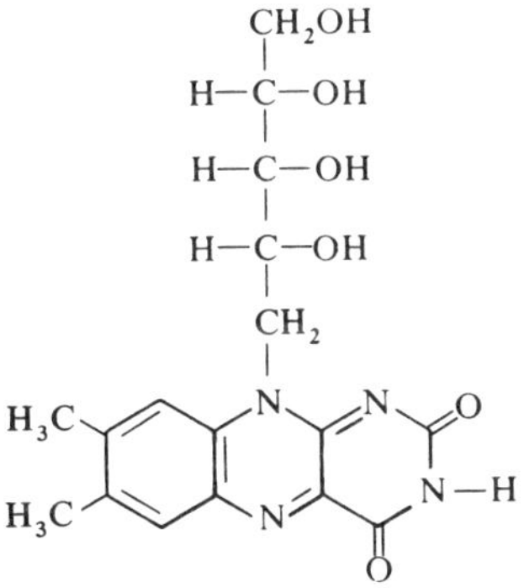

(V) Riboflavine

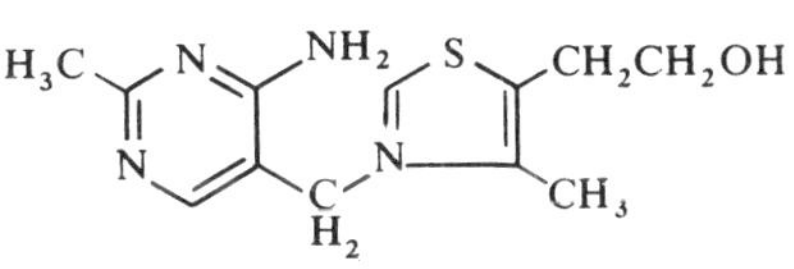

(VI) Thiamine

Riboflavine has also been studied anodically as riboflavone semiquinone by Korshunov et al. (1968), while thiamine (VI) was shown to be oxidized anodically by Tachi and Koide (1951). The half-wave potentials for thiamine were independent of pH and equal to about -0.4 V. The compound was shown to exhibit cathodic behaviour in addition to its oxidation.

An interesting pharmaceutical application, the determination of pyridoxine (vitamin B_6) by oxidation at a carbon paste electrode using linear sweep voltammetry, has been described recently by Soderhjelm and Lindquist (1975). The only interferences occurred from ascorbic acid and ferrous iron, both of which were removed by ion-exchange chromatography prior to analysis. The authors report the voltammetric method to have a better standard deviation than an alternative colorimetric method. The carbon paste electrode was compared to a glassy carbon and found to have superior analytical characteristics in that the residual currents in the region of pyridoxine oxidation were much lower for the former. Peak currents were linear with respect to concentration in the range 10^{-6} to 10^{-4}M. An interesting method for determining the blank was employed, whereby the electrochemically inactive complex of pyridoxine with borate was produced and the

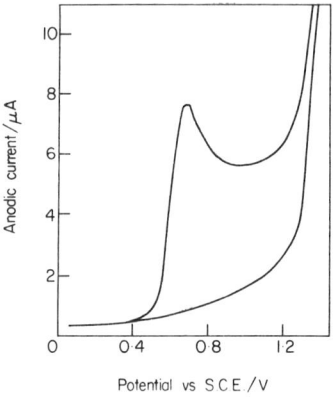

FIG. 4. Voltammogram of 1.0×10^{-4}M pyridoxine in ammonia buffer at pH 9.2. Carbon paste electrode. Lower scan represents a blank solution. (Soderhjelm and Lindquist, 1975.)

resulting solution subjected to the same voltammetric procedure as the samples. A relative standard deviation of $\pm 2\%$ was obtained for tablets containing 2 mg pyridoxine (which contained ascorbic acid and ferrous iron). Figure 4 shows a typical voltammogram for pyridoxine measured at the carbon paste electrode.

Vitamin A has also been determined voltammetrically by Atuma et al. (1974) using anodic oxidation at the carbon paste electrode. Various derivatives of vitamin A were studied including retinyl acetate, retinal, retinol, retinoic acid and retinyl palmitate. Typical anodic polarograms are shown in Figure 5.

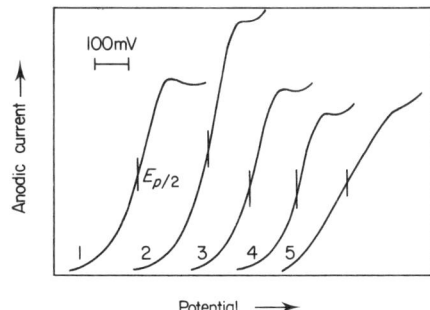

FIG. 5. Typical voltammograms of different vitamin A derivatives: (Atuma et al., 1974).

	$E_{p/s}$ vs s.c.e.
1. Retinyl acetate	0.71
2. Retinal	0.80
3. Retinol	0.67
4. Retinoic acid	0.78
5. Retinyl palmitate	0.75

III. ANALGESICS

Analgesics are an important pharmaceutical class which have been successfully analysed by the use of voltammetric oxidation waves.

The anodic oxidation of morphine (VII) has been used analytically by Rashid and Kalvoda (1971). Solutions of 0.1 mM to 1 mM of morphine tartrate in 1M KOH were determined, the half-wave potential being +0.18 V.

(VII) Morphine

Noninski et al. (1969) determined morphine using a "self-cleaning" rotating silver electrode. Advantages in sensitivity over the dropping mercury electrode were achieved. At a speed of 100 r.p.m. the sensitivity was twice that of the dropping mercury electrode whilst at 1000 r.p.m. sensitivity was improved ×10. Deys (1964) determined morphine at a rotating platinum electrode. 4N H_2SO_4 was used as supporting electrolyte and for the determination a wave was recorded first for 25ml 4N H_2SO_4 from +0.5 V, then after the addition of a known volume of electro oxidizable substance (0.1 ml morphine solution containing 2 mg ml^{-1}) the wave was recorded again. Morphine, pseudomorphine, dihydromorphinone, dihydromorphine, nalorphine and apomorphine all gave an oxidation wave whereas the alkylated hydroxy alkaloids ethylmorphine, diacetylmorphine, benzylmorphine, codeine and thebaine showed no wave.

Hyoscyamine (VIII) and atropine (IX) have been determined amperometrically by Timbekov and Kasymov (1968). The titration was performed in 0.1M K_2SO_4–0.003M H_2SO_4 with 0.002M $KMnO_4$ at +0.4 V with a rotating platinum electrode.

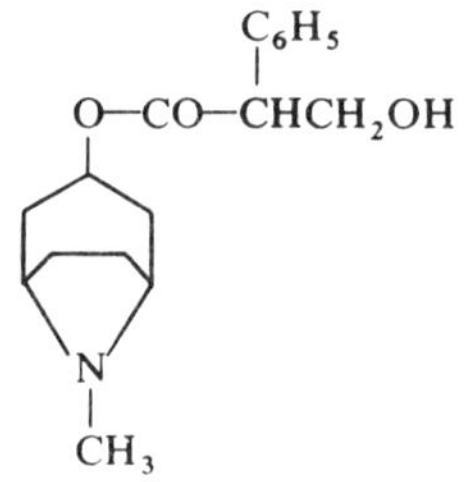

(VIII) Hyoscyamine 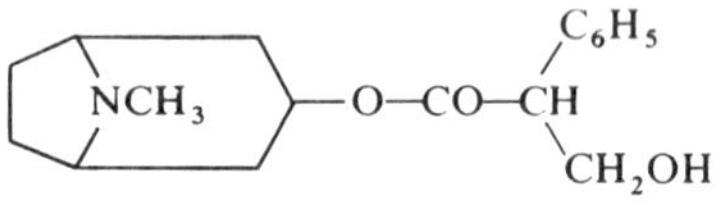   (IX) Atropine

Wasilewska and Szysko (1969) described the amperometric titration of pethidine (X) with tetraphenylborate ($NaBPh_4$). The determination was based on measurement of the anodic oxidation of sodium tetraphenylborate. A d.m.e. was used with external s.c.e. connected via a salt bridge in 1% acetic acid where the anodic wave for $NaBPh_4$ was better developed than in acetate buffer pH 4.6. The half-wave potential was +0.05 V and a potential of +0.12 V was used for the titration. No polarographic interference due to substances in the table mass (e.g. K^+ or NH_4^+, which can be precipitated by $NaBPh_4$, or Cl^-, which can react with Hg) was observed. Deviations from values determined by other methods were ±5%.

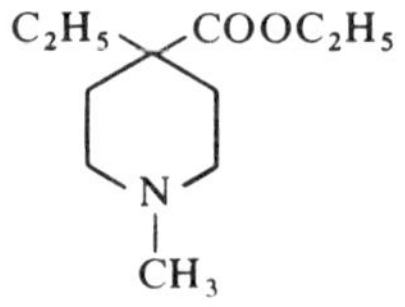

(X) Pethidine

Paracetamol (acetaminophen) (XI) has been determined in dosage forms (Shearer et al. 1972) using peak voltammetry at the glassy carbon electrode. A stability-indicating assay for the compound in both solid and liquid dosage forms was developed, based on oxidation of the compound using an acetate buffer system in methanol as supporting electrolyte. The peak

(XI) Acetaminophen   (XII) p-Aminophenol

potential was +0.5 V whilst the peak for p-aminophenol (XII), the major hydrolytic degradation product, was +0.2 V. Voltammograms are displayed in Fig. 6. It was possible to perform the analysis without separating the two compounds. Most common excipients were shown not to interfere, and of the therapeutically active agents used in the twelve different tablet combinations, only ascorbic acid had any adverse effect on the determinations.

A silicone rubber based graphite electrode was used by Pungor et al. (1971) to determine several analgesics including phenacetin (XIII), p-aminosalicylic acid (XIV), and amidopyrine (XV). Aqueous KCl was used as an electrolyte, and several other drugs including isoproterenal, methyldopa, pheno-

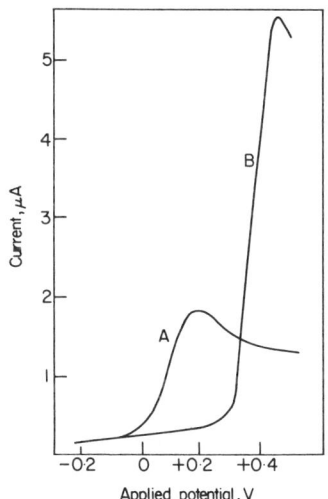

FIG 6. Voltammogram of (A) p-aminophenol (concentration = 0.012 mg/ml^{-1}) and (B) acetaminophen (concentration = 0.1 mg/ml), using a glassy carbon electrode. (Shearer et al., 1972.)

thiazine, chlorpromazine, methophenazine ethanesulphonate, diethazine, trimeprazine maleate and methantheline bromide were similarly determined. Lugovoi and Ryazanov (1967) showed amidopyrine to undergo irreversible oxidation at the platinum disc electrode. Two waves were obtained in an acetate medium of pH 3.5–6. Amidopyrine can be determined by the first wave in solutions of from 1×10^{-5} to 2.3×10^{-3} M with an error of $\pm 3.0\%$. The following substances in 10–20 excess were shown not to interfere: antipyrine, aspirin, caffeine, codeine, phenacetin, salicylic acid, Luminal, Veronal, Novocaine, sugar, starch, ethanol and oxygen. The only interference noted came from analgin.

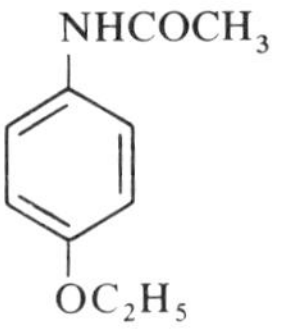

(XIII) Phenacetin

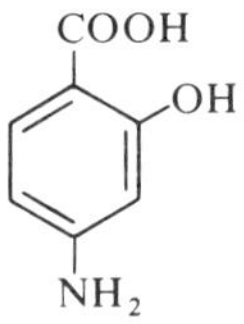

(XIV) p-Aminosalicylic acid

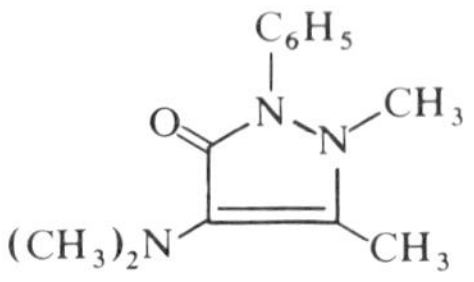

(XV) Amidopyrine

A recent example of oxidation using a graphite electrode in aqueous solutions is the work by Stradins and Gasonov (1973) who performed a voltammetric study of deprotonation during the oxidation of o- (XVI), m- and p-salicylic acid.

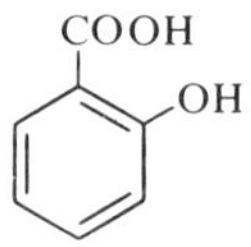

(XVI) o-Salicylic acid

The anodic oxidation of aniline derivatives has been studied by several workers. The kinetics of anodic oxidation of aniline in acetonitrile and water at the rotated ring-disc electrode were examined by Breitenbach and Heckner (1971). Reversible waves with $E_{\frac{1}{2}}$ of $+0.47$ V and $+0.31$ V respectively were obtained. An interesting review by Lutskii et al. (1972) of 46 substituted anilines showed that a linear relationship was found between the potentials for voltammetric oxidation and the π-charge density at the nitrogen atoms of the amines as calculated by the linear combination of atomic orbitals method.

The anodic oxidation of N,N-dimethylaniline in non-aqueous media was investigated by Hand and Nelson (1970). Products formed depended on the concentration of dimethylaniline and the time of electrolysis. At low concentrations (10^{-3}M) the main product was N,N,N',N'-tetramethylbenzidine. At 10^{-2} M, 4,4'-methylene-bis (N,N-dimethylaniline) was the major product, while at higher concentrations and longer electrolysis times the dye Crystal Violet was formed.

IV. CENTRAL NERVOUS SYSTEM AGENTS INCLUDING TRANQUILIZERS AND SEDATIVES

Several drugs in this category are amenable to anodic oxidation. A method of determining l-dopa (XVII) in pharmaceutical dosage forms has been devised by Mason (1973), based on oxidation at the tubular carbon electrode. A comparison with a colorimetric method showed polarography to be faster, simpler and equally precise and accurate. Between 25 and 30 samples could be determined per hour, with a relative standard deviation of less than $\pm 0.9\%$. The method was further developed into the continuous

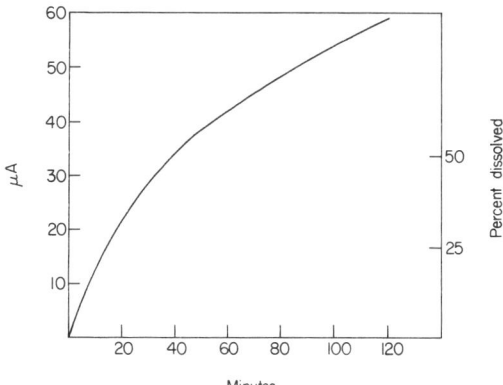

FIG. 7. Recorder trace for continuous dissolution study of l-dopa in tablets, using a tubular carbon electrode (Mason, 1973).

monitoring of dissolution rates of solid dosage forms. Figure 7 shows the dissolution curve as determined by voltammetry. The same technique was used by Stewart et al. (1974) to determine methyldopa (XVIII) in pharmaceutical dosage forms and biological fluids. Limiting currents were linear with respect to concentration in the range 10^{-5}–10^{-3} M.

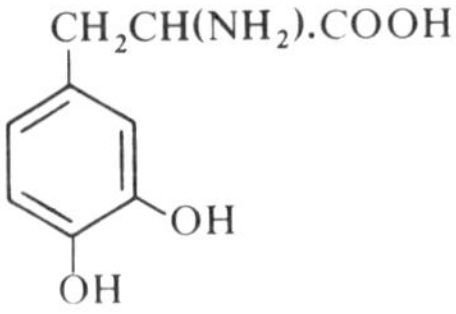

(XVII) L-Dopa — CH₂CH(NH₂).COOH on benzene with OH, OH

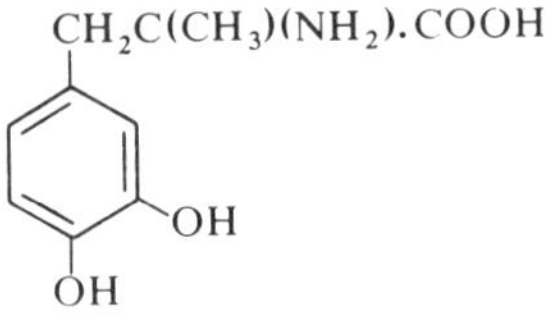

(XVIII) Methyldopa — CH₂C(CH₃)(NH₂).COOH on benzene with OH, OH

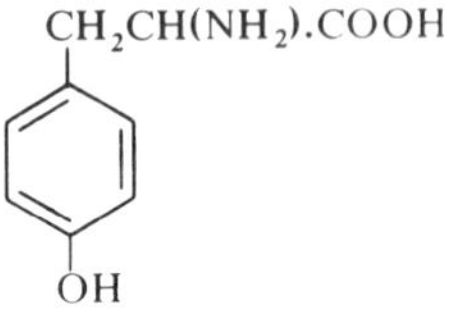

(XIX) Tyrosine — CH₂CH(NH₂).COOH on benzene with OH

Work by Kitagawa and Kanei (1970) and Kitagawa and Tsushima (1971) on the oxidation of tyrosine (XIX) at a carbon paste electrode showed that the compound was irreversibly oxidized in a two-electron step at +1.0 V in sulphuric acid solutions.

Several important amines and quaternary ammonium compounds used as drugs, including amphetamine sulphate, ephedrine sulphate, acetylcholine chloride and hexadecyltrimethyl ammonium bromide, have been determined by Sinsheimer and Hong (1965) using sodium tetraphenylborate as titrant in the amperometric titration of the compounds at a graphite electrode. The method was based on the electrochemical oxidation of sodium tetraphenylborate and was semi-automatic, using a constant rate burette and continuous recording of current. This technique was claimed to be an improvement over the use by Smith *et al.* (1963) of the same titrant in amine determinations at a graphite electrode. Wasilewska and Szyszko (1969) described the amperometric titration of pharmaceutically important organic bases such as promethazine (XX) and chlorpromazine (XXI) with NaBPh₄. The method is identical with that described earlier for pethidine.

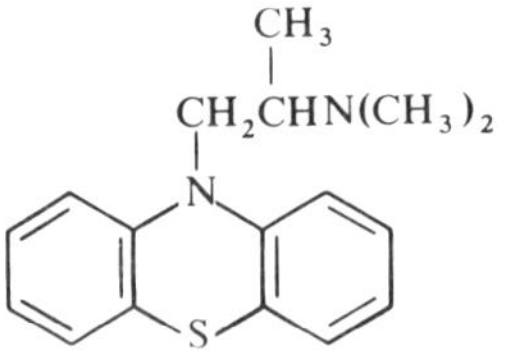

(XX) Promethazine — phenothiazine with N-CH₂CH(CH₃)N(CH₃)₂

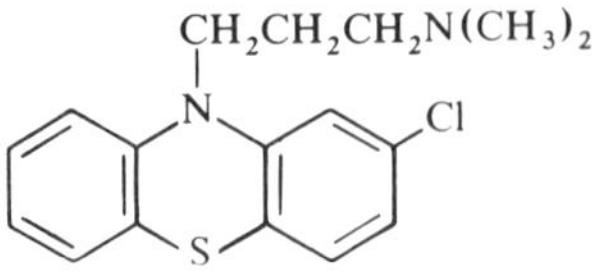

(XXI) Chlorpromazine — phenothiazine with N-CH₂CH₂CH₂N(CH₃)₂ and Cl

Phenothiazine (XXII) gives anodic waves at the dropping mercury electrode in sulphuric acid solutions of greater than 7N and in alkaline solutions at pH 12.5. In 10M H_2SO_4, the wave was well-defined and the limiting current a linear function of the phenothiazine concentration (Kemula and Kalinowski, 1967). Merkle and Discher (1964) have shown that the concentration of sulphuric acid has a differentiating effect on the half-wave potentials of a series of seven phenothiazines corresponding to oxidation of the free radical to the sulphoxide at a rotating platinum microelectrode. The oxidation reactions for phenothiazine derivatives are in general represented by R: → R· + e⁻ and R· + H_2O → S + 2H⁺ + e⁻ in 12N sulphuric acid, and R:→R· + e⁻ and 2R· + H_2O→R: + S + $2H^+$ in 1N sulphuric acid, where R: represents the initial reduced form of the compound, R· represents the free radical obtained upon one-electron oxidation and S represents the corresponding sulphoxide.

Gonzalez and Fernandez-Alonso (1970) studied the behaviour of eighteen phenothiazine derivatives, anodic studies being conducted at the rotating platinum electrode.

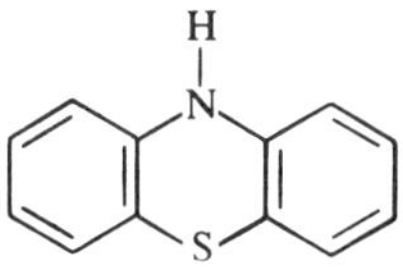

(XXII) Phenothiazine

Several important drugs of the benzodiazepine series and some dibenzocycloheptane derivatives were studied in acetonitrile with 0.1M tetrabutylammonium perchlorate as supporting electrolyte (Volke et al., 1975). Anodic oxidation was performed using rotating disc indicator electrodes. At both Pt and Au electrodes, amitriptyline, dothepin and dithiaden gave well-defined diffusion-controlled waves ($E_{\frac{1}{2}}$ +1.2 → +1.3 V). At Pt electrodes only oxazepam (XXIII), diazepam (XXIV) and flurazepam (XXV) of the benzodiazepines tested gave well-defined oxidation waves which were diffusion controlled and rectilinear up to 1 mM.

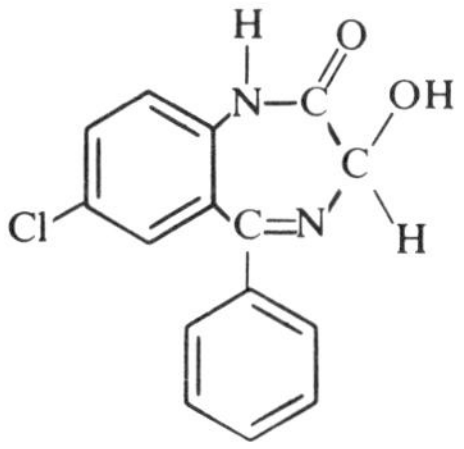

(XXIII) Oxazepam

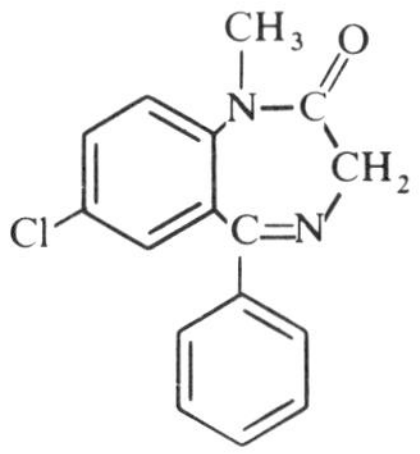

(XXIV) Diazepam

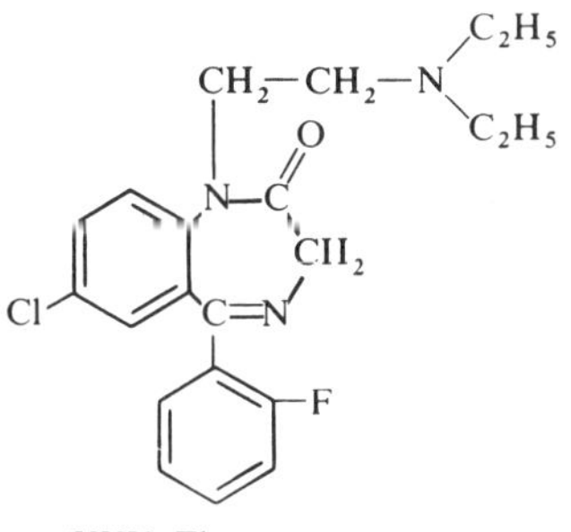

(XXV) Flurazepam

V. ANTITUBERCULINS

The polarographic behaviour of isonicotinic acid hydrazide (XXVI), an important and long established drug, was investigated by Liberti et al. (1952) and Anastasi et al. (1952). The former workers used the oxidation wave in alkaline media for its determination in pharmaceuticals, while the latter workers determined the compound specifically at pH 9.0.

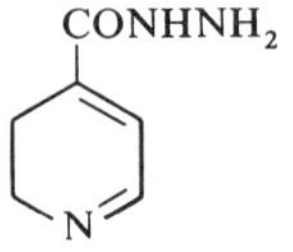

(XXVI) Isonicotinic acid hydrazide

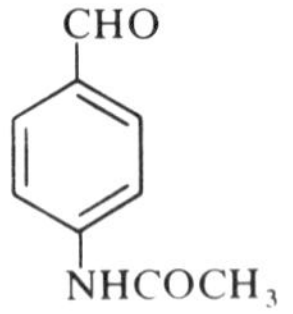

(XXVII) Thiacetazone

Asah (1963) reported that thiacetazone (XXVII) and related compounds produced mercury(I) compounds in alkaline media after anodic oxidation.

Brandys (1966) studied the application of oxidation waves to the determination of drugs derived from isoniazid.

VI. NATURALLY OCCURRING COMPOUNDS

For convenience this section has been sub-divided to describe naturally occurring purines, sulphur compounds and catechols/catecholamines.

A. PURINE AND DERIVATIVES

Much work has recently been described on the anodic oxidation of purine (XXVIII) and related compounds, particularly using voltammetry at the pyrolytic graphite electrode (p.g.e.). The oxidation of purine, (XXVIII), adenine (XXIX), cytosine (XXX) and pyrimidine (XXXI), at both the p.g.e. and mercury electrodes using d.c., a.c. and triangular sweep voltammetry has been reported (Dryhurst and Elving, 1969). Information was obtained on the adsorption of the reactant and product species on the electrode, the reversibility of the electron-transfer step and the accompanying chemical

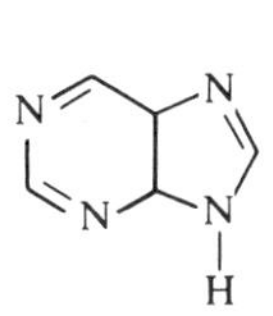

(XXVIII) Purine 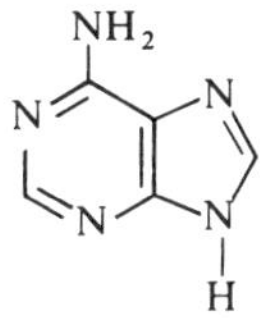   (XXIX) Adenine 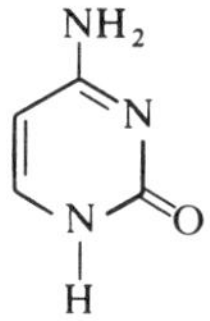   (XXX) Cytosine

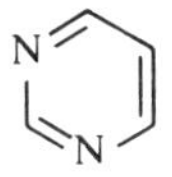

(XXXI) Pyrimidine 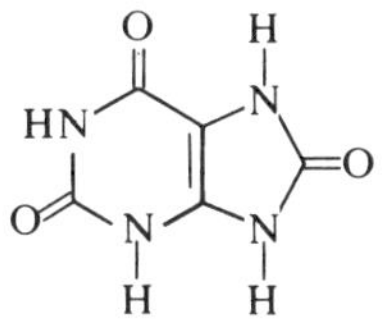   (XXXII) Uric acid

reactions. The oxidation of biologically important purines at the p.g.e. was studied by Dryhurst (1969a, b). Cyclic voltammetry of saturated uric acid (XXXII) in acetate buffer at pH 4.7 showed a reversible electrode reaction at +0.45 V, characteristic of the two-electron oxidation giving a dicarbonium ion.

A single analytically-useful oxidation wave for adenine (XXIX) was obtained by Dryhurst and Elving (1968) using the p.g.e. Macroscale controlled potential electrolysis was performed in aqueous M CH_3COOH (pH 2.3) with exhaustive electrolysis, identification and determination of reaction products and intermediates.

Dryhurst and Pace (1970) studied the electrolytic oxidation of guanine (XXXIII) at the p.g.e. Techniques used included linear sweep and cyclic voltammetry between pH 0 and 12.5 and controlled potential electrolysis

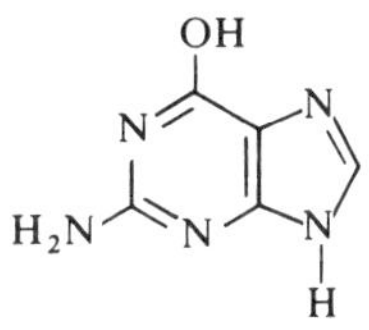

(XXXIII) Guanine 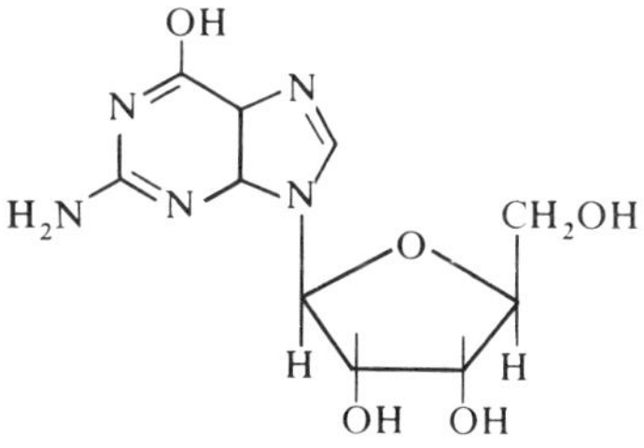   (XXXIV) Guanosine

in aqueous 1M CH_3COOH solution. Guanine appeared to be oxidized to a diol which degraded further to parabanic acid, guanidine and CO_2 and hydrolysed further to oxalyl guanidine and CO_2.

Dryhurst (1971) also determined guanine (XXXIII) in the presence of guanosine (XXXIV) at a p.g.e. Both compounds were oxidized, but adsorption affected the determinations. In the presence of 5 to 15-fold excesses of guanosine, adsorbed guanine was displaced and the process became diffusion controlled, allowing guanine to be determined without interference.

Again using a p.g.e., a direct method of determining allopurinol (XXXV) and uric acid (XXXII) in mixtures has been described (Dryhurst and De, 1972). Uric acid and allopurinol were both electrochemically oxidized at the p.g.e. in aqueous solution, the oxidation of allopurinol occurring at more positive potentials. Uric acid was adsorbed at the p.g.e. producing non-linear calibration curves, an effect that disappeared when the solution was saturated with allopurinol since the latter substance was preferentially adsorbed at the p.g.e., thus displacing adsorbed uric acid from the electrode surface. Allopurinol (0.1–1 mM) could be determined via its polarographic reduction wave at the d.m.e. The solution was then saturated with allopurinol and uric acid (0.05–0.5 mM) determined via its anodic peak at the p.g.e. The procedure is satisfactory in supporting electrolyte systems of pH 0–6.

Dryhurst and Hansen (1971a) studied the voltammetric oxidation of some biologically important xanthines (XXXVII). A two-electron process gives the appropriate uric acid. Dryhurst and Hansen (1971b) also investigated the oxidation of theobromine and caffeine at the p.g.e. Both were

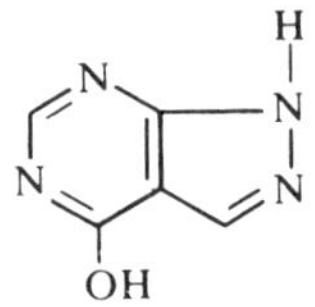

(XXXV) Allopurinol 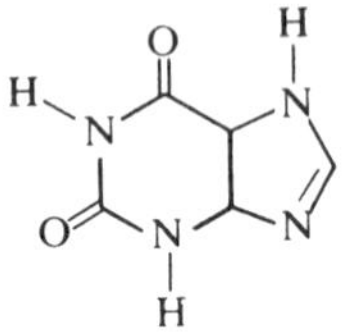 (XXXVI) Xanthine

oxidized in an overall four-electron process. The mechanism proceeds by a two-electron potential-controlling oxidation to give the corresponding uric acid which is further oxidized to a 4,5-diol analogue of uric acid.

Hansen and Dryhurst (1971) studied the electrochemical oxidation of theophylline (XXXVII) at the p.g.e. Products were isolated, identified by u.v. and mass spectrophotometry and oxidation mechanisms proposed.

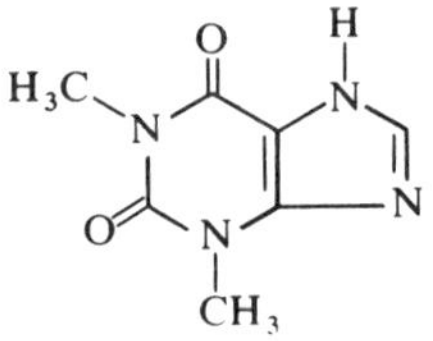

(XXXVII) Theophylline (1,3-dimethylxanthine)

Reduced nicotinamide adenine dinucleotide (NADH) has been determined (Blaedel and Jenkins, 1975) using rotated glassy carbon and platinum electrodes. Well-defined current-potential curves were obtained at micromolar concentrations. Conditions for the quantitative determination of NADH were established. Experimental evidence suggested that the electron transfer involved chemical interaction between the electrode surface and the NADH. The effects of pH, buffer system, electrode material and conditioning upon profile and position of oxidation wave pointed to the critical dependency of the oxidation process on the state of the electrode. NADH could be amperometrically determined at 10 μM concentration with relative standard deviations about 1%.

The electrochemical determination of xanthine oxidase and inhibitors was described by Guilbault et al. (1964). A method based on the catalysed aerobic oxidation of hypoxanthine by xanthine oxidase in pH 7.4 tris buffer was developed. Anodic polarography was used to study the significance of the potentials observed and a procedure described to determine micro amounts of the enzyme and its inhibitors.

An interesting combination of chromatography and voltammetry was recently described by Varadi et al. (1974) who determined guanine, xanthine, hypoxanthine and adenine after separation on a chromatographic column and passage through a voltammetric cell (range $-0.3 \rightarrow +1.5$ V) containing a silicone rubber-based graphite electrode with a Ag/AgCl reference cell. A limit of detection of 0.1×10^{-6} mol was quoted for guanine.

Early work was performed on uracil (XXXVIII) by Manousek and Zuman (1955) who showed that in slightly alkaline solutions the compound gave one anodic wave which did not increase in height at concentrations greater than 10^{-4}M uracil, with borax being the best supporting electrolyte.

(XXXVIII) Uracil

In the field of body fluid analysis, a rapid and accurate analysis of serum uric acid has been developed by Park et al. (1972) using a carbon paste electrode and oxidation at +0.64 V. Good agreement was obtained with the more conventional automated tungstophosphate colorimetric method. It was shown that xanthine and glutathione did not interfere with the analysis but that the presence of ascorbic acid gave slightly high results.

Those purine compounds which contain sulphur will be discussed later in the section on sulphur containing compounds.

B. SULPHUR-CONTAINING COMPOUNDS

Anodic oxidation is of great value in the polarographic determination of sulphur compounds. Methods for many molecules have been reported and early work is well covered by Brezina and Zuman (1958). Important naturally-occurring compounds determined include sulphur-containing purines, cystine–cysteine system, and a variety of miscellaneous sulphur compounds.

(i) *Sulphur-containing purines*

Dryhurst (1969c) reported theoretical studies on the electrolytic oxidation of 6-thiopurine (XXXIX) at the p.g.e. Three well-defined voltammetric waves were produced. Buffers used included aqueous M CH_3COOH (pH 2.3), ammonia buffer (pH 9) and carbonate buffer (pH 9). Electrochemical oxidation was shown to follow a different route to enzymatic oxidation. The same compound was studied by Vachek (1960) who developed a polarographic

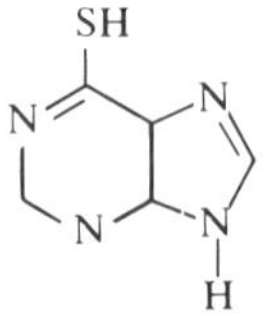

(XXXIX) 6-Thiopurine

method for its estimation, based on anodic oxidation. The compound was dissolved in McIlvane buffer solution at pH 7.1 and subjected to polarography. Hypoxanthine and glucose were shown not to interfere and the method was accurate to within $\pm 2\%$. The $E_{\frac{1}{2}}$ of the anodic wave was -0.26 V. The same author (Vachek, 1965) determined the compound as an impurity in 6-(4-carboxybutyl) thiopurine using this anodic wave. A reduction wave is also given for 6-(4-carboxybutyl)-thiopurine. Dryhurst (1970) has also studied the electrochemical oxidation behaviour of purine-2,6-dithiol at the p.g.e. and has applied such behaviour to determining sulphur-containing purines.

(ii) *Cystine–cysteine*

$$\begin{array}{ll} S-CH_2CH(NH_2).COOH \\ | \\ S-CH_2CH(NH_2).COOH \end{array} \qquad HS.CH_2.CH(NH_2).COOH$$

(XL) Cystine (XLI) Cysteine

Koryta and Pradac (1968a, b) and Pradac and Koryta (1968) studied the system at both platinum and gold electrodes. Three oxidation waves were obtained, with cysteine (XLI) being oxidized to cystine (XL), (in 1N H_2SO_4), or to the adsorbed radical RS· in a one-electron process. Two of the waves increased with solution stirring rate resulting in the disappearance of the 3rd wave. At more negative potentials a cathodic peak was observed due to reduction of oxidation intermediates of cysteine. The peak disappeared above 10^{-4}M cysteine at the platinum and above 10^{-3}M at the gold electrode and was 0.5 V more negative at the platinum electrode.

Ueno (1968) also studied the behaviour of cysteine (XLI), and showed that the compound could be determined polarographically, using differential polarography in 0.5N acetate buffer, pH 5.2.

The pulse polarographic behaviour of the cystine catalytic hydrogen wave was investigated by Gilbert (1969). The wave obtained in the presence of Co(II) in aqueous ammonia was studied. Limits of detection in pure solution of 50 nM and 5 nM were obtained with 10- and 30-mV pulses respectively, with 15 mM Co(II) in a buffer system of NH_3 and NH_4Cl of concentration 0.1 to 0.2M (pH 9 to 10).

(iii) *Miscellaneous naturally occurring sulphur compounds*

Sulphur compounds are important biologically and as such have been analysed extensively in body fluids. Several examples of anodic polarography have been described. Charles and Knevel (1968) have used a.c. polarography to determine nonprotein -SH groups (as in cysteine or glutathione) in biological materials. Ascorbic acid was shown not to interfere and the method was used to detect differences in nonprotein -SH group levels in rat liver caused by stress. The determination of mercapto groups in stomach contents was described by Korobeinik (1968), who used an amperometric titration with 0.001M Hg_2Cl_2 in 0.5M phosphate buffer (pH 7.1). A selection of amperometric titration methods was tested by Kolthoff *et al.* (1965) for determining mercapto groups in serum albumin. Several electrodes were used, the best being a rotating mercury-wire electrode with titration against a mercury-containing titrant. Richmond and Somers (1966) described the amperometric determination of thiols in extracts of biological materials using a rotating platinum electrode. Specifically, dodecanethiol was determined.

The cholinesterase activity of blood was determined using the anodic wave of thiocholine (Lhuguenot *et al.*, 1969) liberated by the hydrolysis of acetylthiocholine perchlorate (0.028M) with cholinesterase in 0.05M phosphate buffer with a rotating platinum electrode. Down to 0.001 i.u. of cholinesterase could be determined with an error of $\pm 7\%$.

Konupcik *et al.* (1960) used the anodic wave of mercaptothiamine at -0.26 V for its determination. A solution in dilute sulphuric acid was used and results were accurate to within $\pm 2\%$ compared with a bromometric method.

Mercaptans have been determined in aqueous solutions by Zhustereva et al. (1972) using an alcoholic/HCl electrolyte system, while Remke (1970) confirmed that an anodic wave (using a d.m.e.) detected in fresh mitochondrial suspension was due to sulphydryl compounds. In fresh, well-developed suspensions the wave was caused by liberation of reduced glutathione from rat liver mitochondria.

Applications of the rotating platinum electrode were described by Bessarabova et al. (1970) and Bessarabova and Songina (1971) who examined several sulphur compounds including thiooxine, cysteine and thiosemicarbazide.

C. CATECHOLS AND CATECHOLAMINES

Early work by Sartori and Cattaneo (1942) on the electrolytic oxidation of adrenaline showed the reaction to be second order up to ca. pH 6 with a lower order above pH 6 with the probable formation of a semiquinone. The polarographic behaviour of epinephrine (XLII) and other polyphenolic amines, particularly norepinephrine (XLIII) and isopropylepinephrine (XLIV) have been investigated by Cantin et al. (1973).

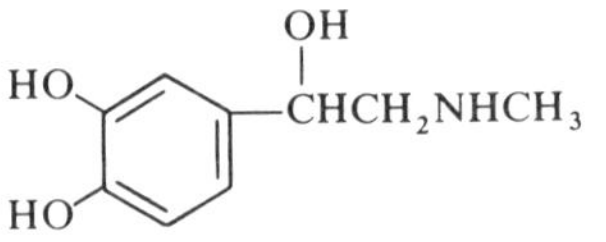

(XLII) Epinephrine

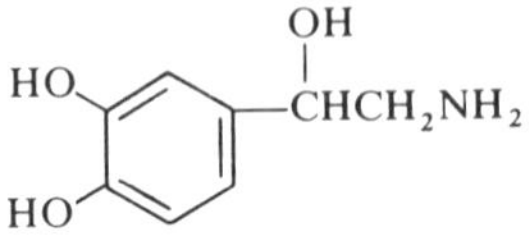

(XLIII) Norepinephrine

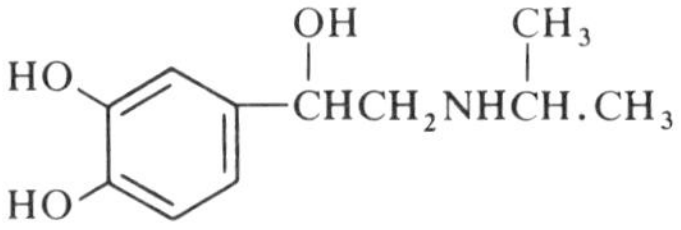

(XLIV) Iso propylepinephrine

Determinations of pharmaceutical samples were made using a rotating wire platinum electrode. $E_{\frac{1}{2}}$ values of $+0.24$ V, $+0.27$ V and $+0.30$ V were obtained for epinephrine, isopropylepinephrine and norepinephrine respectively.

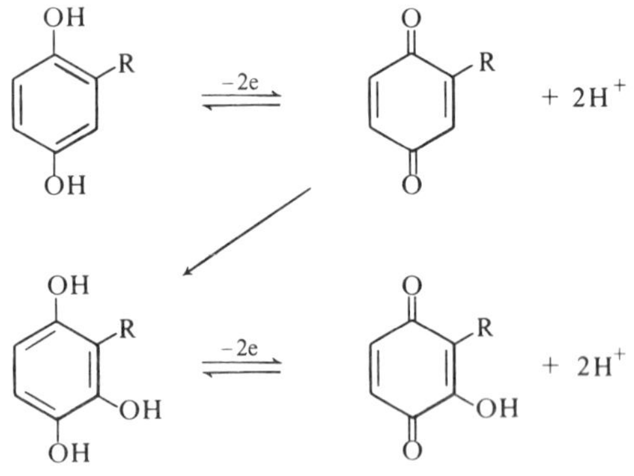

5. OXIDATIVE VOLTAMMETRIC ANALYSIS OF MOLECULES

The electrolytic oxidation pathways of naturally occurring phenolic compounds were studied by Papouchado et al. (1972). A three-electrode, controlled potential system was used, with C-Nujol paste electrodes for voltammetric studies and C cloth for preparative scale electrolysis. Compounds investigated included many 2-substituted hydroquinones, catechols and some p-substituted phenols. The scheme above was proposed as a mechanism for phenolic oxidation. An initial two-electron transfer is followed by a 1,4-nucleophilic addition of water to the quinone followed by a further two-electron oxidation of the resulting trihydroxy compound.

Work by Kajita (1966) on homogentisic acid (XLV) in the presence of erythorbic and ascorbic acids showed that homogentisic acid had no oxidation wave below pH 3.0 but gave a clear wave at pH 6 to 7. A mixture of ascorbic and erythorbic acids could be detected in the presence of homogentisic acid.

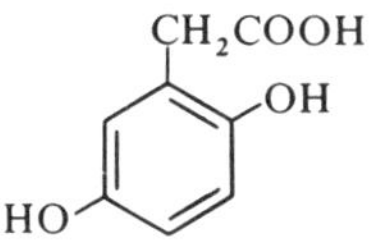

(XLV) Homogentisic acid

Important work on the anodic oxidation of naturally occurring compounds has been performed by Adams (1976) and his co-workers, who have studied phenols, catechols and catecholamines, particularly the latter compounds in mammalian brain. Their efforts were concentrated on the adrenergic transmitters dopamine (DA) (XLVI) and norepinephrine (NE) (XLVII).

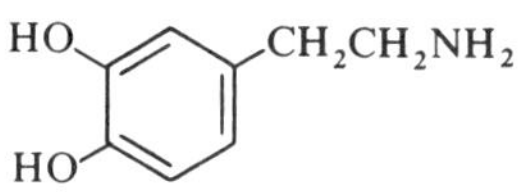

(XLVI) Dopamine (DA) 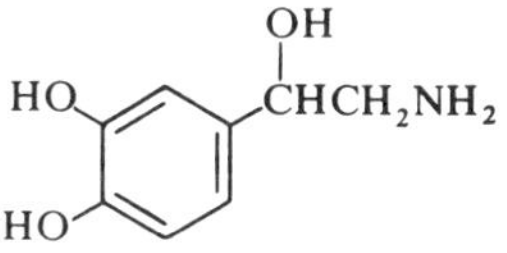   (XLVII) Norepinephrine (NE)

Other compounds studied by Adams' group include homovanillic acid (HVA) (XLVIII), dihydroxyphenylacetic acid (DOPAC) (XLIX) and 5-hydroxy-indole acetic acid (5-HIAA) (L) in cerebral spinal fluid (CSF). An anion exchange resin was used for preliminary separation of the compounds in CSF.

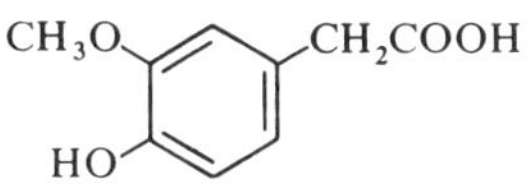

(XLVIII) HVA 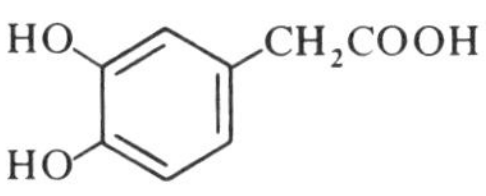   (XLVIX) DOPAC 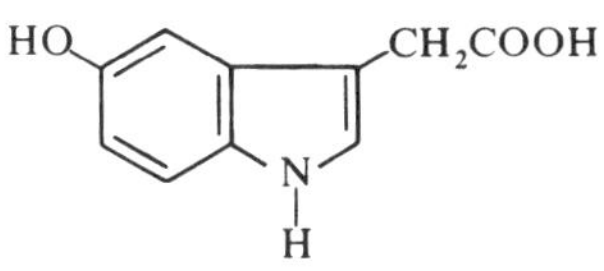   (L) 5-HIAA

Several other workers have used liquid chromatography with electrochemical detection (Kissinger et al., 1974, 1975; Felice and Kissinger, 1976; Riggin et al., 1976) for measuring catecholamines, their metabolites and other phenolic substances in body fluids. Blank and Pike (1976) have used h.p.l.c. with electrochemical detection to measure tyrosine hydroxylase activity via l-dopa separation and analysis.

These determinations following separation and *in vivo* determinations, are detailed in the chapter by Chowdhry (Chapter 6).

Many studies have been carried out on the behaviour of naturally occurring neurotransmitters at the rotating Pt electrode (Bezugly and Beilis, 1965; Barnes and Mann 1967; Lane and Hubbard, 1976).

VII. MISCELLANEOUS COMPOUNDS

Compounds will now be discussed which do not readily fit into any of the preceding categories, and will be classified according to their major functional groups, i.e. nitrogen-, sulphur-, oxygen- and halogen-containing compounds.

A. NITROGEN-CONTAINING COMPOUNDS

Many of these are structurally amenable to electrolytic oxidation. Two important classes are amines and hydroxylamines.

(i) *Amines*

Amines are not reducible at the d.m.e. but can be oxidized at a variety of solid electrodes (Shearer et al., 1972; Wacholz and Pfeifer, 1972).

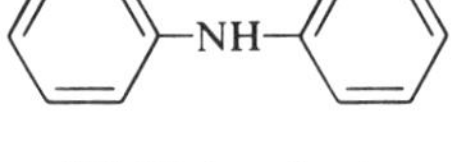

(LI) Diphenylamine

The mechanism of the anodic oxidation of aromatic amines has been studied by Leedy and Adams (1970), who established products and pathways for diphenylamine (LI) and related compounds.

Cyclic voltammetry was used for general qualitative pictures of the electrode reactions while chronoamperometry and chronopotentiometry served for quantitative measurements. A platinum button electrode was used and several systems were studied, including N-methyldiphenylamines, N-phenyl-p-anisidines and N-phenyl-p-aminophenols.

Eisner and Zommer (1971) also studied the anodic oxidation of hydrazine and its derivatives. For example, three oxidation waves were obtained between -0.4 and $+1.0$ V (vs Hg/Hg_2sO_4 electrode) using potential sweep on 5×10^{-3}M dimethyl hydrazine in 0.5N sulphuric acid.

Masui et al. (1968) have studied the anodic oxidation of 17 aliphatic amines by cyclic voltammetry at a glassy carbon electrode in aqueous alkaline solution. Primary amines were reported as showing no wave, with secondary amines showing one and tertiary amines two waves. The postulated mechanism for tertiary amines involved the loss of two electrons, followed by reaction with water to form a secondary amine and an aldehyde, according to the following scheme:

$$R_2\ddot{N}CH_2R' \xrightarrow[slow]{-e^-} [R_2\overset{\cdot+}{N}-CH_2R'] \xrightarrow[fast]{-H^+} R_2N-CHR' \xrightarrow[fast]{-e^-} R_2\overset{+}{N}=CHR'] \xrightarrow[fast]{+H_2O} R_2NH + R'CHO$$

(ii) *Hydroxylamines*

Sternson (1974) has studied some arylhydroxylamines including N-phenylhydroxylamine (LII). They were shown to be very easily oxidized at carbon paste electrodes and yielded well-defined cyclic voltammograms

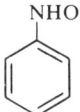

(LII) N-phenylhydroxylamine

in phosphate buffer. Aliphatic hydroxylamines are oxidized at more anodic potentials than aromatic analogues, precluding their analysis in proteinaceous solutions. Aromatics, however, can be determined in human plasma and liver microsomal suspensions, even with large excesses of substrates known to generate such N-hydroxy intermediates, i.e. aromatic nitro and amine compounds. Linear calibration plots were obtained in the concentration range 1×10^{-4} to 2.5×10^{-6}M.

The O-carbamoyl derivatives of N-acyl-N-alkylhydroxylamines undergo oxidation at a platinum microelectrode (Supin et al., 1971). At pH > 2 and low concentrations, the wave height was found to be proportional to concentration. Using the rotating Pt electrode the active centre appeared

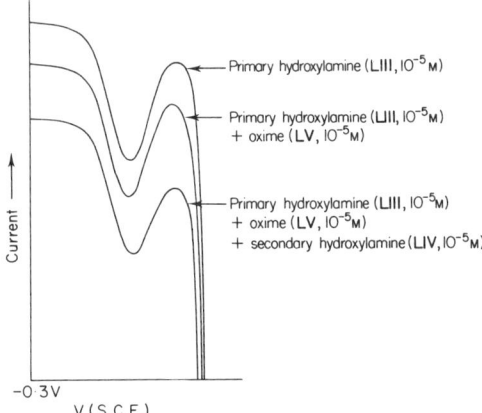

FIG. 8. The effect of secondary hydroxylamine (LIV) and oxime (LV) on the anodic wave of primary hydroxylamine (LIII), all at 10^{-5}M. D.p.p. used with BR buffer, pH 7.0.

to be the nitrogen of the R_2NCO group. It was found that oxidation was facilitated by increasing the basicity of the solution.

Beckett et al. (1977) found that the anodic wave corresponding to oxidation of a primary aliphatic hydroxylamine (LIII) was more suitable for analytical purposes than its corresponding reduction wave. A linear calibration plot was obtained in the range 10^{-4} to 10^{-6}M in BR buffer pH 7.0 containing 10% methanol. This wave was unaffected by equimolar amounts of the potential metabolic interferences, the secondary hydroxylamine (LIV) and the oxime (LV). This is shown in Fig. 8.

(LIII)   (LIV)

(LV)

(iii) Miscellaneous nitrogen compounds

A wide variety of individual compounds have been shown to display anodic behaviour and no single classification is possible. The remainder of this section is therefore devoted to those nitrogen compounds not falling into the two preceding sections. Barbiturates have been shown to give rise to anodic waves corresponding to mercury salt formation at the d.m.e. (Koryta and Zuman, 1952; Zuman et al., 1953) and also anodic waves at solid electrodes (Kato and Dryhurst, 1975).

Passet and Tsivina (1972) studied the benzimidazole group of drugs using voltammetry at a rotating platinum electrode. Benzimidazoles (LVI) and their derivatives were determined in reaction mixtures. Struck and Elving (1964) carried out a polarographic investigation of parabanic acid (imidazolidinetrion) and showed that electrolytic oxidation produced uric acid and that the diffusion current due to parabanic acid was proportional to concentration in acetate/phosphate buffer pH 5.1.

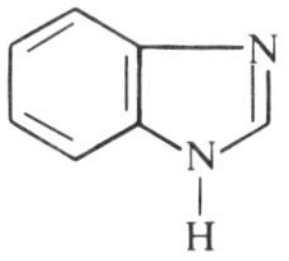

(LVI) Benzimidazole

A rotating platinum electrode was used by Libert and Caullet (1971) to study the electrochemical oxidation of 2,3,4,5 tetraphenylpyrrole (LVII). in nitromethane. Two one-electron waves were obtained.

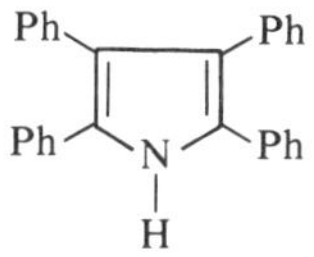

(LVII) Tetraphenylpyrrole 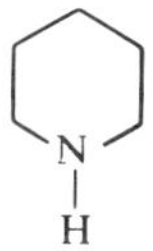   (LVIII) Piperidine

Piperidine (LVIII) undergoes oxidation at mercury and platinum (Barradas et al., 1971). The oxidation in aqueous Na_2SO_4 solution was studied and the diffusion coefficient of piperidine was evaluated. The reaction was found to involve transfer of two electrons with the participation of two OH^- per molecule of piperidine oxidized. Evidence was presented to show that piperdine N-oxide was the most probable product of the oxidation.

The pyridine and quinoline series was studied by Pozdeeva and Novikov (1966, 1969). Semicarbazides and thiosemicarbazides were included in the study.

The irreversible anodic oxidation of carboxylic acid hydrazide was studied by Titov et al. (1970). Results showed the diffusion character and the irreversibility of the oxidation.

Ambrose and Nelson (1968) have studied the anodic oxidation pathways of carbazoles (LIX) and N-substituted derivatives.

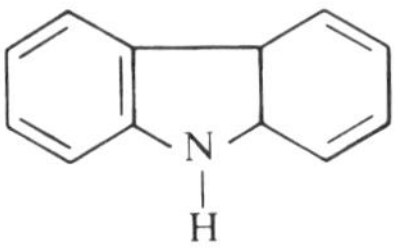

(LIX) Carbazole

For carbazole and the N-alkyl or N-aryl derivatives, ring–ring coupling is the predominant decay pathway of carbazole cations. Coupling rates are rapid. It was shown that N–N coupling is observed initially for carbazole, but is of little consequence in long-term electrolysis. Oxidation products were synthesized by chemical means and compared with species formed electrochemically by matching cyclic polarograms.

B. SULPHUR-CONTAINING COMPOUNDS

In addition to those already discussed, several other sulphur compounds have found application as pharmaceutical products.

Since mercury salt formation is an important oxidation process for many sulphur compounds, most work has been performed at mercury electrodes, with less emphasis on the use of solid electrodes than for organic compounds without sulphur.

(i) Thioamides and thioureas

Both thioamides and thiourea derivatives have been determined by anodic polarography. Early workers included Fedoronko and Zuman (1955) who showed that thiourea derivatives with a hydrazine group also gave anodic waves, and Jensovsky (1956), who studied the single anodic wave due to each of the mercury salts of thiourea and its N-substituted derivatives. The waves were useful analytically and best developed in alkaline solutions.

Reddy and Krishnan (1970) have more recently studied the irreversible oxidation of thiourea using a platinum electrode in solutions of hydrochloric acid.

Several thioamides are on the market as established pharmaceuticals. One such product is ethionamide (2-ethyl-4-thiocarbamidopyridine) (LX). This compound was determined in biological fluids (Kane, 1959), using polarographic reduction. A direct method was used to measure the drug

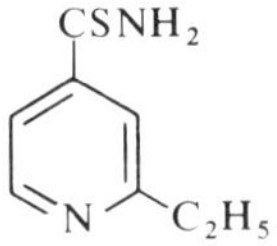

(LX) Ethionamide

in serum and cerebrospinal fluid at levels of 0.1 to 5.0 µg ml^{-1} and in urine at levels above 30 µg ml^{-1}. A sensitivity of about 0.2 µg ml^{-1} was obtained and it was found unnecessary to add a supporting electrolyte/buffering agent to the serum solutions. Urine samples were acidified by adding 5% by volume of glacial acetic acid. The advantage of anodic polarography for measuring ethionamide in body fluids would be the lack of interference from its major metabolites, the corresponding amide and carboxylic acid (Okuda, 1963), neither of which contain sulphur and are thus not amenable to the oxidation process.

The anodic polarography of thioamides has been the subject of investigations in the author's own laboratory (Davidson and Smyth, 1975, Davidson, 1976). Several compounds, including both free bases and hydrochloride salts have been studied, and analytical methods developed for their determination in pharmaceutical formulations. Compounds LXI–LXIV are typical of those studied.

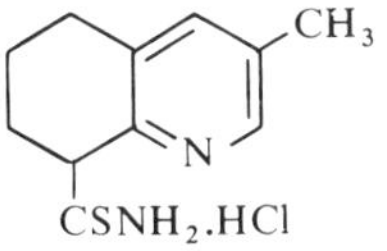

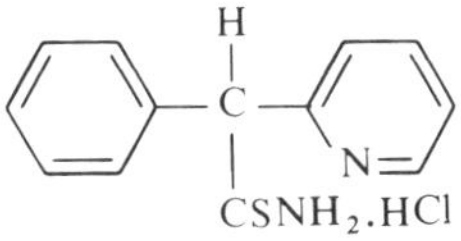

All undergo anodic oxidation at the dropping mercury electrode with a half-wave potential of about -0.25 V for solutions 10^{-4}M in Britton–Robinson buffer pH 9.15.

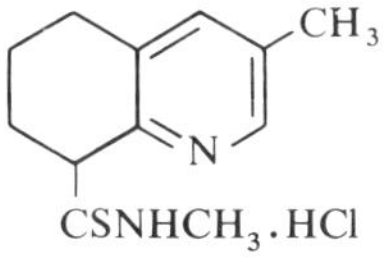

Cathodic stripping voltammetry at the hanging mercury drop electrode has been used to determine compound LXI in blood and urine without the need for prior drug extraction (Davidson and Smyth, 1977), at concentrations down to 2×10^{-8}M. Naturally-occurring sulphur compounds such as glutathione, thiamine, methionine, cystine, cysteine and H$_2$S and anions such as chloride were shown not to interfere. 2 ml blood samples were diluted 1:1 with Britton–Robinson buffer pH 4.78, followed by electrolysis at $+0.05$ V (vs s.c.c.) and subsequent cathodic stripping to -0.75 V (vs s.c.e.). Standard addition was used to determine drug concentration. Cathodic stripping voltammograms for the compound and the effect of glutathione, cysteine and chloride are shown in Figs 9 and 10.

The mechanism of electrolytic plating in this application is thought to

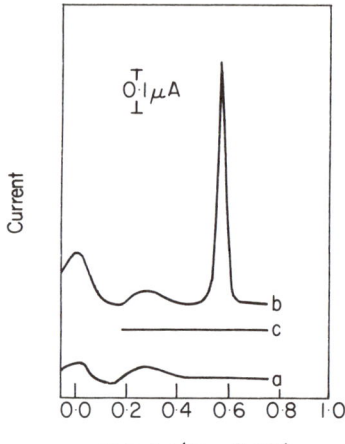

FIG. 9. Cathodic stripping voltammograms for 1:1 horse plasma: pH 4.78 buffer solutions containing (a) no drug, (b) 3.49×10^{-7}M compound 1. Scan (c) represents a post-cathodic stripping scan of solution (b).

involve the production of the corresponding nitrile (LXV) and HgS with deposition of HgS via electrolysis and its subsequent removal in the stripping step, i.e.

$$\text{(LXI)} + Hg^{2+} \longrightarrow \text{(LXV)} + HgS + 2H^+$$

Supportive evidence for this mechanism is to be found in previously documented data on the anodic oxidation of thiobenzamide (Lund, 1960).

Smyth and Osteryoung (1977) have shown that differential pulse polarography can be used to determine thiourea (LXVI), benzyl isothiourea (LXVII) and naphthylthiourea (ANTU.) (LXVII) in a mixture without prior separation using 0.1N sodium hydroxide as supporting electrolyte. It was shown that whereas thiourea and ANTU underwent a two-electron process at the d.m.e. with adsorption, benzyl isothiourea was oxidized in a one-electron process. Limits of detection for ANTU and thiourea were 10^{-7}M, and for benzyl isothiourea were 5×10^{-7}M, using pure solutions. Figure 11 shows the differential pulse polarograms obtained.

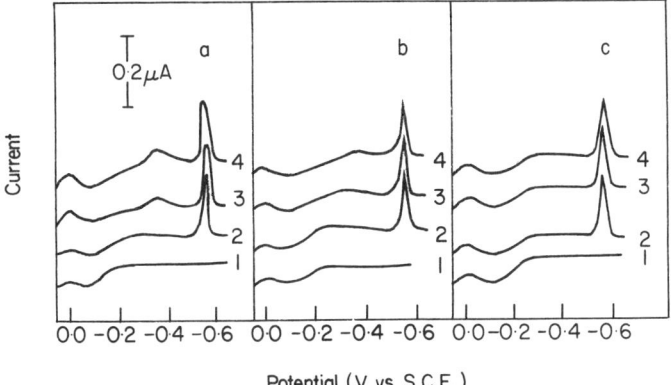

FIG. 10. Effect of possible interferents on stripping voltammograms for 1:1 horse plasma: pH 4.78 buffer solutions, 3.7×10^{-7}M solutions of compound LXI. (a) Additions of (2) 0.0, (3) 12.0 and (4) 24.0 μg of cysteine; (b) additions of (2) 0.0, (3) 3.0 and (4) 6.0 μg of reduced glutathione; (c) additions of (2) 0.0, (3) 15.0 and (4) 30.0 mg of sodium chloride. Scans (1) in each case are for blanks containing no drugs or additives.

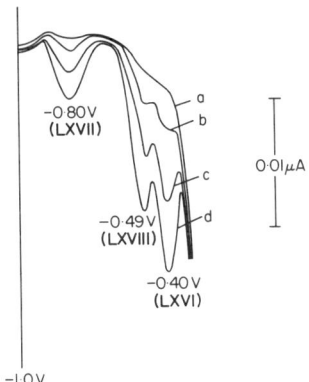

FIG. 11. Dpp determination of thiourea (LXVI), ANTU (LXVII) and benzyl isothiourea (LXVII): a = 1N NaOH; b = 15 ng/ml^{-1} (LXVI), 35 ng/ml^{-1} (LXVII), 27 ng/ml^{-1} (LXVIII); c = 45 ng/ml^{-1} (LXVI), 105 ng/ml^{-1} (LXVII), 81 ng/ml^{-1} (LXVIII); d = 75 ng/ml^{-1} (LXVI), 175 ng/ml^{-1} (LXVII), 135 ng/ml^{-1} (LXVIII). Conditions: scan rate 1 mV/s^{-2}; drop time 2 s; modulation amplitude 100 mV.

$H_2N \cdot CS \cdot NH_2$ $CH_2 \cdot N{=}C(SH){-}NH_2$ $NH \cdot CS \cdot NH_2$

(LXVI) Thiourea   (LXVII) Benzyl isothiourea   (LXVIII) Naphthylthiourea

(ii) *Miscellaneous sulphur compounds.*

A recent method of determining low levels of the sodium salt of 2-mercaptopyridine-N-oxide (LXIX) (down to 8×10^{-10}M) has been described by Csejka *et al.* (1975) using differential pulse cathodic stripping voltammetry, after deposition onto a hanging mercury drop electrode using anodic oxidation.

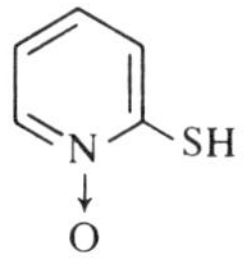

(LXIX) 2-mercaptopyridine-N-oxide

Several other sulphur compounds have been determined anodically, including diethyl dithiocarbamate (Brainina and Krapivkina, 1969; Bessarabova *et al.*, 1969), the 5,5'-disubstituted 2-thiobarbiturates (Smyth, 1970), thiobarbituric acid (Takamura and Sakamoto, 1972) and 2-mercaptopropionyl glycine (Pantani *et al.*, 1973). A solid graphite electrode was applied to the measurement of mercaptides from dimercaptothiopyrones (Arishkevich *et al.*, 1971).

Amperometric titration has been used to rapidly determine dithiocarbamates. (Ikeda and Satake, 1976). The titrant used was silver nitrate with measurement of the oxidation current at $+0.7$ V using a rotating platinum wire electrode. Results were precise, with errors between 0.2 and 0.4%.

The anodic polarography of thiosemicarbazides was studied by Lugovoi *et al.* (1973), while Supin *et al.* (1972) studied benzoxazolinethione.

C. OXYGEN-CONTAINING COMPOUNDS

Many drugs contain phenolic groups and several references are available on the anodic oxidation of these groups, in addition to those already mentioned as naturally-occurring compounds.

The rotating graphite electrode was used by Gorokhovskii *et al.* (1973) to investigate phenol oxidation while Lutskii *et al.* (1972) have studied the oxidation of 22 substituted phenols, with 1, 2 or 3 substituents selected from OH, OCH_3, Cl and NO_2. They were compared in terms of voltammetric oxidation potentials. The latter values were correlated with the ionization potential and with the π-electronic charge at the O of the OH group.

Parker and Ronlan (1971) showed that 2,6-di-t-butyl-*p*-cresol underwent a 2-electron oxidation during constant current coulometry in CH_3CN containing 2,6-lutidine.

Pelcere *et al.* (1969) determined both 2-phenyl-1,3-indandione (LXX) and 2-methoxy-2-phenyl-1,3-indandione (omephin) (LXXI) by dissolving in

50 ml ethanol, 20 ml 0.01M KOH and 20 ml M LiCl and making up to 200 ml with ethanol. The former compound was determined in tablets at an $E_{\frac{1}{2}}$ of $+0.39$ V with an error less than 2% at a concentration of 10^{-4} to 10^{-3}M.

(LXX) 2-Phenyl-1,3-indandione

(LXXI) Omephine

Sundholm (1971) studied the electrochemical oxidation of some aliphatic alcohols in acetonitrile, propylene carbonate, dichloromethane and sulpholane. Anodic oxidation was measured experimentally by linear sweep cyclic voltammetry at a platinum wire electrode. All alcohols studied gave a single well-distinguished, irreversible oxidation peak in CH_3CN, propylene carbonate and CH_2Cl_2 whereas no oxidation peaks were detected in sulpholane. Simonson and Murray (1975) investigated the anodic oxidation of hydroxide ion in acetonitrile and dimethylsulphoxide using a platinum electrode. Two anodic waves and a connected cathodic wave in acetonitrile correspond to surface oxide formation and stripping. The surface waves disappeared in dimethyl sulphoxide.

Other recent work of interest is that by Manousek and Volke (1973) who studied the anodic oxidation of aromatic aldehydes in alkaline solution, and Zuman (1974) who elucidated the mechanism of the process yielding anodic waves of aldehydes by oxidation in alkaline media.

D. HALOGEN-CONTAINING COMPOUNDS

The anodic oxidation of Cl^- is well-documented and analytical details are described by Heyrovsky and Zuman (1968). The determination is of importance to pharmaceutical analysts, many drugs being formulated as their hydrochloride salts. The chloride moiety is therefore a useful functional group to consider for analytical purposes. The electrolysis process corresponds to a reaction in which dissolved mercury ions are precipitated at the electrode surface by the approaching chloride ions to form calomel.

Dryhurst and Elving (1966) have reported work on the voltammetric determination of iodide and bromide ion at the rotating pyrolytic graphite electrode. The p.g.e. was rotated at 250 r.p.m. and was resurfaced before each run. In 0.5M H_2SO_4 and 0.5M K_2SO_4–0.5M H_2SO_4 supporting electrolytes, iodide showed two reproducible oxidation waves. The limit of detection of iodide was 0.1 mM. Bromide ion gave one analytically useful anodic wave, the limit of detection being 0.1 mM. No oxidation waves were observed for Cl^-. The accuracy of the iodide determination from 0.01 to 1.0 mM was about $\pm 4.5\%$ in acidic K_2SO_4 and somewhat less in neutral K_2SO_4. Bromide was determined with an accuracy of $\pm 3.4\%$.

The same authors (Dryhurst and Elving, 1967) studied the electrolytic oxidation of halides at the pyrolytic graphite electrode in aqueous and acetonitrile solutions, using single sweep and cyclic voltammetry. In aqueous solutions, iodide was oxidized in two well-defined steps. Bromide gave a single anodic wave in aqueous solution. No anodic wave was seen for chloride at the p.g.e. in aqueous solution. In CH_3CN, iodide gave three anodic waves.

Canterford et al. (1973) discussed the advantages of rapid d.c. polarography in the presence of analytically undesirable phenomena associated with anodic waves for analytical methods based on Hg compound formation. The work was particularly applied to chloride and iodide determinations. It was shown that at normal mercury drop times, abnormal phenomena associated with the presence of reaction products on the electrode surface may give rise to adverse effects such as maxima, erratic drop behaviour, or non-linear calibration curves, thus preventing collection of useful analytical data. By using short, controlled drop times of say 0.16 sec, well-defined waves could be produced and limiting current concentration plots were linear over a wide concentration range. Canterford and Buchanan (1973) studied the application of differential pulse polarography to anodic electrode processes involving mercury compound formation, particularly in chloride and iodide analysis.

An interesting method of chloride, bromide and iodide determination in algae was described by Calzalari et al. (1969). After sample preparation chloride and iodide were determined anodically between -0.25 to $+0.40$ V and -0.45 to $+0.20$ V respectively at concentration levels of 1 to 5 mM in 0.1M KCl solution at pH 3 to 4. Bromide was oxidized and measured by cathodic reduction.

VIII. GENERAL REVIEWS ON ANODIC OXIDATION

Several reviews on electrochemical oxidation, mainly concerning solid electrodes, have been published. The monograph by Adams (1969) is concerned with the use of solid electrodes in anodic oxidation with the mercury electrode being of limited use in this context.

Petrii and his co-workers have produced several reviews. Petrii and Lohaniai (1968) reviewed the adsorption and electrooxidation of simple organic substances on a rhodium electrode. The oxidation of methanol and formaldehyde were discussed in detail. Similar reviews were compiled by Petrii et al. (1968), Petrii and Podlovchenkco (1968) and Petrii (1969) concerning the platinum group metals, the former review dealing with methanol, alcohols, aldehydes and formic acid, the latter dealing mainly with methanol and methane.

Weinberg and Weinberg (1968) reviewed the whole subject of electrochemical oxidation of organic compounds and considered a wide range of compounds, aromatics, olefines, acetylenes, amines, amino acids, quaternary ammonium salts, phenols, esters, amides, lactams, halides, organometallics and carboxylic acids. A review of electrochemical oxidation was also undertaken by Andrews (1972).

Finally, a good text by Brezina and Volke (1968) contains data on the general polarography of organic compounds used in medicine and pharmacy.

REFERENCES

Adams, R. N. (1969) "Electrochemistry at Solid Electrodes". Marce Dekker, New York.
Adams, R. N. (1976). *Anal. Chem.* **48**, 1128A.
Ambrose, J. F. and Nelson, R. F. (1968) *J. Electrochem. Soc.* **115**(11), 1159.
Anastasi, A., Mecarelli, E. and Novacic, L. (1952). *Microchem. Microchim. Acta,* **40**, 113.
Andrews, J. E. L. (1972). M.Sc. Thesis, University of London.
Arishkevich, A. M., Usatenko, Y. I., Shidlovskaya, A. I., Kroik, A. A. and Moroz, A. A. (1971). *Khim. Tekhnol. (Karkov)*, **17**, 172.
Asah, Y. (1963). *Chem. Pharm. Bull.,* Japan, **11**(7), 930.
Atuma, S. A. and Lindquist, J. (1973). *Analyst,* **98**, 886.
Atuma, S. A., Lindquist, J. and Lundstrom, K. (1974). *Analyst,* **99**, 683.
Barnes, K. K. and Mann, C. K. (1967) *J. Org. Chem.* **32**, 1474.
Barradas, R. G., Giordano, M. C. and Sheffield, W. H. (1971). *Electrochim. Acta,* **16**(8), 1235.
Beckett, A. H., Rahman, N. M. and Franklin Smyth, W. (1977). *Anal. Chim. Acta* **92**, 353.
Bessarabova, I. M. and Songina, O. A. (1971). *Izv. Akad. Nauk. Kaz. Ssr., Ser, Khim.* **21**(6), 58.
Bessarabova, I. M., Zakharov, V. A., Songina, O. A. and Kufeld, G. R. (1969). *Izv. Akad. Nauk. Kaz, Ssr, Ser, Khim.* **21**(6), 58.
Bessarabova, I. M., Zakharov, V. A. and Songina, O. A. (1970). *Khim. Khim. Tekhnol (Alma-ATA),* No. 1, 24.
Bezugly, V. B. and Beilis, Y. I. (1965). *Zh. Anal. Chim.* **20**, 1000.
Blaedel, W. J. and Jenkins, R. A. (1975). *Anal. Chem.* **47**, 1337.
Blank, C. L. and Pike, R. (1976). *Life Sci.* **18**, 859.
Brainina, Kh. Z. and Krapivkina, T. A. (1969). *Anal. Lett.* **2**(5), 269
Brandys, J. (1966). *Dissnes Pharm. Warsz.* **18**, 319.
Breitenbach, M. and Heckner, K. H. (1971). *J. Electroanal. Chem. Interfacial Electrochem.,* **33**(1), 45.
Brezina, M. and Volke, J. (1968). Beckman Reports (1–2), 11.
Brezina, M. and Zuman, P. (1958). "Polarography in Medicine, Biochemistry and Pharmacy". Interscience, Chichester and New York.
Brezina, M., Koryta, J., Loucka, T. Marsikova, D., Pradac, J. (1972). *J. Electroanal. Chem. Interfacial Electrochem.* **40**(1), 13.

Calzalari, C., Gabriella, L. F. and Marletta, G. P. (1969). *Analyst*, **94**, 774.
Canterford, D. R. and Buchanan, A. S. (1973). *J. Electroanal. Chem. Interfacial Electrochem.* **4**(2), 291.
Canterford, D. R., Buchanan, A. S. and Bond, A. M. (1973). *Anal. Chem.* **45**(3), 1327.
Cantin, D., Alary, J. and Coeur, A. (1973). *Analusis*, **2**(9), 654.
Cescon, P. and Montalti, M. (1970) *Annalt Chem.* **60**(1), 63.
Charles, R. and Knevel, A. M. (1968). *Anal Biochem.* **22**, 179.
Cospito, M., Raspi, G. and Lucarini, L. (1969). *Anal. Chim. Acta*, **47**, 388.
Cozzi, D., Raspi, G. and Nucci, L. (1963). *J. Electroanal. Chem.* **6**, 275.
Csejka, D. A., Nakos, S. T. and Du Bord, E. N. (1975). *Anal. Chem.* **47**, 322.
Davidson, I. E. (1976). *Proc. Anal. Div. Chem. Soc.*, **13**(8), 229.
Davidson, I. E. and Smyth, W. F. (1975). Paper presented at 35th International Congress of Pharmaceutical Sciences, Dublin, 1st–5th September 1975.
Davidson, I. E. and Smyth, W. F. (1977). *Anal. Chem.* **49**, 1196.
Deys, H. P. (1964). *Pharm. Weekbl.* **99**(28), 737.
Dryhurst, G. (1969a). *J. Electrochem. Soc.* **116**(10), 1411.
Dryhurst, G. (1969b). *Anal. Chim. Acta*, **47**(2), 275.
Dryhurst, G. (1969c). *J. Electrochem. Soc.* **116**(8), 1097.
Dryhurst, G. (1970). *J. Electrochem. Soc.* **117**(9), 1113.
Dryhurst, G. (1971). *Anal. Chim. Acta*, **57**, 137.
Dryhurst, G. and De, P. K. (1972). *Anal. Chim. Acta*, **58**(1), 183.
Dryhurst, G. and Elving, P. J. (1966). *J. Electroanal. Chem.* **12**(5/6), 416.
Dryhurst, G. and Elving, P. J. (1967). *Anal. Chem.* **39**(6), 606.
Dryhurst, G. and Elving, P. J. (1968). *J. Electrochem. Soc.* **115**(10), 1014.
Dryhurst, G. and Elving, P. J. (1969). *Talanta*, **16**(7), 855.
Dryhurst, G. and Hansen, B. H. (1971a). *J. Electroanal. Chem. Interfacial Electrochem.* **30**(3), 417.
Dryhurst, G. and Hansen, B. H. (1971b). *J. Electroanal. Chem. Interfacial Electrochem.* **30**(3), 407.
Dryhurst, G. and Pace, G. F. (1970). *J. Electrochem. Soc.* **17**(10), 1259.
Eisner, U. and Zommer, N. (1971). *J. Electroanal. Chem. Interfacial Electrochem.* **30**(3), 433.
El-Sourady, H. A. (1971). *Egypt. Pharm. J.* **53**, 51.
Fedoronko, M. and Zuman, P. (1955). *Chem. Listy*, **49**, 1484.
Felice, L. J. and Kissinger, P. T. (1976). *Anal. Chem.* **48**, 795.
Freude, F., Kaniss, N. and Wunderlich, H. (1953). *Pharmazie*, **8**, 631.
Gilbert, D. D. (1969). *Anal. Chem.* **41**, 1567.
Gonzalez, J., Fernandez-Alonso, J. I. (1970). *An. Quim. Farm.* **66**(12), 931.
Gorokhovskii, V. M., Kuzovenko, N. M., Khaikim, M. S., Fedorina, L. G. and Vakatova, N. I. (1973). *Ref. ZH. Khim.* Abstr. No. No. 13B1540.
Grundova, K., Davidek, J., Velisek, J. and Janicek, G. (1973). *Lebensm-Wiss. Technol.* **6**(1), 11.
Guilbault, G. G., Kramer, D. N. and Cannon, P. L. (1964). *Anal. Chem.* **36**, 606.
Hand, R. and Nelson, R. F. (1970). *J. Electrochem. Soc.* **117**(11), 1353.
Hansen, B. H. and Dryhurst, G. (1971). *J. Electroanal. Chem.* **32**, 405.
Hawley, M. D., Tatawawadi, S. V., Piekarski, S. and Adams, R. N. (1967) *J. Amer. Chem. Soc.* **89**, 447.
Heyrovsky, J. and Zuman (1968). "Practical Polarography", p. 102. Academic Press, London.
Ikeda, S. and Satake, H. (1976). *Bunseki Kagaku*, **25**(a), 611.
Jensovsky, L. (1956). *Collect. Czech. Chem. Commun.* **21**, 459.
Kajita, T. (1966). *Nippon Shokuhin Kogyo Gakkaishi*, **13**(7), 288.
Kajita, T. and Senda, M. (1970). *Bunseki Kagaku*, **19**(3), 330.

Kajita, T., Yamamoto, Y., and Senda, M. (1973). *Bunseki Kagaku*, **22**(8), 1051.
Kane, P. O. (1959). *Nature, Lond.* **183**. 1674.
Kato, S. and Dryhurst, G. (1975) *J. Electroanal. Chem.* **62**, 415.
Kemula, W. and Kalinowski, M. K. (1967). *Z. Anal. Chem.* **224**, 383.
Kissinger, P. T., Felice, L. J., Riggins, R. M., Pachla, L. A. and Wenke, D. C. (1974). *Clin. Chem.* **20**, 992.
Kissinger, P. T., Riggin, R. M., Alcorn, R. L. and Rau, L. (1975). *Biochem. Med.*, **13**, 299.
Kitagawa, T. and Kanei, Y. (1970). *Japan Analyst*, **19**. 642.
Kitagawa, T. and Tsushima, S. (1971). *Bunseki Kagaku* **20**(12), 1561.
Kolthoff, I. M., Shore, W. S., Tan, B. H. and Matsuoka, M. (1965). *Anal. Biochem.* **12**, 497.
Konupcik, M., Liska, M. and Kupcik, F. (1960). *Ceskosl. Farm.* **9**(10), 502.
Korobeinik, F. G. (1968). *Lab. Delo.* **11**, 692.
Koryta, J. and Pradac, J. (1968a). *J. Electroanal. Chem. Interfacial Electrochem.* **17**(1/2), 185.
Koryta, J. and Pradac, J. (1968b). *J. Electroanal. Chem. Interfacial Electrochem.* **17**(1/2), 177.
Koryta, J. and Zuman, P. (1952). *Chem. Listy*, **46**, 389.
Kruze, I. E. (1969). *Farmatsiya (Moscow)*, **18**(4), 59.
Lane, R. F. and Hubbard, A. T. (1976). *Anal. Chem.* **48**, 1287.
Leedy, D. W. and Adams, R. N. (1976). *J. Amer. Chem. Soc.* **92**(6), 1646.
Lhuguenot, J. C., Coq, H. M. and Baron, C. (1969). *Bull. Soc. Chim. Biol.* **51**(2), 424.
Libert, M. and Caullet, C. (1971). *Bull. Soc. Chim. Fr.* No. 5, 1947.
Liberti, A., Cervone, E. and Cattaneo, C. (1952). *Giorn. Biochem.* **1**, 440.
Lindquist, J. (1975). *Analyst*, **100**, 339.
Lindquist, J. and Farroha, S. M. (1975). *Analyst*, **100**, 377.
Lucarini, L., Cospito, M. and Raspi, G. (1970). *Farmaco, Ed. Prat.* **25**(1), 39.
Lugovoi, S. V. and Ryazonov, I. P. (1967). *Zh. Analit. Khim.* **22**(7), 1093.
Lugovoi, S. V., Chernova, T. N. and Chistota, V. D. (1973). *Zh. Analit. Khim.* **28**(5), 991.
Lund, H. (1960). *Coll. Czech. Chem. Commun.* **25**, 3313.
Lutskii, A. E., Beilis, Y. I. and Fedorchenko, V. I. (1972). *Zh. Obshch. Khim.* **42**(11), 2535.
Malik, W. V. and Singh, K. L. (1968). *Indian J. Technol.* **6**(11), 344.
Manousek, O. and Volke, J. (1973). *J. Electroanal. Chem. Interfacial. Electrochem.* **43**(3), 365.
Manousek, O. and Zuman, P. (1955). *Chem. Listy*, **49**, 668.
Mason, W. D. (1973). *J. Pharm. Sci.* **62**, 6, 999.
Mason, W. D., Gardner, T. D. and Stewart, J. T. (1972). *J. Pharm. Sci.* **61**(8), 1301.
Masui, M. Sayo, H. and Tsuda, Y. (1968). *J. Chem. Soc. B*, **9**, 973.
Merkle, F. H. and Discher, C. A. (1964). *Anal. Chem.* **36**, 1639.
Niederstebruch, A. and Hinsch, I. (1967). *Fette Seifen Anstr-Mittel*, **69**(8), 559.
Noninski, Kh. I., Dryanovska-Noninska, L. and Iliev, L. St. (1969). *Farmatsiya, Sof.* **19**(3), 24.
Okuda, Y. (1963). *Rev. Polarography (Japan)*, **11**, 197.
Owen, R. S. and Franklin Smyth, W. (1975). *J. Fd. Technol.* **10**, 263.
Pachla, L. A. and Kissinger, P. T. (1976) *Anal. Chem.* **48**, 364.
Pantani, F., Legittimo, P. and Ciantelli, G. (1973). *Bull. Chim. Farm.* **112**(1), 29.
Papouchado, L., Petrie, G. and Adams, R. N. (1972). *J. Electroanal. Chem. Interfacial Electrochem.* **38**(2), 389.
Park, G., Adams, R. N, and White, W. R. (1972). *Anal. Lett.* **5**, 007.
Parker, V. D. and Ronlan, A. (1971). *J. Electroanal. Chem. Interfacial Electrochem.* **30**(3), 502.
Passet, B. V. and Tsivina, N. S. (1972). *Khim-Farm. Zh.* **6**(4), 56.
Pelcere, I., Priede, V., Karklins, A. and Veiss, A. (1969). *Latv. Psr. Zinat. Akad. Vestis. Kim. Ser.* No. 6, 690.
Perone, S. P. and Kretlow, W. J. (1966). *Anal. Chem.* **38**, 1760.
Petrii, O. A. (1969). *Progr. Elektrochim. Org. Soedin*, **1**, 278.

Petrii, O. A. and Lohaniai, N. (1968). *Electrokhimiya*, **4**(6), 656.
Petrii, O. A. and Podlovchenko, B. I. (1968) *Topl. Elem.* 169.
Petrii, O. A., Podlovchenko, B. I. and Frumkin, A. N. (1968) *Sovrem. Probl. Fiz. Khim.* **2**, 196.
Pozdeeva, A. G. and Novikov, E. G. (1966). *Zh. Prikl. Khim.* **39**(12), 2669.
Pozdeeva, A. G. and Novikov, E. G. (1969). *Zh. Prikl. Khim.* **42**(11), 2626.
Pradac, J. and Koryta, J. (1968). *J. Electroanal. Chem. Interfacial Electrochem.* **17**(1/2), 167.
Pungor, E., Feher, Z, and Nagy, G. (1971). *Magy. Kem. Foly*, **77**, 298.
Rashid, A. and Kalvoda, R. (1971). *Cesk. Farm.* **20**(4), 143.
Reddy, S. J. and Krishnan, V. R. (1970). *J. Electroanal. Chem. Interfacial Electrochem.* **27**(3), 473.
Remke, H. (1970). *Acta Biol. Med. Ger.* **24**(1–2), 13.
Richmond, D. V. and Somers, E. (1966). *Chem. Ind.*, **1**, 18.
Riggin, R. M., Alcorn, R. L. and Kissinger, P. T. (1976). *Clin. Chem.*, **22**, 782.
Sartori, G. and Cattaneo, C. (1942). *Gazz. Chim. Ital.* **72**, 525.
Schmandke, H. (1965). *Int. Z. Vitam-Forsch*, **35**, 237.
Schmandke, H. and Crohlke, H. (1965). *Clin. Chim. Acta*, **11**, 491.
Schubert, E. and Roland, U. (1968). *Die Nahrung* **12**(7), 715.
Shearer, C. M., Christenson, K., Mukherji, A. and Papariello, G. J. (1972). *J. Pharm. Sci.* **61**(10), 1627.
Shiozaki, K., Fukui, K., and Kitagawa, T. (1971). *Japan Analyst*, **20**(4), 438.
Sinsheimer, J. E. and Hong, D. (1965). *J. Pharm. Sci.*, **54**(5), 805
Simonson, L. A. and Murray, R. A. (1975). *Anal. Chem.* **47**, 290.
Smith, E., Worrell, L. F., and Sinsheimer, J. E. (1963). *Anal. Chem.* **35**, 58.
Smith, L. I., Kolthoff, I. M., Wawzonek, S. and Ruoff, P. M. (1941). *J. Amer. Chem. Soc.* **63**, 1018.
Smith, L. I., Spillane, I. J. and Kolthoff, I. M. (1942). *J. Amer. Chem. Soc.* **64**, 646.
Smyth, M. R. and Osteryoung, J. G. (1977). *Anal. Chem.* **49**, 2310.
Smyth, W. F. (1970). Proc. Anal. Chem. Conf., 3rd, **2**, 123.
Soderhjelm, P. and Lindquist, J. (1975). *Analyst*, **100**, 349.
Sternson, L. A. (1974). *Anal. Chem.* **46**, 2228.
Stewart, J. T., Hoo, C. L. and Mason, W. D. (1974). *J. Pharm. Sci.* **63**, 954.
Stradins, J. and Gasonov, B. R. (1973). *Latv. Psr. Zinat. Akad. Vestis. Kim, Ser.* No. 3, 370.
Struck, W. A. and Elving, P. J. (1964). *Anal. Chem.* **36**, 1374.
Sundholm, G. (1971). *Acta. Chem. Scand.* **25**(8), 3188.
Supin, G. S., Baskakov, V. A. Konstantinova, N. V. and Kolobanova, L. P. (1971), *Zh. Obshch. Khim.* **41**(3), 502.
Supin, G. S., Poznanskaya, N. L., Shvetsov-Shilovskii, N. I., Melnikov, N. N. and Goloskova, A. V., (1972). *Zh. Obshch. Khim.* **42**(6), 1190.
Tachi, I. and Koide, S. (1951). Sbornik Mexinarod Polarogr. Szedu Praze, 1st Congr. Pt. 1, Proc. 450, 469.
Takamura, K. and Sakamoto, M. (1972). *Tokyo Yakka Daigaku Kenkyu Nempo*, No. 22, 177.
Timbekov, E. Kh. and Kasymov, M. K. (1968). *Referat. Zh., Khim.* **19**GD(2), Abstr. No. 2G177.
Titov, E. V., Popova, I. V. and Lagutskaya, L. I. (1970). *Teor. Eksp. Khim.* **6**(6). 789.
Ueno, Y. (1968). *Experientia*, **24**(9), 970.
Vachek, J. (1960). *Ceskosl. Farm.* **9**(3), 126.
Vachek, J. (1965). *Ceskosl. Farm.* **14**(5), 216.
Varadi, M., Feher, Z. and Pungor, E. (1974). *J. Chromatogr.* **90**, 259.
Volke, J., El-Laithy, M. M. and Volkova, V. (1975). *J. Electroanal. Chem.* **60**(2), 239.
Volova, I. G., Kuznetsova, M. A. and Brainina, Kh. Z. (1972). *Ref. Zh. Khim.* Abstr. No. 11R11.
Wacholz, E. and Pfeifer, S. (1972). *Pharmazie* **27**, 97.

Wasilewska, L. and Szyszko, E. (1969). *Diss. Pharm. Pharmacol.* **21**(6), 591.
Weinberg, N. L. and Weinberg, H. R. (1968). *Chem. Rev.* **68**, 449.
Zhustereva, S. S., Krunchak, V. G., Lomonova, M. A., Mikhailova, V. P. and Khvorostin, Ya. S. (1972). *Tr. Vses. Nauch-Issled. Inst. Tsellyul. Bum. Prom.* **60**, 170.
Zuman, P. (1974) *Proc. Soc. Anal. Chem.* December, 338.
Zuman, P., Koryta, J. and Kalvoda, R. (1953). *Chem. Listy*, **47**, 345.

APPLICATIONS

In the Basic Medical Sciences

Chapter 6

SOME RECENT APPLICATIONS OF ORGANIC VOLTAMMETRY IN THE BASIC MEDICAL SCIENCES

BABUR Z. CHOWDHRY

Department of Chemistry, Chelsea College, University of London, Manresa Road, London, England

I. INTRODUCTION

The application of voltammetric methods of analysis to problems of clinical interest has for long been practised in many Eastern European countries. The review by Brezina and Zuman (1958) highlighted the applications of polarography in the related fields of medicine, biochemistry and pharmacy. Although further reviews on this subject have been written by Elving (1966), Milazzo *et al.* (1971) and Brezina and Volke (1975), many clinicians in Western Europe have yet to fully realize the usefulness of voltammetric methods of analysis. It is the intention of this chapter, therefore, to give a review of the diverse ways in which voltammetric techniques have been applied in the basic medical sciences for a readership in these disciplines.

II. CLINICAL MEDICINE

A. CLINICAL DIAGNOSIS

(i) *The Brdicka reaction*

Since the polarographic behaviour of proteins was first observed in 1928, much effort has been expended in both a study of the electrochemical reactions of proteins (Homolka, 1971; Lewitova, 1974; Kolthoff *et al.*, 1974; Anisimova *et al.*, 1975; Kuznetsov and Shumakovich, 1975; Kuznetsov *et al.*, 1977; Senda *et al.*, 1976) and in the application of these reactions to diagnosis in clinical medicine.

The basic reaction used is the Brdicka reaction in which catalytic hydrogen evolution is observed as a characteristic "double-wave" following polarography at the d.m.e. In most cases polarography is carried out on a sample of the body fluid (usually serum) mixed with 10^{-3} M Co(II) or Co(III) in 0.1M ammoniacal buffer. The process is believed to be dependent on the presence of —SH and —S—S— groups in the protein with the Co(II)-protein complex acting as catalyst in hydrogen evolution.

There are three modifications of the basic reaction: (1) the denaturation reaction in which 0.1 ml serum is mixed with 1.4 ml 0.9% NaCl and 1.5 ml 0.25M NaOH. After a selected time, 0.1 ml of this mixture is diluted with 4 ml 10^{-3} M Co(III) and subjected to polarography from -0.8 V to -1.8 V; (2) the filtrate reaction illustrated below:

$$0.4 \text{ ml fresh serum} + 1 \text{ ml } 0.1\text{M KOH} \xrightarrow[45 \text{ min}]{20°C} \xrightarrow[\text{salicyclic acid}]{1 \text{ ml } 20\% \text{ sulpho-}} \xrightarrow{\text{mix}} \xrightarrow[10 \text{ min}]{\text{stand}}$$

$$\xrightarrow[\text{sample}]{\text{filter whole}} \text{filtrate (0.4 ml)} \xrightarrow[\text{Co(III) solution}]{4 \text{ ml Brdicka}} \text{polarography from } -0.8 \text{ V to } -1.8 \text{ V}$$

(3) that in which electrophoresis precedes polarographic examination.

Examples of some of the diseased states that have been diagnosed polarographically are given in Table I.

TABLE I

Brdicka Basic Reaction
Inflammatory rheumatic disease
Diseases reflected by changes in composition of cerebrospinal fluid

Electrophoretic-Polarographic Test
Lipoid nephrosis
Hepatic parenchyma
Infectious hepatitis
Cirrhosis
Myeloma
Nephritis
Nephrosis
Cystopyelitis

Brdicka Filtrate Reaction
Hepatic cirrhosis
Beta-2-myeloma
Carcinoma of digestive tract and pancreas
Dupuytrens contractive
Intestinal tuberculosis and malabsorption
Chronic rheumatism

The Brdicka denaturation test (Vishnyak and Krusnova, 1965) showed differences in the polarographic behaviour of sera taken from patients suffering from different forms of schizophrenia. As the condition of the patients improved it was found that the height of polarographic waves produced by protein-free filtrates decreased to normal values while in Brdicka's denaturation test the height of the waves increased.

Ueno et al. (1968), in a polarographic and paper electrophoretic study of sera and cerebrospinal fluids (CSF) taken from normal and mental patients (schizophrenic, epileptic and general paretic), found that there were very significant changes in the polarographic behaviour of the latter fluids.

Matyus and Scheda (1969), using a Brdicka reaction, found that in endogenous psychoses the wave heights were relatively low, whereas in other states, e.g. senile psychoses, arteriosclerotic dementias and acute softenings, the wave heights were higher than normal. They attributed these findings to low and high concentrations of sulphur-containing amino acids respectively. Relatively high wave heights were also found in patients suffering from tumours and meningitis.

In an interesting paper, Bukaresti et al. (1969) have suggested the use of the Brdicka filtrate reaction in the assessment of decompensated heart patients (valvular diseases, cardiosclerotics and cor pulmonale). The polarographic tests were claimed to be a much better guide than the classical dysproteinemic liver tests in such conditions and even some complications such as rheumatic fever did not present analytical problems.

Polarographic tests have also been used to detect acute disorders of brain circulation (Anisimova, 1971). In this examination, the blood sera of 230 patients subsequent to stroke were tested. Patients with haemorrages in the parenchymatous and subarachnoidal space showed differences in the Brdicka "double wave" after the second day whereas patients suffering from

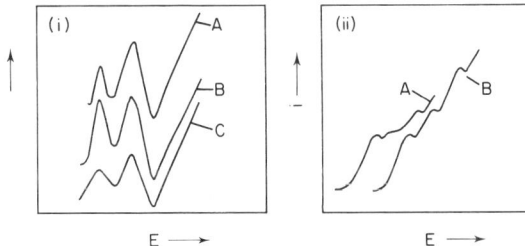

FIG. 1. (i) Polarographic analysis of spinal fluid by the Brdicka reaction of patients suffering from (A) meningoencephalitis and (B) inflammation of tumour of the brain. (C) is normal. (After Moysenko, 1972.) (ii) Oscillopolarograms of albumin-free filtrate of blood serum (A) from healthy patients, (B) from patient suffering from tumour of the brain. (After Burgman, 1973.)

ischemic strokes showed no change until the fourth day. Anisimova (1971) recommended that polarography could be used in conjunction with other methods in establishing the exact nature of the stroke.

A study of 196 patients (130 with meningoencephalitis and 66 with benign tumours of the brain) revealed that the polarographic behaviour of the sera of the two groups of patients differed significantly. In inflammatory and neoplastic processes a pronounced shift in the potential of the first and second wave to more negative values occurred (Moysenko, 1972; Moysenko and Vlasenko, 1973). In tumours of the brain a greater increase in the height of the first differential wave than of the second was observed while in meningoencephalitis the reverse occurred [Fig. 1(i)]. Samples of both healthy and diseased subjects were also analysed by the sulphosalicyclate filtrate test for albumin extracted from spinal cord.

Burgman (1973) has claimed that oscillopolarographic recording of blood serum free of albumin allows a distinction between healthy patients and those suffering from brain tumours [Fig. 1(ii)].

With the aid of the d.m.e. (vs Ag anode) polarograms recorded from -1.2 V to more negative potentials have allowed the diagnosis of cystinurea (Kuzel, 1973). The test involved mixing 0.05 ml urine and 6.0 ml ammonia/nickel sulphate solution (9×10^{-2}M NH_4Cl, 1.17×10^{-1}M NH_3, 2×10^{-3} $NiCl_2$; 1.25×10^{-5} M carbowax, 1/40 saturated sulphite); the mixture was allowed to stand for one hour and then subjected to polarography. The method was found to be simple and was not disrupted by sulphur-containing proteins nor by amino acids in the urine.

An electrophoretic polarographic analysis of blood serum has provided more helpful criteria in the assessment of rheumatism than the more classical procedures (e.g. examination of albumin levels by electrophoresis—Stas, 1973). Changes in the sera of alcoholics as well as in those of patients with diseases of the thyroid were also noted.

Diseases such as cholecystitis, chronic cholecystitis of an inflammatory nature and cirrhosis of the liver have been distinguished by examination of both the height and shape of waves obtained by Brdicka reactions (Mansurova, 1974).

(ii) *Non-Brdicka reactions*

Carruthers (1968) has found significant changes in the levels of reducible compounds present in normal patients and those with malignant growths. High activity of glutathionase has also been reported as being a distinctive feature of some chemically induced rat hematomas (e.g. using 3-methyl-4-dimethylamino azobenzene) and the cysteine liberated in such cases may be determined polarographically (Fiala *et al.*, 1972).

Charles and Knevel (1968) have used a.c. polarography to determine non-protein sulphydryl compounds in liver and brain tissue whereas Remke (1970) attributed the anodic wave observed in ageing mitochondrial suspensions to the release of free-SH groups into the protoplasm.

The presence of fructose in urine has been demonstrated by Blinkov (1965) using polarography. Moreover a correlation was observed between the height and shape of the two waves found in urine and the severity of the liver involvement in infectious hepatitis. Sushkin (1969) has claimed that different types of fructosuria may be identified by polarography and that (if required) the presence of various antituberculosis agents could be monitored simultaneously using the same technique.

Hata et al, (1971) and Hata (1973), has shown that the i-E curves obtained following polarography of tissue extracts and urine in animals suffering from cancer differ from those of normal animals. A.c. polarography of tissue extracts from mice and rats [using a supporting electrolyte of Na_2HPO_4 (4.22×10^{-2}M) and KH_2PO_4 (2.57×10^{-2}M); NaCl was added in some instances] gave a wave at -1.7 V and usually an additional one at -1.4 V (with reference to s.c.e.). A third wave at -1.9 V existed in tissue extracts of malignant mammalian tumours (spleen and liver) and in non-cancerous tissue (confirmed histologically) of animals with implanted tumours. Moreover the third wave was present in the urine of non-cancerous humans but was never present in the tissue extracts of animals without cancer or in the urine of patients suffering from cancer. It was postulated that substances producing the wave at -1.9 V were eliminated by healthy individuals but retained by cancerous individuals and that the substances responsible could have some relationship with the exposed carbohydrate on the surface of structurally altered carcinogenic cells (a heat stable alcohol precipitate compound was postulated).

Oxygen measurements by polarographic techniques have been used in the investigation of mitochondrial activity in relation to certain disease states and/or drug activity. The reader is referred to the works of Schreiber et al. (1970), De Montalvo et al. (1972), Matsubara and Tochino (1969, 1970), Bouhnik et al. (1972), and Saito et al. (1972).

B. STEROIDS, HORMONES AND VITAMINS

Voltammetric analysis has been successfully employed in clinical medicine for the determination of many drugs and their degradation products in formulations (Franklin Smyth et al., 1979) and for drugs and their metabolites in body fluids (Smyth and Franklin Smyth, 1978). Other chapters in this volume deal with the determination of psychotropic (Brooks, Chapter 3) and antibiotic (Browne, Chapter 4) drugs. It is thought necessary to give here, therefore, a brief account of the polarographic methods available for the determination of other groups of compounds which are important in clinical medicine, i.e. steroids, hormones and vitamins.

The most common naturally occurring steroids which can be determined by polarographic methods of analysis are those containing a keto group, e.g. 3 keto-steroids. Well-defined d.p.p. waves in 0.1M TEAP/50% CH_3OH (Smyth, 1977) are found when unsaturation is present $\alpha\beta$ to the keto group, with a reduction potential of approximately -1.8 V, e.g. testosterone, cortisone.

Since the reduction potentials of keto steroids occur at such negative values, it is common to employ derivatization procedures in order to obtain lower limits of detection (i.e. generally $<5 \times 10^{-6}M$). Buecher and Franke (1965a, b), for instance, used a hydrolytic procedure to convert 17-oxosteroids to their β-vinyl hydrazone derivatives by heating in 15% HCl at 100°C for 10 minutes. Starka and Brabencova (1960), on the other hand, used HIO_4 oxidation to determine acetaldehydrogenic steroids in the presence of formaldehydrogenic steroids in urine. Polarographic methods have also been described for the determination of corticosteroids (Hake, 1966) and tocopheronolactone (Schmandke and Crohlke, 1965) in urine following a t.l.c. separation step.

The cardiovascular drugs digoxin (I) and digitoxin (II) also contain the basic steroid-type nucleus but in this case, the reduction occurs at

I: R = OH
II: R = H

the $\diagup\text{C}=\text{C}\diagdown$ bond which is conjugated to the keto group in the five membered ring. Digoxin and digitoxin give rise to waves at -2.28 and -2.32 V vs s.c.e. respectively in isopropanol/0.01M TBAI. The limit of detection for these compounds using d.p.p. was quoted to be $2.5 \times 10^{-6}M$ (Kadish and Spiehler, 1975). This assay may prove useful for formulation analysis; however much greater sensitivity and selectivity can be obtained using radioimmunoassay techniques (Butler, 1976).

Norethisterone (17α-etinyl-17β-hydroxyl-4-estren-3-one) is a pharmaceutical compound used in oral contraceptives. It has been rapidly assayed in tablets by d.p.p. without the need for prior extraction. The procedure involved solubilization in 40% methanolic solutions in the presence of

0.2 M TMAB, degassing with nitrogen for thirty to forty minutes followed by polarography at the d.m.e. (Opheim, 1977). Both E_p and i_p were studied as a function of pH. One peak was obtained in the pH ranges 2.9–6.1 and 9.0–11.6 whilst two peaks were obtained between pH 7.1 and 8.1. A linear peak current–concentration relationship was found in the range 7×10^{-6}M– 7×10^{-4}M.

Four progestrogens (norethisterone, norethisterone acetate, dimethisterone and norgesterel) have been assayed polarographically both in the pure form and in pharmaceutical formulations by Chatten et al. (1977). The steroids were dissolved in DMF (norethisterone and norgesterel) or 95% ethanol (norethisterone acetate and dimethisterone), diluted with Sörensens phosphate buffer, pH 6.0, and subjected to d.p.p. The reduction at the d.m.e. corresponded to a one-electron process for all the steroids and peak potentials did not differ appreciably ($E_p = -1.42 \rightarrow -1.48$ V). Results for assays of formulations agreed reasonably well with the analytical data supplied by the manufacturers. Both diffusion and kinetic factors were found to be important in the reduction processes of the compounds studied.

The hormones vasopressin and oxytocin can also be determined using polarographic methods of analysis. The polarographic behaviour of vasopressin has been studied by Krupicka and Zaoral (1969) whereas Boto and Williams (1976) have determined oxytocin in the presence of vasopressin in 1.1 mM hexamino-cobaltichloride base electrolyte, pH 9.0. This method could determine 0.2–2.5 µg l^{-1} oxytocin in pharmaceutical preparations. Oxytocin has also been determined by d.c. polarography in 1M KCl and the mechanisms shown to involve reduction of the disulphide bond (Rishpon and Miller, 1975).

The polarographic analysis of vitamins has been dealt with elsewhere (Heyrovsky and Zuman, 1968; Lindquist and Farroha, 1975; Gulaid, 1976, Davidson, this volume, Chapter 5).

C. RESPIRATORY PHYSIOLOGY

Voltammetric measurements of oxygen principally by platinum electrodes have been widely used in respiratory physiology. Since this subject is beyond the scope of this chapter the reader is referred to:

(a) Determination of the oxygen content of blood [Solymar et al. (1971), Volter et al. (1972), Borgström et al. (1974), Gelshuen et al. (1971)].
(b) Measurement of the oxygen flux across tear epithelial interfaces of young adult rabbit cornea in vivo (Ausberger and Hill, 1972).
(c) Oxygen uptake of seven day chick neural retina cells using Clark electrodes [Kraul and Richmond (1971), Flower (1976)].

(d) Application of oscillopolarography to measure the difference in respiratory condition of a single neuron in the resting state and after electrical stimulation (Kogan et al. 1974).

(e) Measurement of pO_2 [Weiss and Cohen (1974), Garbus et al. (1975), Longo, (1976)] or consumption of oxygen [Lehmenkühler, (1976), Kunke et al., (1972)] following exposure to CO/CO_2.

(f) Comparison of the respiratory activity of bone and marrow cells by different methods (Gesinski et al. 1968).

(g) Measurement of oxygen in the brain (Halsey and McFarland, 1974).

(h) Determination of pO_2 in carotid body tissue (Acker and Lubbers, (1976).

(i) Determination of the dissociation curve of oxyhaemoglobin (Grinberg, 1977).

(j) The use of such electrodes in routine clinical diagnosis [Fenner (1974), Soutter et al. (1976), Conway (1976), Parker (1977)].

(k) Errors in pO_2 blood measurement by polarography (Akhmetov, 1977).

(l) Reviews of methodology and instrumentation [Bielowski (1971); Payne and Hill (1975); Kessler (1971)].

(m) Determination of compounds of biological significance [Weitzmann (1976), Kumar and Christian (1977)] and in microbiological analysis (Harrison, 1974).

The investigation of blood flow by voltammetry is also beyond the direct scope of this chapter and the reader is referred to the following:

(1) Local blood flow and distribution studied in pig liver (Aune, 1972).

(2) Local and average total blood flow in anaesthetized and unanaesthetized animals (Fein et al., 1975)

(3) Evaluation of subendo- and subepicardial blood flow under different physiological conditions (Moggio and Hammond, 1976).

(4) Evaluation of the functional state of the brain during hypoxia [Shakhnovich et al. (1971), Cherynakov (1966)]

(5) Local microflow of blood within small tissue volumes (Stossek et al. 1974)

(6) Problem of intercompartmental diffusion in cerebral blood flow measurements (Halsey et al., 1977).

For a review of this subject by other techniques the paper by Sandler and Tator (1976) should be consulted.

III. CHEMICAL CARCINOGENESIS

In recent years various workers have attempted to derive relationships between the chemical nature of carcinogens and the biological effects they cause (Brulé et al., 1973; Miller and Miller, 1974; Ferguson, 1975; Howe, 1975; Heidelberger, 1975, 1976; Calvin, 1975; Elashoff and Beal, 1976; Walters, 1977). Voltammetric techniques have been employed in this context and have also been used for the determination of carcinogens and their metabolites in a variety of materials.

A. THEORETICAL ASPECTS

The following general relationships have been postulated:

(a) The more positive the reduction potential (i.e. the greater the electron affinity) is for an aldehyde or ketone, the greater is the inhibitory effect on the proliferation of *E. coli* (Hata, 1970).
(b) The more positive are the reduction potentials of polycyclic aromatic hydrocarbons in DMF/DMSO (e.g. benzpyrenes), the more carcinogenic they are (Podany and Vachalkova, 1973). The situation for structurally similar polycyclic aromatic heterocycles was somewhat more complicated (Podany et al., 1975).
(c) The carcinogenicity of aromatic azo dyes is partially related to their ease of reduction and also to products that are formed on electroreduction (Carruthers, 1976; Franklin Smyth and Hassanzadeh, 1976).

B. ANALYTICAL ASPECTS

Nitroquinoline and 1-oxide derivatives are amenable to polarographic analysis and the method may be used to distinguish closely related compounds. Tachibana et al. (1967) examined 4-nitro-quinoline-1-oxide (4NQO; III), 4-hydroxyamino-quinoline-1-oxide (4HAQO; IV) and 4-amino-quinoline-1-oxide (V) polarographically; 4NQO is converted into 4HAQO *in vivo* and both compounds are carcinogenic. All three compounds show $E_{\frac{1}{2}}$ values which are pH dependent. The reduction processes for the compounds at pH 3.78 were formulated in the following manner, yielding 4-amino-quinoline (VI) as the final product:

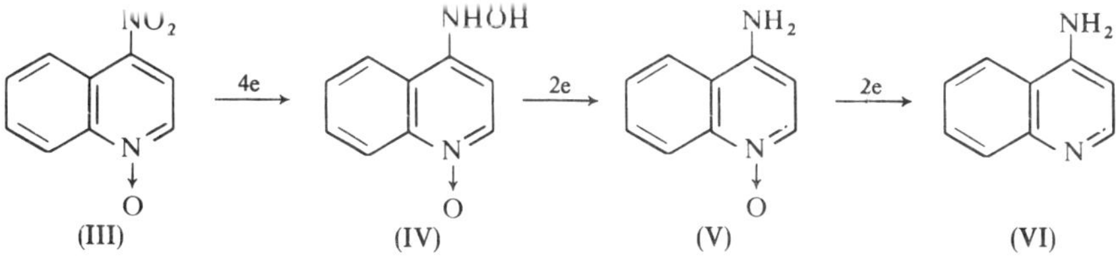

Chodkowski et al. (1974) found that acridine derivatives unsubstituted by a nitro group were reduced in two one-electron waves in alkaline media but in acidic media only one reduction step was observed. In Britton–Robinson buffer, pH 3.35, the hydrochlorides of 1-nitro-9-amino, 9-ethyl-, 9-propyl- and 9-butyl-amino-acridine were reduced in three steps. The definition of the waves was found to be a function of the chain length of the compounds.

The detection of aerial 3,4-benzpyrene using a polarographic finish has been investigated by Mulik et al. (1975). The results, however, were indecisive and analysis by other methods such as g.l.c., spectrofluorimetry or luminescence spectrometry utilizing the Shpolski effect at 77 K (Farooq and Kirkbright, 1976) are recommended. Polynuclear aromatic hydrocarbons can also be analysed in air using concerted solvent partition, column chromatography and high pressure liquid chromatography (Bartle et al., 1976; Fox, 1976), attaining much lower levels than by using polarography.

Franklin Smyth et al. (1975) have made a combined spectral and polarographic study of different structural classes of N-nitroso compounds over a wide range of pH. They suggested the structures that were most likely to predominate at physiological pH's, postulated mechanisms of reduction and derived optimum conditions for their determination (Table II).

TABLE II. *Pulse polarography in Britton–Robinson buffer of certain carcinogenic nitroso compounds* (after Smyth et al., 1975)

Compound	E_p	
	pH 2	pH 10
N-nitroso-piperidine	−0.75	−1.35
Dimethyl-N-nitrosamine	−0.87	−1.55
Dibenzyl-N-nitrosamine	—	−1.35
N-nitroso-sarcosinamide	—	−1.35
Diphenyl-N-nitrosamine	−0.6	−0.95
N-methyl-N-nitrosourea	−0.65	—
N-methyl-N-nitrosoaniline	−0.65	−1.2
N-methyl-N-nitroso-N'-nitroguanidine	Multiple	Multiple
p-nitrosophenol	Multiple	Multiple
p-nitrophenol	Multiple	Multiple

Since the *in vivo* levels of many of these carcinogens are extremely low (e.g. N-nitroso-pyrrolidine occurs in fried bacon at levels of 100–200 µg kg^{-1} —Walters, 1977) and since the chemiluminescent (thermal energy) detector for g.l.c. is more sensitive, selective and specific for nitroso compounds, polarography would be recommended as a tool to investigate their metabolism where ultimate sensitivity is not required.

The nitroso compounds, N-nitroso-pyrrolidine, N-nitroso-proline and N-nitroso-4-hydroxyproline have been determined by differential pulse polarography (Hasebe and Osteryoung, 1975) at levels of the order of 10^{-7} M. In acidic media N-nitroso-pyrrolidine and its derivatives gave one irreversible reduction wave whereas at pH 6 only N-nitroso-pyrrolidine was reduced.

Mitomycin B (VII) and mitromycin C (VIII) are clinically important antineoplastic agents. Their electrochemical behaviour has been reported by Rao et al. (1977a, b), using polarography and single and multisweep voltammetry at the h.m.d.e.

VII: $R^1 = -OCH_3$
$R^2 = -OH$
$R^3 = -CH_3$

VIII: $R^1 = -NH_2$
$R^2 = -OCH_3$
$R^3 = -H$

In 0.1M phosphate buffer/0.1M HCl (pH 6–8) both compounds exhibited reversible behaviour. The voltammetric data allowed the authors to postulate possible mechanisms of *in vivo* activation of the mitomycins.

The application of polarography to studies of the mechanism of action and metabolism of various foreign compounds has been dealt with in several papers (Bendirdjian et al., 1975; Chien, 1976; Franklin Smyth and Smyth, 1976; Malfoy et al., 1977). Such studies are of real *in vivo* relevance and parallels between *in vivo* biological reactions and those occurring at the electrode surface have been highlighted by Elving (1974).

IV. NEUROPHYSIOLOGY

The main adrenergic neurotransmitters (Fig. 2) almost all contain either amino or phenolic groups and can thus be oxidized at a variety of solid electrodes. In certain cases it has been possible to follow the levels of these compounds continuously *in vivo* using miniaturized electrodes (McCreery et al., 1974b). Drawbacks of these assays include the uncertainty of electrode area, recognition of the chemical species giving rise to a particular oxidation process, limited sensitivity in the presence of interferences in the body fluid and susceptibility to electrode poisoning. With recent advances in instrumentation, however, some of these problems are likely to be overcome in the near future (Adams, 1976).

Apart from direct determinations, neurotransmitters can be extracted from brain materials, subjected to separation procedures (e.g. h.p.l.c.) and analysed voltammetrically (Kissinger, 1977).

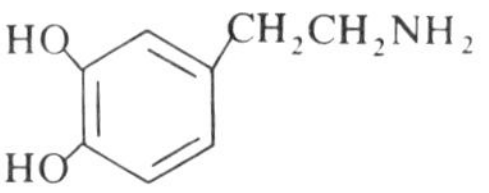

(IX)

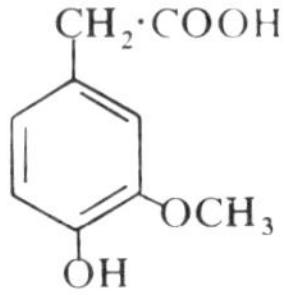

(XII)

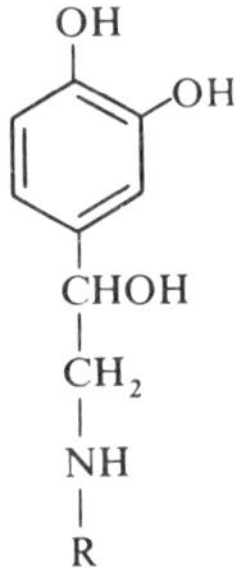

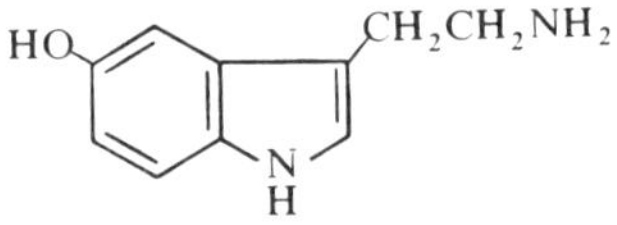

(XIII)

(X) (a) R = CH_3
(b) R = H
(c) R = 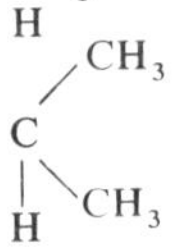

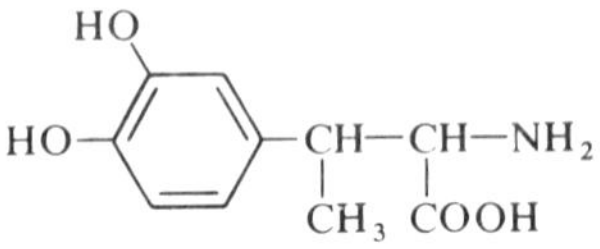

(XIV)

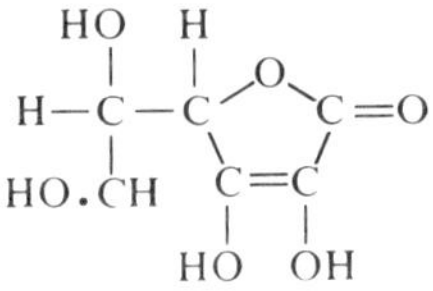

(XI)

FIG. 2. Structures of compounds of neurophysiological significance detailed in the text. (IX) Dopamine (DA); (X) (a) = epinephrine, (b) = norepinephrine, (c) = isopropylepinephrine; (XI) ascorbic acid; (XII) homovanillic acid (HVA); (XIII) serotonin (5HT); (XIV) α-methyl dopa.

A. SOME DIRECT DETERMINATIONS *IN VIVO*

6-Hydroxydopamine (6-OHDA; XV) has been implicated in the etiology of schizophrenia and its importance as a chemical sympathectomy agent is well known. Its *in vivo* metabolic fate, however, remains largely unknown. Significant quantitative differences between the cyclic voltammetry of 6-OHDA *in vitro* and in rat caudate nucleus was accounted for by the oxidation of 6-OHDA in brain tissue (McCreery *et al.*, 1974a). After injection of 6-OHQ

(XVI), two new peaks appeared at +0.08 V and −0.32 V (vs Ag/AgCl) due possibly to the oxidation of 6-OHDA and reduction of 6-OHQ respectively.

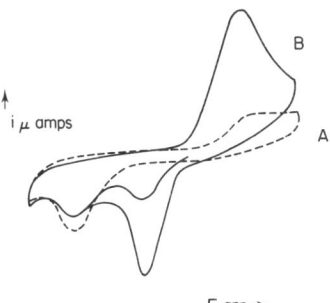

(XV) +0.08 V   (XVI) −0.32 V

The injection of 6-OHDA or 6-OHQ into caudate nucleus has been followed as a function of time. The concentration decay rates for 6-OHDA and 6-OHQ were similar but both faster than DA (half life of the substances being 25 and 15 minutes respectively). The authors concluded that the oxidation-reduction behaviour of injected 6-OHDA and 6-OHQ was suggestive of a redox buffer system such that the percentage of the oxidized form in the mixture was about 40% (Fig. 3). Kissinger et al. (1973) have

FIG. 3. Cyclic voltammograms resulting from electrodes implanted in the rat caudate nucleus. (A) Before drug injection, (B) after injection of 6-hydroxydopamine. (After McCreery et al., 1974a).

implanted small active surface diameter carbon paste electrodes in the caudate nucleus of anaesthetized rats (cerebral cortex; hippocampus) and recorded cyclic voltagrams between −0.2 to +0.6 V. An oxidation process was observed which was irreversible in nature. The nature of the species involved was not known but postulated to be ascorbic acid XI ($F_{\frac{1}{2}}$ of 0.37 V vs Ag/AgCl).

Electrical stimulation of the dopaminergic substantia nigra can be brought about by applying a potential to a micro carbon electrode stereotaxically placed in a lateral ventricle. This has been shown to cause a release into the CSF of the dopamine metabolite homovanillic acid (HVA; XII) (Wightman et al. (1976).) Using a baseline of ascorbic acid and HVA and applying a triangular wave potential sweep from 0·0 to 1.4 V (which was then returned to −0.2 V) permitted the determination of HVA (at +0.8 V vs a micro saturated s.c.e.). Oxidation and reduction of dihydroxyphenylacetic acid (DOPAC) and its *o*-quinone at more negative potentials were also observed. Two important points are of note. Firstly, repeated stimulation after a given interval produced a reproducible response. Secondly, independent chemical identification of HVA, DOPAC or 5-HIAA by high pressure liquid chromatography was achieved. Lane *et al.* (1976) have used "high purity" iodine treated platinum electrodes for direct *in vivo* detection of brain catecholamines. Electrodes were stereotaxically implanted either into the head of the caudate nucleus or into the lateral ventricle of adult male albino rats anaesthetized with sodium pentobarbital. An Ag/AgCl (1M NaCl) reference electrode in Teflon tubing was positioned in the contra-lateral cortex via a stainless steel guide cannula which itself acted as the auxiliary electrode. Low levels of dopamine (DA) of the order of 8×10^{-7}M could be determined directly by this method. How far these concentrations reflect the availability of the molecule has yet to be ascertained.

Fike and Curran (1977) have designed a new micrometer based thin layer electrochemical cell for use in *in vivo* studies. A pretreated Pt electrode and a saturated calomel electrode have been used for the determination of some catecholamines (epinephrine, norepinephrine, dopamine and dopa) in 1M H_2SO_4 by linear sweep voltammetry with current integration. The cavity volume of the cell was 1 µl and concentrations of catecholamines in the concentration range 0.1 mM–5.0 mM were detectable.

B. SOME DETERMINATIONS FOLLOWING SEPARATION PROCEDURES

The enzyme catalysed *o*-methylation of norepinphrine (NE) (X, b) by catechol-O-methyl transferase has been followed voltammetrically (Sternson *et al.*, 1976). The reaction is important since it is responsible for the extra neuronal (hepatic) inactivation of NE via a nucleophilic substitution reaction (methylation of a phenol function) to give normetenephrine (NME). In an hepatic incubation media NE is converted to NME by the enzyme in the presence of Mg(II) and S-adenosyl methionine. Both NE and NME are then combined with bis-(2-ethyl hexyl) hydrogen phosphte at pH 7.4 to form an ion pair which can be quantitatively extracted with an immiscible organic solvent using partition chromatography. NE and NME can be monitored simultaneously in 0.5M H_2SO_4 by d.p.p. at carbon paste electrodes (NE, $E_p = +0.55$ V and NME, $E_p = +0.75$ V vs s.c.e.) and also by c.v. (NE $E_{p/2} = +0.49$ V and NME $E_{p/2} = +0.70$ V). The advantages of the assay are primarily that no time consuming pre-separations are required (unlike spectrophotometric, fluorescence and radiochemical techniques) and that the sensitivities (about 10^{-6}M) are those which are required for biological investigations.

Lankelma and Poppe (1976) have designed a coulometric detector, (employing a glassy carbon electrode of large surface area) for use in conjunction with an h.p.l.c. column. The system was used for the separation and detection of biogenic amines and neuroleptic compounds for which detection limits in the picogram region were claimed.

Epinephrine, norepinephrine and isopropyl epinephrine have been estimated in phosphate buffer pH 7.0 using a revolving platinum electrode and scanning between -0.2 and $+1.2$ V (vs s.c.e.) (Cantin et al., 1973). At concentrations of 5×10^{-4}M the $E_{\frac{1}{2}}$ values were $+0.27$, $+0.24$ and $+0.30$ V respectively; calibration plots were linear in the range 1×10^{-3} to 8×10^{-5}M.

Sasa and Blank (1977) have determined serotonin (5HT) and dopamine (DA) in both whole brain and cerebellum of mice. H.p.l.c. with electrochemical detection at the carbon paste electrode using a working potential of $+0.6$ V (vs s.c.e.) was employed. As standards, dopamine (HCl derivative) and serotonin (as creatinine sulphate monohydrate) were dissolved in 0.01M HCl and applied to the h.l.p.c. column [cation exchange column utilizing Dupont Zipax SCS Strong Cation Exchange Resin and 3,4-dihydroxy benzylamine (DHBA) as the internal standard]. 5HT and DA were extracted from whole brain by the addition of 10^{-1}M EDTA, 10^{-5}M DHBA and 2.5×10^{-2}M HCl to brain samples, followed by ultrasonic homogenization and extraction of the compounds with butanol and/or heptane. The extraction from cerebellum utilized EDTA and ascorbic acid.

Variables in experimental technique were all thoroughly investigated in order to optmize analytical results (these included tissue homogenization methodology, methods of extraction from brain tissue, volumes of solution injected into the h.p.l.c. column, pH of solutions, temperature and time of shaking of brain samples).

Possible interferences such as ascorbic acid and biogenic amines related to DA were not found to interfere. The results shown in Table III were obtained.

TABLE III

Compound	Whole brain (kEM) ng (g wet tissue)$^{-1}$	Cerebellum ng (g wet tissue)$^{-1}$
DA	805 ± 12	240 ± 14
5HT	973 ± 16	19 ± 2

Liquid chromatography was used by Kissinger et al. (1973) to separate DA and NE, with subsequent electrochemical detection. The column effluent was allowed to flow through a thin channel past a small carbon paste electrode surface. At a fixed applied potential the detector measures current for the oxidation of the electroactive species as they elute from a cation exchange resin. Samples of the order of 20–200 pg of injected DA and NE have been measured with a relative error of $\pm 10\%$. The method has been used to provide mapping of catecholamine distribution in mammalian brain, by taking 0.5–50 mg samples as needle punches from thin brain slices.

Riggin and Kissinger (1977) have suggested a simple and sensitive assay for the determination of tetrahydroisoquinoline (THIO), tetrahydropapaveroline (THP) and salsolinol (SAL) in urine and brain tissue. This involved the combination of high performance cation exchange chromatography (pellicular Vidac SCX stationary phase in glass column; mobile phase of McIlvaine buffer pH 3.8) and amperometric detection at a carbon paste electrode set at $+0.72$ V (vs Ag/AgCl). A mixture of NE, DA, SAI, THP and epinephrine (E) could be resolved using these conditions. For urine samples (4 ml) the pretreatment involved the use of acid hydrolysis and protein removal. The compounds were extracted with ethyl acetate followed by hexane, adsorbed on alumina, eluted with acetic acid and the eluate applied to the cation exchange column. A detection limit of 2 ng ml^{-1} was quoted for both THP and SAL and little interference from other compounds encountered. A detection limit of 2 ng g^{-1} was quoted for whole brain using a similar procedure. These authors were also the first to be able to detect, *in vivo*, 6,7-dihydroxy tetrahydroisoquinoline following the injection of L-DOPA.

Thrivikraman et al. (1974) have described a method for the determination of ascorbic acid in brain tissue following h.p.l.c. separation.

Recently Lane and Hubbard (1976) have drawn attention to the use of differential double pulse voltammetry (a technique utilizing two simultaneously varying unequal square wave potential pulses as an alternative to the linear d.c. ramp) at chemically modified platinum electrodes for preliminary investigations towards *in vivo* detection of catecholamines. The advantage of the technique is that "pure" electrodes are used (having undergone chemical modification of their surface by aqueous iodide) and there is no progressive build up of insulating films on the electrode surface as there is with conventional d.p.p. Dopamine and norepinephrine (in the form of hydrocholride and bitartrate monohydrate salts respectively) gave good responses in a supporting electrolyte of 0.1M phsophate buffer pH 7.4 and 0.9% NaCl. Under the conditions used Δi_p was linear with concentration for the above compounds.

V. VIROLOGY

Due to their structural "simplicity", viruses are of value in helping to elucidate many facets of molecular biology (e.g. morphogenesis; Watson, 1975). They are also important since they may be (potentially) pathogenic in certain cases (*Lancet*, 1976).

Viruses represent stable supramolecular complexes containing nucleic acids and proteins (though carbohydrate and/or lipid may also be present) organized into a characteristic three dimensional arrangement. Although of large particle weight and diverse size, shape and chemical composition (Andrewes and Pereira, 1972; Colter and Paranchych, 1967), they can be isolated in homogeneous form and may be crystallized.

Ruttkay-Nedecky (1964a, b; Ruttkay-Nedecky and Anderlova, 1967; Ruttkay-Nedecky and Bezuch, 1971a, b, 1973) has examined the polarographic behaviour of tobacco mosaic virus (T.M.V.). T.M.V. is a rodlike helical plant virus containing ribonucleic acid (6000 nucleotides) and a protein shell capsid (MW = 3.4×10^{-7}); it has a diameter of about 180 Å with a complete length of 3000 Å. The structural aspects of T.M.V. have recently been reviewed by Holmes *et al.* (1975). It has 2150 identical proteins, each of molecular weight 17,000. T.M.V. gives rise to a catalytic double wave because of the presence of protein in the virus. Using a supporting electrolyte of 0.001M $Co(NH_3)_6Cl_3$, 0.1M NH_3 and 0.1M NH_4Cl, this occurs at -1.45 to -1.50 V (vs Hg anode). Ruttkay-Nedecky and Bezuch (1973) have postulated that these polarographic peaks originate from a single cysteinyl residue in position 27 of the protein and have also shown that the virus can be identified by the appearance of an additional wave following a phenol extraction and partial denaturation with concentrated urea. This additional wave disappears on renaturation dialysis (pH 5–6) and the original polarographic behaviour is restored.

The polarographic studies have been supported by electron microscopy and amperometric titration of sulphydryl groups. Studies with cytochrome c and use of T.M.V. mutants with known amino acid sequence in their polypeptide units also lend credence to the cysteine hypothesis. Finally, studies on the mechanism of protein behaviour at various electrodes (Kolthoff *et al.*, 1974) have indicated that under the conditions used, the sulphydryl group in cysteine may play a central role in the polarographic behaviour of T.M.V. virus.

Godschalk and Veldstra (1965) have examined turnip yellow mosaic virus (T.Y.M.V.). Polarography was used to assess the reaction between thiol groups and four different mercury compounds (methyl, ethyl, n-propyl and n-butyl mercuric nitrate) at the d.m.e. using a base electrolyte of 0.1M borate buffer (pH 8.2) and 0.00036% Triton X100 (vs Ag/AgCl). A structural analysis of the virus using spectrophotometric and ultracentrifugal techniques was made. It was concluded from the study that a linear relationship existed between the extent of degradation and the number of sulphydryl groups substituted by mercurials and that substitution and degradation occurred at a rate which was a function of the number of carbon atoms of the mercurial compound. Moreover treatment of T.Y.M.V. with *p*-chloro-mercuribenzoate resulted in the disintegration of the virus protein with concomitant release of RNA.

Polarography, therefore, offers a means of identifying certain viruses (Gurken-virus has also been examined) and could possibly be used to show the presence or absence of non-viral protein in viral preparations in instances of contamination. Various voltammetric procedures might also give simultaneous determination of the nucleic acid content of viruses since methods for the microdetermination of nucleic acids have been reported (Palacek and Pechan, 1971).

Irtiakov (1975) has recommended that the polarography method may be used in the examination of cases of clinical viral hepatitis (this includes diagnosis and prognosis as well as for evaluation of the efficiency of hormonal therapy).

VI. IMMUNOLOGY

Immunoglobulins form part of the immune system of mammals which is presently an area of immense scientific interest to medical scientists (Roitt, 1971; Nisoroff *et al.*, 1975). The topics of transplantation surgery (tissue rejection problems) and autoimmune diseases are of particular importance. Immunoglobulins themselves are used in radioimmunoassay methods for the estimation and detection of both natural and synthetic substances in animal tissues and body fluids (Simmons and Ewing, 1974; Pasternak, 1975; Clausen, 1971).

Poljak *et al.* (1975, 1976) has reviewed the structure of immunoglobulins. The majority of electrochemical studies of immunoglobulins come from the laboratories of Zikan (1966; Zikan and Sterzl, 1967), Churchich and Mottda (1962) and Fontaine *et al.* (1973, 1974). Fontaine *et al.* (1973) examined human myeloma IgG (Fig. 4) prepared by electrophoretic and chromatographic methods. Fragments of the immunoglobulins were also prepared using enzyme degradation combined with chromatographic separation (ion exchange and sephadex gel filtration). Total reduction of inter- and intra-chain disulphide bonds, alkylation of sulphydryl groups and blocking of lysine amino acid residues using succinylation, acetylation and acidic TNBS

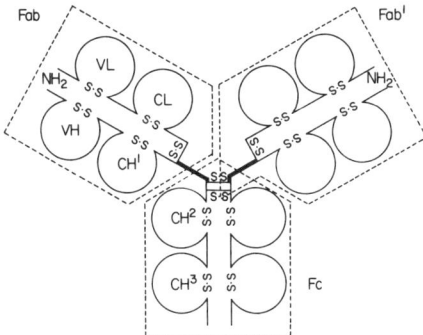

FIG. 4. Structure of a human IgG molecule (after Poljak 1975). (Fab, Fab' and Fc are the major fragments; VL, CH^1, and CH^2 and CH^3 are different homology regions.)

were also undertaken. The purity of the sample was ascertained by immunoelectrophoretic techniques (specific antiimmunoglobulin sera).

Coulometry, cyclic voltammetry and sinusoidal polarography (5–20 µM samples) were the voltammetric techniques employed. The effect of pH, temperature and concentration on polarographic behaviour, together with electrode mechanisms were studied. Some typical data for IgG is shown in Table IV.

TABLE IV. *Polarographic data for IgG* (after Fontaine et al., 1973)

Immunoglobulin	Wave	$E_{\frac{1}{2}}$ (vs s.c.e.)	pH	Electrode Mechanism
IgG_1	1	−0.28	3–9	Adsorption
IgG_2	1	−0.28	3–9	Adsorption
IgG_3	1a	−0.28	3–6.5	Adsorption
	1b	−0.36	7–9	Adsorption
	(plus another wave variable with pH)			
IgG_4	1	−0.28	3–9	Adsorption
	2	−0.48	3–7	Diffusion
$(Fab')_2^a$	1	−0.28	3–9	Adsorption
	2	−0.48	3–7	Diffusion

[a] $(Fab')_2$ This is a fragment of IgG immunoglobulin.

The derivatized forms of IgG differed in electrochemical behaviour from the underivatized forms as did different concentrations of $(Fab')_2$ fragments (Fig. 5). Lastly denatured forms of immunoglobulins (e.g. in 4M urea) could be distinguished from unmodified forms.

As for IgC, the interchain disulphide bonds played an important role in the electrochemical reduction mechanism. The diverse immunoglobulin forms of IgA were separated using polyacrylamide gel electrophoresis. Monomeric forms of IgA gave two waves in acidic media (pH 4.8), an adsorption wave at -0.27 V (vs s.c.e.) and a diffusion controlled wave at -0.44 V (vs s.c.e.). A peak at -0.28 V (vs s.c.e.; pH 7.70) was also observed. At pH 4.8 the dimer gave four waves (-0.20, -0.22, -0.30 and -0.44 V vs s.c.e.) and at pH 7.70 three waves (-0.15, -0.25 and -0.35 V vs s.c.e.).

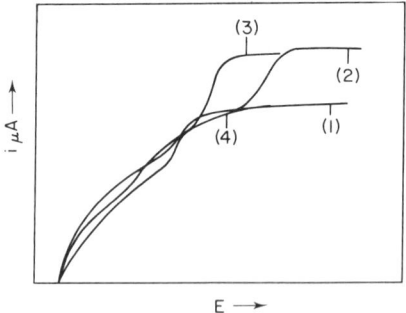

FIG. 5. Polarography of IgG and related structures (after Fontaine et al., 1973). (1) (Fab')$_2$ (1 M KCl, 0.05M sodium acetate pH 3.5); (2) higher concentration of (Fab')$_2$ in same buffer; (3) IgG-succinylated derivative (1M KCl, 0.05 M sodium acetate pH 4.0); (4) IgG (M KCl, 0.05M sodium phosphate, pH 7.0).

The value of polarography in immunology therefore is that micromolar quantities of immunoglobulin (and/or hapten) can be estimated and identified in short periods of time by a non-biological method. Substantial changes in electrochemical behaviour are also apparent upon derivatization and/or degradation.

VII. CELLS, ORGANELLES AND MEMBRANES

Water soluble, non-particular Mg(II) activated ATPase derived from *Streptococcus faecalis* membrane ghosts gave a d.c. polarographic wave (Redwood and Godschalk, 1972) which was perturbed by hen egg yolk phosphatidyl vesicles in succinate–$MgCl_2$–NaOH buffer at pH 6.3 in a ratio of 1:5 (enzyme:vesicles). Matrix rank analysis on phase sensitive a.c. polarograms of the mixture (Fig. 6b) indicated the presence of three components suggesting the formation of a vesicle-ATPase complex. The enzyme also gave an a.c. polarographic wave at -1.1 V which disappeared on the action of detergent (cetyl-dimethyl benzoyl ammonium bromide). In the presence of vesicles a maximum at -1.07 V was also obtained.

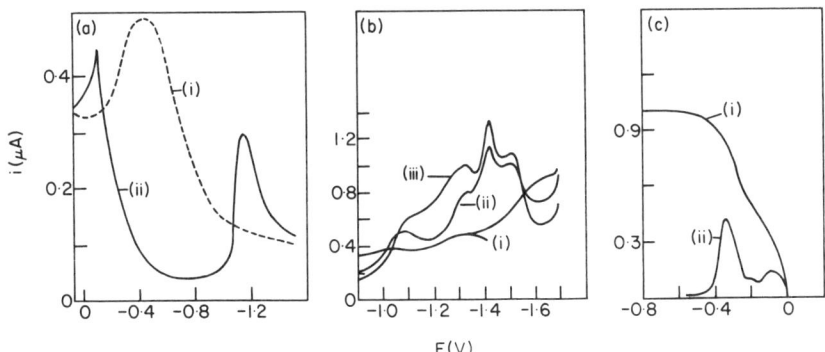

FIG. 6. (a) A.c. polarograms of (i) background solution, (ii) background solution + ATPase (Bach et al., 1973). (b) Phase-sensitive a.c. polarograms of (i) ATPase, (ii) liposomes, (iii) ATPase + liposomes (Redwood et al., 1972). (c) Polarography of human spermatozoa; (i) d.c. polarogram in KNO_3 pH 7.4, (ii) a.c. polarogram in KNO_3, pH 7.4 (Valasquez et al., 1972).

There are many different kinds of membrane-bound ATPases which are thought to be involved in the movement of ions across membranes. Bach et al. (1973) studied Na, K dependent ATPase of rabbit kidney using d.c. and a.c. polarography. Results revealed that the adsorption of the enzyme took plase in the voltage range -0.1 V to -1.2 V (vs Ag/AgCl in 0.1N NaCl) (Fig. 6a) and this was shown to be dependent on enzyme concentration. Adsorption studies at various pH's indicated an isoelectric point of pH 4.0 to pH 5.5.

Many other authors have pointed to the value of voltammetric techniques in the study of artificial and natural membranes (Gerber et al., 1965; Chien, 1972; Masters and Miller, 1974; Lyons et al., 1974; Brezina et al., 1974).

Valasquez and Rosado (1972) have carried out conductimetric, amperometric and polarographic experiments (the latter involving d.c. and a.c. measurements at the d.m.e. in 1M KNO_3, pH 7.4) on whole human sperm cells. Spermatazoa in 1M KNO_3 (37°C) yielded an anodic polarographic wave at -0.30 V (vs s.c.e.) (Fig. 6c) due to free —SH groups at the spermatozoan surface. A.c. signals could also be obtained in 0.1M sodium acetate and 0.25M sucrose at pH 7.0. Where a net negative charge was found to be due to free-SH and free sialic acid groupings comparison of polarographic behaviour of N-acetyl-neuramminic acid, sialic acid and pyruvate under the same conditions as those used with spermatozoa (0.1M sodium acetate) supported the conclusions of the authors as to the cause of spermatozoan polarographic behaviour. The charge on the spermatozoa was calculated by conductimetry (anodic and cathodic) in 0.25 M sucrose at -2.20 and $+0.1$ V. The addition of $CaCl_2$ allowed the residual charge at neutral pH on the cell surface to be calculated.

Red blood cells (rbc) of "normal" human adults also produce a characteristic polarographic wave in phosphate buffer (Allan, 1971). The rbc obtained gave a half wave potential at -0.22 V (± 0.01 V, vs s.c.e.; reaction: $RSH + Hg \rightarrow RSHg + H^+ + e^-$). The sulphydryl groups on the surface of the cell were thought to be responsible for the polarographic behaviour as evidenced by (a) lack of response using rotating platinum or glassy carbon electrodes and (b) the fact that N-ethyl-maleimide treatment removed the wave at the d.m.e. Allan used the Δi_d-time data for the cells to show that there was a rapid decrease in metabolic viability during the first 8–10 days followed by a considerably slower decrease. This indicated that experiments with stored cells should always be carried out at a fixed time after collection of cells.

VIII. MISCELLANEOUS COMPOUNDS

The polarographic behaviour of miscellaneous compounds is listed in Table V.

IX. CONCLUSIONS

This review has tried to show the diverse ways in which voltammetric techniques are of use in the basic medical sciences. It has not been possible, however, to cover every subject or to cite all the compounds that are amenable to voltammetric methods of analysis, e.g. blood flow proteins and amino acids (Zuman, 1974), nucleotides (Elving et al., 1976), porphyrins (Wilson and Neri, 1973) or metal ions (Brooks and Mark, 1975). The author has included little comparison with other techniques other than those which

TABLE V. *Polarographic behaviour of Miscellaneous Compounds*

Molecule	Sample	Indicator Electrode	$E_{\frac{1}{2}}$ (V vs s.c.e.)	Supporting Electrolyte/ Components of Assay Medium	Notes	Reference
Ubiquinone	Rat liver tissue		-0.38 V	0.4M $(C_2H_5)_4$ NOH pH 7.6 (C_2H_5OH used for solubilization)	Linear calibration plot	Fomenko and Zaslavskii (1974)
NADH	Aqueous solution	Glassy C electrode	$+0.6$ V^{a,p} (E_p)	Pyrophosphate buffer, 0.1M, pH 8.8	Linearity 0–0.1 mM	Blaedel and Jenkins (1974) (See also Steinbrecht and Kunz, 1972)
Glycine		Dropping Cu amalgam electrode	-0.28 V^{an}	0.5M $NaClO_4$, pH 8.0	Linearity 10^{-2}–10^{-4} M	Mendez *et al.* (1975)
Human Methaemoglobin	Aqueous solution	Vibrating dropping Hg electrode and d.m.e.	-0.75 V^k -1.04 V^k	Tris-hydroxy methyl amino methane (0.05M), cacodylic acid (0.05M) and perchloric acid (0.1M), pH 6.8 adjusted with NaOH	Plus other ill-defined waves (controlled potential electrolysis also carried out)	Betso and Cover (1972)
Purine bases Guanine (I'); Adenine (II'); Xanthine (III'); Hypoxanthine (IV'); in the form of admixtures	Aqueous solution	Silicone rubber-based graphite electrode	I' 0.81 V^p II' 1.04 V^p III' 1.12 V^p IV' 0.96 V^p	III' and IV' phosphate buffer, pH 8.3; II' and I' acetate buffer, pH 4.8	Detection limit 10^{-10} M. Technique is very specific and is preceded by h.p.l.c. Previously used techniques such as u.v. do not give good resolution	Varadi *et al.* (1974)

TABLE V—continued

Molecule	Sample	Indicator Electrode	$E_{\frac{1}{2}}$ (V vs s.c.e)	Supporting Electrolyte/Components of Assay Medium	Notes	Reference
Adenosine-5'-monophosphate (I') and Adenosine-5'-O-monothiophosphate (II')	Aqueous solution	Silicone rubber-based graphite electrode		0.25M Na_2SO_4, 3×10^{-3}M $CuSO_4$, 10^{-3}M KCl, 4×10^{-4}M tri-n-butyl phosphate, pH 5.0, 0.05M H_2SO_4	I', 10^{-6}M– 5×10^{-8}M II', 5×10^{-7}M– 10^{-8}M Simultaneous detection [I'] ≥ 10[II']	Weinhold and Sohr (1977)
Xanthine in xanthosine	Aqueous solution	Pyrolytic graphite electrode linear sweep voltammetry	1.11 V^p	McIlvaine Buffer, pH 6–8	Reproducibility 5–10%, E_p indicated shows presence of xanthine. (u.v. not applicable, chromatographic methods are time consuming)	Owens and Dryhurst (1977)
Cysteamine (I')–cystamine (II')		d.c., a.c., d.p.p. at d.m.e.		in presence of Triton X100	Cysteamine anodic wave [l.o.d. (I') ≡ 6 mg l^{-1} in presence of 11 mg l^{-1} of (II')]	Mairesse-Ducarmois et al. (1977b)

Substance	Medium	Electrode	Conditions	Observations	Reference
tRNAΦ				a.c. adsorption −1.110 V (vs s.c.e.) a.c. desorption or desorption plus reduction dependent on pH (pH 7, −1.320 V; pH 5.5, −1.370 V)	D.c. and a.c. with phase sensitive demodulation d.c. wave kinetically limited; peak or wave obtained in an acid medium corresponds to the reduction of reducible bases (adenines and cytosines) in the destabilized zones to tRNA Φ.	Reynaud (1976)
3-keto sugars (e.g., 3-ketolactose, methyl-β-3-keto-D-gluco pyranoside)	Aqueous solution	d.m.e.	$-1.50 \rightarrow 1.57$ V	LiCl (0.1M) + Na$_2$CO$_3$ (0.15M), pH 11.25	Interferences by traces of proteins and 5-methyl-phenazinium methyl sulphate	VanBeeumen and Deley (1971)
Bacterial pigments (β-carotene, prodigiosin, violacein and "red pigment")	Aqueous solution	d.m.e. (vs Hg pool)		McIlvaine Buffer	Differential cathode ray polarography Levels of 5×10^{-5} ml^{-1} detectable. At pH 8.0 violacein and prodigiosin can be distinguished	Whitnach and Soli (1966)

TABLE V—continued

Molecule	Sample	Indicator Electrode	$E_{\frac{1}{2}}$ (V vs s.c.e.)	Supporting Electrolyte/ Components of Assay Medium	Notes	Reference
Lipoic acid	Aqueous solution	D.c., a.c., normal pulse polarography (n.p.p.)	a.c. -0.615 V^p n.p.p. -0.590 V d.c. -0.59 V	pH 7.1 buffer (5% C_2H_5OH)	Concentration and pH (1.5–12.0) effects studied	Mairesse-Ducarmois et al. (1977a)
Proteins (bovine serum albumin, deoxyribonuclease I, Ex pancrease, lysozyme, pancreatic ribonuclease, trypsin)	Aqueous solution	d.m.e. vs Hg pool	-1.4 V (bovine serum albumin)	6×10^{-4} M $[Co(NH_3)_6]Cl_3$, 1M NH_4Cl, 1M NH_4OH, Triton X100 (2×10^{-5}%)	Concentration range 0.05–10 µg ml^{-1}; few interferences	Palecek and Pechan (1971) See also Horn (1975)
Trypsin		d.m.e.		0.5M Tris HCl, pH 8.2, 0.025M $CaCl_2$, 0.04% gelatin DMF	Enzyme activity can be calculated. N-tosyl-l-arginine-p-nitroanilide (substrate). p-nitroaniline is product	Bartik (1974)
Alcohol	Blood and urine	Differential amperometry at tubular carbon electrodes	$+0.07$ V		Redox reactions carried out, using 2,6-dichlorophenolindophenol; sample volume 10–20 µl	Smith and Olson (1975)

Enzyme	Source	Technique	Potential	Buffer	Comments	Reference
Glycogen phosphorylase and pyridoxal 5'-phosphate		d.m.e.		0.1M KCl, pH 4.9	Information concerning linkage between enzyme and cofactor obtained	Scheller and Will (1973)
Superoxide dismutase	Bovine	d.m.e.	-1.0 V	Sodium borate 0.025M, pH 9.90 plus 2×10^{-4} triphenylphosphine	Enzyme concentration of 10^{-10}–1.5×10^{-8} M detectable in body fluids. Effect of ionic strength, pH and external agents on enzymic activity investigated	Rigo et al. (1975a, b)
Aniline hydroxylase	Pure and microsomal liver suspension	Chronoamperometric stationary carbon paste (graphite–Nujol, 1:1)	$+0.25$ V	0.1M phosphate buffer, pH 7.4	Incubation mixture consists of (1) aniline HCl, (2) microsomal liver suspension (3) NADPH generating system. Incubation is for ten minutes. Assay dependent upon detection of levels of p-amino phenol	Sternson and Hes (1975)

TABLE V—continued

Molecule	Sample	Indicator Electrode	$E_{\frac{1}{2}}$ (V vs s.c.e.)	Supporting Electrolyte/ Components of Assay Medium	Notes	Reference
Prothrombin	Bovine	Pt electrode	Cathodic peak +0.2 V	Saline, 0.15 N NaCl	Information on electrosorption. Important in prothrombin-thrombin conversion in blood coagulation. 0.1–1.0 unit ml	Ramasamy et al. (1973a)
Thrombin	Bovine	Pt electrode	More −ve potential than prothrombin	Saline, 0.15N NaCl		
Fibrinogen	Human plasma	Potentiodynamic, steady state potentiostatic Pt electrode	$-0.8 \to -1.0$ V	Saline	Concentration range 0.007–0.07 mg ml^{-1}	Ramasamy et al. (1973b)

[a] reference electrode Ag/AgCl; [an] anodic wave; [k] kinetically controlled; [p] peak potential quoted.

deal specifically with the analysis of a certain molecule. Many of the applications cited herein are basically qualitative in nature, e.g. detection of diseased states, and a critical assessment of their applicability to clinical medicine could best be served by a greater collaboration between clinicians and electrochemists.

ACKNOWLEDGEMENTS

The author wishes to thank the Science Research Council for financial support. The editor Dr W. Franklin Smyth has been of great help and has given invaluable advice in the course of preparation of this manuscript.

REFERENCES

Acker, H. and Lubbers, D. W. (1976). *Pflugers Arch.* **366**, 241.
Adams, R. N. (1976). *Analyt. Chem.* **48**, 1128A.
Akhmetov, A. G., Stakhov, A. A. and Usmanova, G. Ya. (1977). *Med. Tekh.* **1**, 37.
Allan, M. J. (1971). *Coll. Czech. Chem. Commun.* **36**, 658.
Andrewes, C. and Pereira, H. G. (1972). *In* "Viruses of Vertebrates", 3rd Edn. Bailliére Tindall, London.
Anisimova, L. M. (1971) *Zh. Neuropatol. Psikhiatr. Im. SS Korsakova*, **71**, 933.
Anisimova, L. M., Kolesova, T. S. and Flerov, V. N. (1975). *Uch. Tr. Gov'k. Gos. Med. Inst.* **49**, 53.
Aune, S. (1972). *Microvas. Res.* **4**, 463.
Ausberger, A. R. and Hill, R. M. (1972). *Arch. Ophthal.* **88**, 305.
Bach, D., Britten, J. S. and Blank, M. (1973). *J. Membr. Biol.* **11**, 227.
Bartik, M. (1974). *Clin. Chim. Acta.* **56**, 23.
Bartle, K. D., Lee, M. L. and Novotny, M. (1976). *Proc. Analyt. Div. Chem. Soc.* **13**, 304.
Bendirdjian, J. P., Foucher, B., Rollin, P. and Fillastre, J. P. (1975). *C.R. Acad. Sci. Paris (D)*, **280**, 1489.
Betso, S. R. and Cover, R. E. (1972). *J. Chem. Soc. Chem. Commun.* **10**, 261.
Bielowski, J. (1971). *Post Biochem.* **17**, 565.
Blaedel, W. J. and Jenkins, R. A. (1974). *Analyt. Chem.* **46**, 1952.
Blinkov, I. L. (1965). *Veotnik. Akad. Med. Nauk. SSSR.* **20**, 36.
Borgstrom, K., Hagerdal, M., Lewis, L. and Ponten, U. (1974). *Scand. J. Clin. Lab. Invest.* **34**, 375.
Boto, K. G. and Williams, L. F. G. (1976). *Anal. Chim. Acta.* **85**, 179.
Bouhnik, J., Michell, O. and Michell, R. (1972). *Israel J. Med. Sci.* **8**, 1885.
Brezina, M. and Volke, J. (1975). *Prog. Med. Chem.* **12**, 247.
Brezina, M. and Zuman, P. (1958). "Polarography in Medicine, Biochemistry and Pharmacy". Interscience, New York and London.
Brezina, M., Hofmanova, M. and Koryta, J. (1974). *Biophys, Chem.* **2**, 264.
Brooks, E. E. and Mark, H. B. Jr (1975). *Rev. Analyt, Chem.* **3**, 2.
Brule, G., Eckhardt, S. J., Hall, T. C. and Winkler, A. (1973). *In* "Drug Therapy of Cancer", World Health Organization.

Buecher, H. and Franke, R. (1965a). *Abhandl. Deut. Akad. Wiss. Berl. Kl. Med.* **1**, 93.
Buecher, H. and Franke, R. (1965b) *Acta, Biol. Med. Germ.* **14**, 1.
Bukaresti, I., Hadnagy, C. S., Brassai, Z., Csiki, J. N., Fagaresan, M. and Siko, G. (1969), *Rev. Roum. Med. Int.* **6**, 241.
Burgman, G. P. (1973). *Lab. Delo.* **7**, 402.
Butler, V. P. (1976). *Metd. Dev. Biochem.*, **5**, 71.
Calvin, M. (1975). *Naturwissenschaften*, **62**, 405.
Cantin, D., Alary, J. and Coeur, A. (1973). *Analusis*, **2**, 654.
Carruthers, C. (1968). *Proc. Soc. Exp. Biol. Med.*, **127**, 1214.
Carruthers, C. (1976). *Anal. Chim. Acta.* **86**, 273.
Charles, R. and Knevel, A. M. (1968). *Analyt. Biochem.* **22**, 179.
Chatten, L. G., Yadov, R. N., Binnington, S. and Moskalyk, R. E. (1977) *Analyst*, **102**, 323.
Cherynyakov, I. N. (1966). *Biofizika*, **11**, 188.
Chien, Y. W. (1972). *Diss. Abst. Int. B*, **33**, 1643.
Chien, Y. W. (1976). *J. Pharm. Sci.* **65**, 471.
Chodowski, J., Kuvak, W. and Walczak, E. (1974). *Rocz. Chem.* **44**, 1603.
Churchich, J. E. and Mottda, H. (1962). *Annales. Assoc. Quim. Arg.* **50**, 19.
Clausen, J. (1971). *In* "Immunochemical Techniques for the Identification and Estimation of Macromolecules" (Laboratory Techniques in Biochemistry and Molecular Biology), (T. S. Work and E. Work. ed.) North-Holland, Amsterdam.
Colter, J. S. and Paranchych, W. (1967). *In* "The Molecular Biology of Viruses". Academic Press, New York and London.
Conway, M. (1976). *Pediatrics*, **57**, 244.
deMontalvo, A., Piciolella, E., Frigola, A., Pepe, G. and Guerritore, D. (1972). *Bull. Soc. Hal. Biol. Sper.* **48**, 233.
Elashoff, R. M. and Beal, S. (1976). *Ann. Rev. Biophys. Bioeng.* **5**, 562.
Elving, P. J. (1966). *Abhandl. Deut. Akad. Wiss. Berlin Kl. Med.* **4**, 635.
Elving, P. J. (1974). Electroanalytical chemistry. *In* "Advances in Analytical Chemistry and Instrumentation" (H. W. Nurnberg, ed.). Interscience, New York and London.
Elving, P. J., Schmaker, C. O. and Santhanam, K. S. V. (1976). *C.R.C. Crit. Rev. Anal. Chem.* **6**, 1.
Farooq, R. and Kirkbright, G. F. (1976). *Analyst*, **101**, 566.
Fein, J. M., Willis, J., Hamilton, J. and Parkhurst, J. (1975). *Stroke*, **6**, 42.
Fenner, A. (1974). *Biotelemetry*, **1**, 227.
Ferguson, L. N. (1975). *Chem. Soc. Rev.* **4**, 289.
Fiala, S., Fiala, A. E. and Dixon, B. (1972). *J. Natl. Cancer Inst.* **48**, 1393.
Fike, R. R. and Curran, D. J. (1977). *Anal. Chem.* **49**(8), 1205.
Flower, R. W. (1976). *Adv. Exp. Med. Biol.* **75**, 417.
Fomenko, B. S. and Zaslavskii, Yu, A. (1974). *Vopr. Med. Khim.* **20**, 554.
Fontaine, M., Rivat, C., Roparitz, C. and Caullet, C. (1973). *Bull. Soc. Chim. Franc.* **6**, 1873.
Fontaine, M., Rivat, C., Roparitz, D. and Caullet, C. (1974). *Bull. Soc. Chim. Franc.* **11**, 1513.
Fox, M. A. (1976). *Analyt. Chem.* **48**, 993.
Franklin Smyth, W. and Smyth, M. R. (1976). *Proc. Analyt. Div. Chem. Soc.* **13**, 223.
Franklin Smyth, W. and Hassanzadeh, H. (1976). *Z. Anal. Chem.*, **280**, 299.
Franklin Smyth, W., Hill, D. E. and Rendell, T. C. (1979). Recent applications of polarography to drug analysis, *In* "Polarography Fifty Years On" (B. Fleet, ed.). Macmillan, London and Basingstoke. To be published.
Franklin Smyth, W., Watkiss, P., Burmicz, J. S. and Hanley, H. O. (1975). *Anal. Chim. Acta*, **78**, 81.
Garbus, J., Miller, R. M., Caponite, M. J. and Myers, C. S. (1975). *Anal. Biochem*, **67**, 669.

Gelshuen, G. G., Nisnevich, E. D., Meltina, R. A., Kagan, L. Z. and Lapikova, I. I. (1971). *Lab. Delo.* **7**, 414.
Gerber, H., Hugl, F. and Scharch, H. (1965). *Chimia*, **19**, 503.
Gesinki, R. M., Morrison, J. H. and Toepper, J. R. (1968). *J. Appl. Physiol.* **24**, 751.
Godschalk, W. and Veldstra, H. (1965). *Arch. Biochem. Biophys.* **111**, 161.
Grinberg, L. N. (1977). *Lab. Delo.* **1**, 23.
Gulaid, A. (1976). M.Sc. Dissertation, Chelsea College, University of London.
Hake, J. (1966). *J. Electroanal. Chem.* **11**, 31.
Halsey, J. H., Jr, Capra, N. F. and McFarland, R. (1977). *Stroke*, **8**(3), 351.
Halsey, J. H. and McFarland, S. (1974). *Stroke*, **5**, 219.
Harrison, D. E. F. (1974). *In* "Measurement Oxygen Proceedings Inter Disciplinary Symposium" (H. Degn, I. Balser and R. Brook, eds), p. 53. Elsevier, Amsterdam.
Hasebe, K. and Osteryoung, J. G. (1975). *Analyt. Chem.*, **47**, 2412.
Hata, S-I. (1970). *Bioenergetics*, **1**, 325.
Hata, S-I. (1973). *Bioenergetics*, **5**, 140.
Hata, S-I., Egyud, L. G. and Szert-Gyorgi, A. (1971). *Proc. Natl. Acad. Sci.*, **68**, 2292.
Heidelberger, C. (1975). *Ann. Rev. Biochem.*, **44**, 79.
Heidelberger, C. (1976). *In* "Carcinogenesis—A Comprehensive Survey": Vol. 1. "Polynuclear Aromatic Hydrocarbons, Chemistry Metabolism" (R. Frendenthal and P. W. Jones, eds). North-Holland Elsevier, Amsterdam.
Heyrovsky, J. and Zuman, P. (1968). *In* "Practical Polarography". Plenum Press, New York.
Holmes, K. C., Stubbs, G. J., Mandelkov, E. and Gallwitz, U. (1975). *Nature*, **254**, 192.
Homolka, J. (1971). *Meth. Biochem. Anal.* **19**, 435.
Horn, G. (1975). *Hung. Sci. Instrum.* **33**, 7.
Howe, J. R. (1975). *Lab. Practice*, **24**, 457.
Irtiakov, U. R. (1975) *Vrach. Delo.* **8**, 146.
Kadish, K. M. and Spiehler, V. R. (1975). *Analyt. Chem.*, **47**, 1714.
Kessler, M. (ed.). (1971). *In* "Oxygen Supply Workshop; Oxygen, Hydrogen, Enzyme Polarography". University Park Press, Baltimore, Md. (Published 1975).
Kissinger, P. T., Hart, J. B. and Adams, R. N. (1973). *Brain Res.* **55**, 209.
Kissinger, P. T., Refshange, C., Dreiling, R. and Adams, R. N. (1973). *Anal. Letters* **6**, 465.
Kissinger, P. T. (1977). *Analyt. Chem.* **49**, 447A.
Kogan, A. B., Zagyshin, S. C. and Zagyshina, L. D. (1974). *Biol. Nauki. Nauch. Eng. Dok. Unz.* **17**, 58.
Kolthoff, I. M., Yamashita, K. and Hie, T. B. (1974). *Proc. Nat. Acad. Sci.* **71**, 2072.
Kraul, K. and Richmond, J. E. (1971). *Comp. Biochem. Physiol.* **39**, 649.
Krupicka, J. and Zaoral, M. (1969). *Coll. Czech. Chem. Commun.* **34**, 678.
Kumar, A. and Christian, G. D. (1977). *Clin. Chim. Acta*, **74**, 101.
Kunke, S., Erdman, W. and Metzger, H. (1972). *J. Appl. Physiol.* **32**, 436.
Kuzel, K. (1973). *Clin. Chim. Acta*, **48**, 377.
Kuznetsov, B. A. and Shumakovich, G. P. (1975). *Bioelectrochem. Bioenerg.* **2**, 35.
Kuznetsov, B. A., Mestechkina, N. M. and Shumakovich, G. P. (1977). *Bioelectrochem. Bioenerg.* **4**, 1.
Lancet (1976) **7951**, 129.
Lane, R. F. and Hubbard, A. T. (1976). *Analyt. Chem.* **48**, 1287.
Lane, R. F., Hubbard, A. T., Fukunaga, K. and Blanchard, R. J. (1976). *Brain Res.* **114**, 346.
Lankelma, J. and Poppe, H. (1976). *J. Chromat.* **125**, 375.
Lehmenkuhler, A. (1976). *Adv. Exp. Med. Biol.* **75**, 3.
Lewitova, A. (1974). *Coll. Czech. Chem. Commun.* **39**, 369.
Lindquist, J. and Farroha, S. M. (1975). *Analyst*, **100**, 377.

Longo, L. D. (1976). *Science*, **194**, 523.
Lyons, J. M., Raisson, J. K. and Kumamoto. (1974). *Meth. Enz.* **32B**, 258.
Mairesse-Ducarmois, C. A., Patriarche, G. J. and Vandenbalck, J. L. (1977a). *Anal. Chim. Acta*, **88**, 47.
Mairesse-Ducarmois, C. A., Vandenbalck, J. L. and Patriarche, G. J. (1977b). *Anal. Chim. Acta*, **90**, 103.
Malfoy, B., Sequaris, J. M., Valenta, P. and Nurnberg, H. W. (1977). *J. Electroanal. Chem.* **75**, 455.
Mansurova, I. D. (1974). *Lab. Delo.* **7**, 714.
Masters, B. and Miller, I. R. (1974). *Bioelectrochem. Bioenerg.* **1**, 446.
Matsubara, T. and Tochino, Y. (1969). *J. Biochem. (Tokyo)*, **66**, 397.
Matsubara, T. and Tochino, Y. (1970). *J. Biochem. (Tokyo)*, **68**, 731.
Matyus, L. and Scheda, V. (1969). *Idiggovogy Szemle*, **22**, 410.
McCreery, R. L., Dreiling, R. and Adams, R. N. (1974a). *Brain Res.* **73**, 15.
McCreery, R. L., Dreiling, R. and Adams, R. N. (1974b). *Brain Res.* **73**, 25.
Mendez, J. H., Perez, A. S. and Conde, F. L. (1975). *J. Electroanal. Chem.* **66**, 53.
Milazzo, G., Jones, P. E. and Rampazzo, L. (1971). *Experientia*, **18**, suppl.
Miller, E. C. and Miller, J. A. (1974). *In* "The Molecular Biology of Cancer" (H. Busch, ed.). Academic Press, New York and London.
Moggio, R. N. and Hammond, G. L. (1976). *Annal. Surg.* **183**, 282.
Moysenko, L. E. (1972). *Neuropatol. Psikhiatr. IM. SS Korsakova*, **72**(6), 854.
Moysenko, L. E. and Vlasenko, S. N. (1973). *Vrach. Delo.* **1**, 139.
Mulik, M. C., Guyer, M. E., Semenuik, G. M. and Sawieki, E. (1975). *Anal. Letters*, **8**, 571.
Nisoroff, A., Hopper, J. E. and Spring, S. B. (1975). *In* "The Antibody Molecule". Academic Press, New York and London.
Opheim, L. N. (1977). *Anal. Chim. Acta*, **89**, 225.
Owens, J. L. and Dryhurst, G. (1977). *Anal. Chim. Acta*, **89**, 93.
Palecek, E. and Pechan, Z. (1971). *Analyt. Biochem.* **42**, 59.
Pasternak, C. A. (1975). *In* "Radioimmunoassay in Clinical Biochemistry". Heyden, London.
Payne, J. P. and Hill, D. W. (eds) (1975). "Oxygen Measurements in Biology and Medicine". Butterworths, London.
Podany, V. and Vachalkova, A. (1973). *Neoplasma*, **20**, 631.
Podany, V. Vachalkova, A., Miertus, S. and Bahna, L. (1975). *Neoplasma*, **22**, 469.
Poljak, R. J. (1975). *Nature, Lond.* **256**, 373.
Poljak, R. J., Arnzel, L. M. and Phizackerley, R. I. (1976). *Prog. Biophys. Mol. Biol.* **31**, 67.
Ramasamy, N., Parameshwarans, Redner, A. and Srinwasan, S. (1973a). *Trans. Soc. Advan. Electrochem. Sci. Technol.* **8**, 50.
Ramasamy, N., Ranganathan, M., Dulc, L., Srinvasan, S. and Sawyer, P. N. (1973b). *J. Electrochem. Soc.* **120**, 354.
Rao, G. M., Begleiter, A., Lown, J. W. and Plambeck, J. A. (1977a). *J. Electrochem. Soc.* **124**, 195.
Rao, G. M., Lown, J. W. and Plambeck, J. A. (1977b). *J. Electrochem. Soc.* **124**, 199.
Redwood, W. R. and Godschalk, W. (1972). *Biochim. Biophys. Acta*, **274**, 575.
Remke, H. (1970). *Acta. Biol. Med. Germ.* **24**, 13.
Reynaud, J. A. (1976). *Bioelectrochem. Bioenerg.* **3**, 561.
Riggin, R. M. and Kissinger, P. T. (1977). *Analyt. Chem.* **49**, 530.
Rigo, A., Rotillo, G., Viglino, P. V. and Tomat, R. (1975a). *FEBS Lett.* **50**, 86.
Rigo, A., Viglino, P. and Retilio, G. (1975b). *Analyt. Biochem.* **68**, 1.
Rishpon, J. and Miller, I. R. (1975). *Bioelectrochem. Bioenerg.* **2**, 215.
Roitt, I. M. (1971). "Essential Immunology". Blackwell Scientific, Oxford.
Ruttkay-Nedecky, G. (1964a) *Abhandl. Deut. Akad. Wiss. Berl. Kl. Chem. Geol. Biol.* **1**, 217.

Ruttkay-Nedecky, G. (1964b). *Coll. Czech. Chem. Commun.* **29**, 1809.
Ruttkay-Nedecky, G. and Anderlova, A. (1967). *Nature, Lond.* **213**, 564.
Ruttkay-Nedecky, G. and Bezuch, B. (1971a). *J. Molec: Biol.* **55**, 101.
Ruttkay-Nedecky, G. and Bezuch, B. (1971b). *Experientia Suppl.* **18**, 553.
Ruttkay-Nedecky, G. and Bezuch, B. (1973). *Stud. Biophys.* **40**, 211.
Saito, H., Tanabe, T. and Yoshio, H. (1972). *Pharmacol. Res. Commun.* **4**, 17.
Sandler, A. N. and Tator, C. H. (1976). *Brain Res.* **118**, 181.
Sasa, S. and Blank, C. L. (1977). *Analyt. Chem.* **49**, 354.
Scheller, F. and Will, H. (1973). *FEBS Lett.* **29**, 47.
Schmandke, H. and Crohlke, H. (1965). *Clin. Chim. Acta*, **11**, 491.
Schreiber, J. R., Balcavaga, W. X. and Morris, H. P. (1970) *Cancer Res.* **309**, 2497.
Senda, M., Ikeda, T. and Kinoshita, H. (1976). *Bioelectrochem. Bioenerg.* **3**, 253.
Shakhnovich, A. R., Fedorov, S. N., Milovanova, L. S. and Semenov, L. G. (1971). *Biofizika*, **16**, 915.
Simmons, I. L. and Ewing, G. W. (1974). *In* "Methods in Radioimmunoassay, Toxicology and Related Areas" (Prog. Anal. Chem. Series). Plenum Press, New York and London.
Smith, M. D. and Olson, C. L. (1975). *Analyt. Chem.* **47**, 1074.
Smyth, M. R. (1977). Private Communication.
Smyth, M. R. and Franklin Smyth, W. (1978). *Analyst*, **103**, 529.
Solymar, M., Rucklidge, M. A. and Prys-Roberts, C. (1971). *J. Appl. Physiol.* **30**, 272.
Soutter, L. P., Conway, M. J. and Parker, D. (1975). *Biochem. Eng.* 257.
Starka, L. and Brabencova, H. (1960). *Clin. Chim. Acta*, **5**, 423.
Stas, J. (1973). *Pol. Tig. Lek.* **28**, 955.
Steinbrecht, I. and Kunz, W. (1972). *Acta Biol. Med. Ger.* **29**(4/5), 495.
Sternson, L. A. and Hes, J. (1975). *Analyt. Biochem.* **67**, 74.
Sternson, L. A., Sternson, A. W. and Bannister, S. J. (1976). *Analyt. Biochem.* **75**, 142.
Stossek, K. D., Lubbers and Cottin, N. (1974). *Pfluggers Arch.* **348**, 225.
Sushkin, A. G. (1969). *Probl. Tuberk.* **47**, 55.
Tachibana, M., Sawaki, S. and Kawazoe. (1967). *Chem. Pharm. Bull. (Japan)*, **15**, 1112.
Thrivikraman, K. V., Refshange, C. and Adams, R. N. (1974). *Life Sci.* **15**, 1335.
Ueno, Y., Kuraishi, F., Uematsu, S. and Tsuruoka, T. (1968). *Folia. Psychiatr. Neurol. Japan*, **22**, 167.
Valasquez, A. and Rosado, A. (1972). *Fert. Ster.* **23**, 562.
VanBeeumen, J. and Deley, J. (1971). *Analyt. Biochem.* **44**, 254.
Varadi, M., Fehrer, Z. and Pungor, E. (1974). *J. Chromat.* **90**, 259.
Vishnyak, Y. I. and Krusnova, A. I. (1965). *Zh. Neuropatol. Psikhiat. Im. SS Korsakova*, **65**, 251.
Volter, F., Vauzelle, D., Lautier, A. and Laurent, D. (1972). *Bull. Physio-Pathol. Resp.* **8**, 947.
Walters, C. L. (1977). *Chem. Brit.* **13**, 140.
Watson, J. D. (1975). *In* "Molecular Biology of the Gene", 3rd Edn, Benjamin, New York.
Weinhold, J. and Sohr, H. (1977). *Anal. Chim. Acta*. **89**, 927.
Weiss, H. R. and Cohen, J. A. (1974). *Environ. Physiol. Biochem.* **4**, 31.
Weitzmann, P. D. J. (1976). *Biochem. Soc. Trans.* **4**, 724.
Whitnach, G. C. and Soli, G. (1966). *J. Electroanal. Chem.* **12**, 60.
Wightman, R. M., Strope, E., Plotsky, P. M. and Adams, R. N. (1976). *Nature, Lond,* **262**, 145.
Wilson, G. S. and Neri, B. P. (1973). *Ann. N.Y. Acad. Sci.* **206**, 568.
Zikan, J. (1966). *Coll. Czech. Chem. Commun.* **31**, 4260.
Zikan, J. and Sterzl, J. (1967). *Nature, Lond.* **214**, 1225.
Zuman, P. (1974). *In* "Experimental Methods in Biophysical Chemistry" (C. Nicolan, ed.), p. 393. Wiley, Chichester and New York.

APPLICATIONS

In Environmental Science

Chapter 7

ELECTROANALYSIS OF TRACE FOREIGN ORGANIC MATERIALS IN THE AQUEOUS ENVIRONMENT

B. J. BIRCH
Unilever Research, Port Sunlight Laboratory,
Merseyside, England

and

J. P. HART
Orthopaedic Department, Charing Cross Hospital,
Hammersmith, London, England

I. INTRODUCTION

The term "foreign" in the title of this article needs defining at the outset. It is taken not to mean the substances arising from the natural decay of leaves, plankton, etc., but rather from the products of man's activity upon this planet. Excellent, detailed reviews of the possible effects of these substances on the environment are available (Middleton and Rosen, 1956: Henderson et al., 1960; Ryckman, 1966); suffice it to say, therefore, that concern may arise from:

(i) Damage to the aesthetic quality of waters, both from the viewpoints of consumption and of leisure, from the formation of foams, odours, etc.
(ii) The complexing of trace heavy metals by organic species, thus altering their toxicity.
(iii) The presence of a cumulative residue of non-degradable, poisonous material. Classically, this is exemplified by the persistence of DDT in the environment.

(iv) The reduction of the dissolved oxygen content of waters by biologically oxidizable substances, with subsequent damage to the natural life of those waters.

This article is concerned with a discussion of applications of electroanalytical methods to such "foreign" molecules in the aqueous environment, brief experimental methods, $E_{\frac{1}{2}}$ values and references to accessible work; electroanalytical instrumentation and methodology have been comprehensively dealt with elsewhere in this volume.

II. APPLICATIONS

Although polarographic methods have been applied to the determination of macro amounts of organic materials for many years, only comparatively recently has attention been directed to the analysis of trace organic molecules in waters (Whitnack, 1961; Ballinger, 1963; Siegerman et al., 1972; Ishu, 1972, 1973; Maienthal and Taylor, 1973).

A. CARBONYL COMPOUNDS

Although the carbonyl group is electro-reducible, the little work that has been carried out on carbonyl compounds by direct polarographic methods has mainly been confined to simple aldehydes. Kuchumova et al. (1963) have determined formaldehyde, acetaldehyde, furfural and butyraldehyde at the µg ml^{-1} level in waste water. LiOH was used as supporting electrolyte with 5% citric acid added to mask the effect of iron. Formaldehyde gave a reduction wave in the -1.6 to -1.8 V region, acetaldehyde and butyraldehyde in the -1.8 to -2.0 V region and furfural between -1.4 and -1.6 V. Similar work was carried out by Dyatlovitskaya and Berezouskii (1962) and by Bodyu and Feldman (1963) for the analysis of furfural in waste waters. Their methods had minor variations in the supporting electrolyte and pH conditions used, but the $E_{\frac{1}{2}}$ values and detection limits of all three procedures were very similar. Melcer and Melcerova (1971) have determined furfural using a derivative polarographic method. A sensitivity of 0.3 µg ml^{-1} was claimed, using 2M NaOH as the preferred supporting electrolyte. Similar work on formaldehyde and furfural has been reported by Ponomarev et al. (1974) and Zhantalai et al. (1976).

The use of derivatization of the carbonyl group has been exploited by Afghan et al. (1975) in a systematic study of these compounds in various supporting electrolytes, using twin cell potential sweep voltammetry. The formation of the semicarbazone was found to be most satisfactory. The formation and subsequent reduction of this species proceeds as shown in Chapter 1, pp. 31, 32.

Using a citrate buffer, with EDTA added to complex interfering heavy metals, a sensitivity of 0.25 ng ml^{-1} was claimed for the determination of carbonyl compounds present in natural waters and industrial effluents.

B. SEQUESTERING AGENTS

Materials of the amino polycarboxylic acid type can be determined polarographically, by observing the reduction of a heavy metal–ligand complex, in the presence of excess metal ions. The reduction potential of

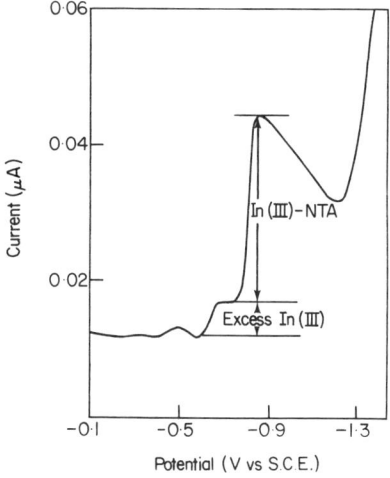

FIG. 1. Activated sludge-treated synthetic sewage with 0.0257 µg ml^{-1} of NTA (plus 0.0257 µg ml^{-1} from NTA–^{14}C to give a total of 0.0514 µg ml^{-1}). It was necessary to add approximately a 30-fold molar excess of In(III) to total NTA before an excess In(III) wave appeared.

this complex is shifted to more negative values, and the limiting current of this reduction is used for the quantitative determination of the ligand. Haberman (1971) demonstrated how nitrilotriacetic acid (NTA), which is a detergent builder with the formula $N(CH_3COOH)_3$, could be determined using In(III). After an excess of In(III) had been added to the NTA in the aqueous solution, the free metal gave a polarographic wave and the limiting current of the complexed In(III) could be found by difference as shown in Fig. 1. Other workers have shown that NTA can be determined using metals other than In(III). Afghan et al. (1972), Hoover (1973) and Taylor et al. (1972) all used Bi(III), Asplund and Wanninen (1971) and Wernet and Wahl (1970) used Cd(II), and Pb (II) was used by Afghan and Goulden (1971). Studies on metal–NTA complexes have shown that the optimum pH for trivalent metal ion complexation is pH 2 and for divalent metal ions pH 7. This has been verified by Afghan et al. (1972), who showed that larger cathodic currents were produced for the reduction of trivalent metal–NTA complexes in acidic media, whereas divalent metal–NTA complexes gave larger reduction currents

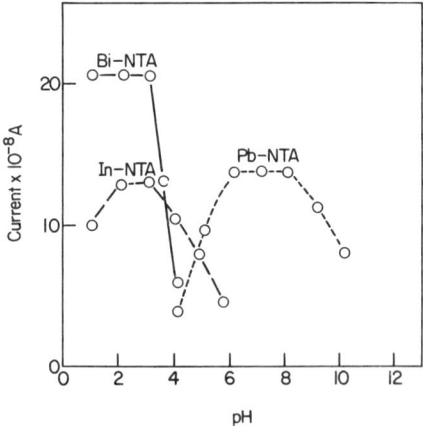

FIG. 2. Effect of pH on peak height of bismuth-NTA complex, lead-NTA complex and indium-NTA complex.

in neutral pH regions (Fig. 2). The polarographic current results from the reduction of complexed metal as shown below.

$$M^{n+}(NTA) + ne^- \rightarrow M^0 + NTA$$

Using an anion exchange column to concentrate NTA, Haberman (1971) was able to determine 0.025 p.p.m. when In(III) was used as the complexing metal. However, Afghan et al. (1972) has found that, without a preconcentration step, 0.01 µg ml^{-1} of NTA can be detected in natural waters and sewage using Bi to complex the NTA. Afghan's method involves automation of part of the procedure and 15 samples per hour have been analysed. The supporting electrolyte used in the In(III)–NTA analyses was the eluent from the anionic exchange column and consisted of 1M NaCl in 0.1M acetate buffer, pH 4.7. This pH, as indicated by Afghan's work, would not appear to give the optimum conditions for polarographic measurements, and a solution pH between 2 to 3 may have improved sensitivity.

C. SIMPLE AROMATIC COMPOUNDS

Direct polarographic methods for nitro-compounds in waste water and effluents have been developed. For example, nitrobenzene in industrial wastes was determined by Dyatlovitskaya et al. (1963) after separating nitrophenols by distillation. The supporting electrolyte consisted of aqueous ethanol containing hydrochloric acid. Fleszar (1964) has determined 0.005 mg l^{-1} of nitrochlorobenzenes in water following extraction with activated charcoal and polarography of the eluted acetone solution in a pyridinium hydrochloride supporting electrolyte. Nitrophenols, nitrocresols and nitrotoluenes, in addition to nitrochlorobenzenes present in waste water, have been polarographically analysed by Zaitsev and Dichenskii (1966). The supporting electrolyte used was 0.25M NaOH in 3:1 CH$_3$OH:H$_2$O, and $E_{\frac{1}{2}}$ values in the region -0.70 to -1.20 V were obtained. A very sensitive method for dinitro-o-cresol in water has been described by Supin et al. (1971). Extraction of the acidified water sample with petroleum ether was followed by linear sweep voltammetry (l.s.v.) in an aqueous borate buffer. This method offered a sensitivity of 0.004 mg l^{-1}. Direct methods for the determination of phenol in waste water have been developed using noble metals as the indicating electrodes (Ginsburg and Frishman, 1959; Strafelda, 1970). The methods could be used for determining less than 0.01 mg l^{-1} of phenol.

Indirect procedures have been described for the determination of benzene and phenol in waste water (Adamovsky, 1966; Avedouard et al., 1975). In the method for benzene, a stream of nitrogen was passed through the water which was subsequently directed into the nitration mixture (ammonium nitrate–sulphuric acid–acetic acid). The resulting m-dinitro-benzene was then subjected to polarographic analysis. For the analysis of phenol, the water sample was firstly made alkaline and concentrated by evaporation. The liquid was then acidified and the phenol extracted into diethyl ether. After evaporating to dryness, the residue was nitrated with hot concentrated nitric and sulphuric acids. Pulse polarography was carried out after adjustment of the nitration mixture to pH 4.9 with a Britton–Robinson buffer. The sensitivity was 1 ng ml^{-1}, whereas for benzene the author quoted a value of 1 μg ml^{-1}.

D. HERBICIDES AND PESTICIDES

Herbicides and pesticides are used throughout the world for the protection of crops, and this has led to the development of trace methods of analysis in surface and natural waters. Comparatively few compounds have been determined in water samples by polarographic methods, although the technique is perhaps one of the most direct and sensitive for them.

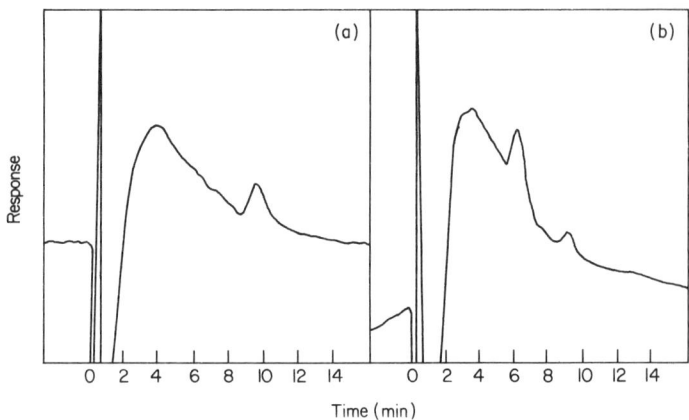

FIG. 3. Chromatograms of (a) blank pond water and (b) pond water fortified with 0.001 µg ml^{-1} of ametryne (equivalent of 3 ng of ametryne).

McKone et al. (1972) have compared gas chromatographic, polarographic and spectrophotometric methods for the determination of the triazine herbicides terbutryne, ametryne and atrazine in pond and canal water. The polarographic method involved extraction with CH_2Cl_2, taking the residue up in 50% $CH_3OH/0.01N\ H_2SO_4$ and running the linear sweep voltammogram from -0.8 V to the decay of the supporting electrolyte. In this medium, atrazine gave rise to a wave at -1.05 V (vs Hg anode) and could be determined down to 0.01 µg ml^{-1} whereas terbutryne and ametryne were both reduced at -1.45 V and could be determined down to 0.005 µg ml^{-1}. The

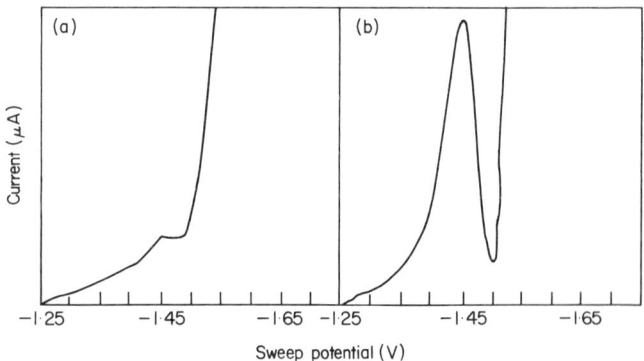

FIG. 4. Polarograms of (a) blank pond water and (b) pond water fortified with 0.01 µg ml^{-1} of ametryne.

polarographic method matched the g.l.c. procedure in terms of time of analysis and ease of calculation of the results but suffered a little in terms of sensitivity and selectivity (Figs. 3 and 4).

In contrast to herbicides mentioned above, the herbicide glyphosate (N-phosphonomethyl-glycine)(I) was found to be more conveniently analysed by polarography than by g.l.c. (Bronstad and Friestad 1976). For the latter method a lengthy four-stage clean-up and two-stage derivatization procedure was necessary, whereas ion exchange followed by nitrosation only was required for polarography. The eluate from ion exchange was treated with 50% sulphuric acid and potassium bromide and sodium nitrite solutions. After fifteen minutes, ammonium sulphamate was added to destroy excess nitrite and polarography was carried out after de-aerating with nitrogen.

$$\text{OH}-\underset{\text{(I)}}{\overset{\overset{\text{O}}{\|}}{\text{C}}-\text{CH}_2-\underset{\text{H}}{\text{N}}-\text{CH}_2-\overset{\overset{\text{O}}{\|}}{\underset{\text{OH}}{\text{P}}}-\text{OH}} \qquad \text{HO}-\underset{\text{(II)}}{\overset{\overset{\text{O}}{\|}}{\text{C}}-\text{CH}_2-\underset{\text{NO}}{\text{N}}-\text{CH}_2-\overset{\overset{\text{O}}{\|}}{\underset{\text{OH}}{\text{P}}}-\text{OH}}$$

The N-nitroso (II) derivative gave a reduction peak at -0.78 V, which could be used to monitor between 35 and 210 μg l^{-1} of glyphosate, in natural waters (Fig. 5). It was suggested that one analyst could analyse twenty samples a day, using the above procedure.

Sobina et al. (1976) have determined fenitrothion (III) in drainage run-offs from fields. The method involved extraction with either butanol or isobutanol, which was subsequently made 0.1M in LiCl. In this medium fenitrothion had an $E_{\frac{1}{2}}$ value of -0.95 V (vs Hg pool) and oscillopolarography could determine concentrations of fenitrothion down to 0.02 mg l^{-1}.

$$\underset{\text{CH}_3\text{O}}{\overset{\text{CH}_3\text{O}}{\diagdown}}\overset{\overset{\text{S}}{\|}}{\text{P}}-\text{O}-\hspace{-2pt}\left\langle\right\rangle\hspace{-2pt}-\text{NO}_2$$
$$\text{CH}_3$$
(III)

In a method for the determination of picloram (potassium-4-amino-3,5,6- trichloropicolinate) (IV) in water (Filimonova and Gorbunova, 1973), filtering was the only pretreatment required. The filtrate was adjusted to pH 1 with 10N sulphuric acid and, after suitable dilution, polarography was

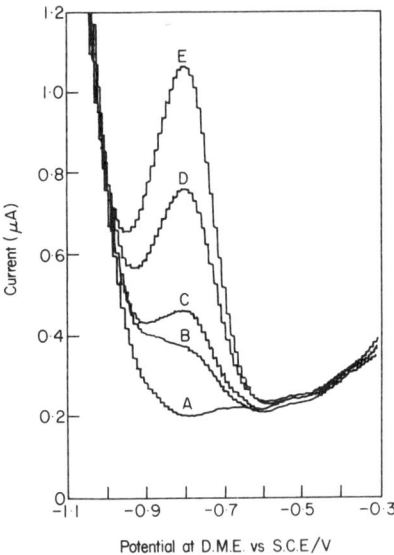

FIG. 5. Current-potential graphs for glyphosate nitrosamine after anion exchange treatment and nitrosation of various amounts of glyphosate added to 1 litre of local tap water. A: 0 (reagent blank); B: 35; C: 70; D: 140; and E: 210 µg added.

carried out. The method was based on the fact that when the amino group is protonated, the reduction of this species to ammonia and 1,3,5-trichloropicolinic acid at -1.42 V gave a measure of the original chloramp. The reactions involved can be written as:

(IV)

The range of linearity of the calibration plot of i_{lim} vs conc. in 0.1 HCl or 0.1 H_2SO_4 was 9.6×10^{-7} to 1.8×10^{-4} M with a sensitivity of 0.2 mg l^{-1} in water.

The herbicide, N,N-diethyl-2,6-dinitro-4-trifluoromethylaniline has been electrochemically reduced in an alkaline supporting electrolyte, consisting of 0.05N tetraethyl ammonium hydroxide in 60% methanol (Filminonova et al., 1975). The compound was first extracted into hexane, and the solvent evaporated to a minimum volume before the addition of supporting electrolyte. Reduction was found to occur in two four-electron steps, at -0.61 V and -0.85 V, corresponding to reduction of nitro groups to hydroxylamines. Using a calibration graph between 0.01 and 0.1 mg l^{-1} of herbicide could be analysed.

The insecticide nemagon, 1,2-dibromo-3-chloropropane, has been polarographically investigated in four different supporting electrolytes (Vovik and Kozlova, 1972; Novik 1971; Novik and Pavlovskii, 1976). The supporting electrolytes examined were 1M KCl, aqueous alcohol-1M KCl, 1M KCl in 30% isopropyl alcohol and 1M Na$_2$SO$_3$. Of these, the most suitable was found to be the latter. Recoveries were made on samples of tap water spiked with 19 to 88 mg l^{-1} of nemagon and the coefficient of variation was found to be 21.2% and 7.2% at the lower and upper concentration levels respectively. The recommended solvent for extraction of the water sample was petroleum ether (1:5).

Perhaps the most well-known of all pesticides is DDT (dichlorodiphenyltrichloroethane), and it is not surprising that polarographic methods have been developed for determinations of this compound in water. The most common procedure developed involved extraction from the water sample with carbon tetrachloride, followed by evaporation, and nitration of the residue (Davidek and Janicek, 1961; Ivanchenko, 1973). After suitable dilution of the nitration mixture with water and methanol, the tetranitro-derivative was determined from its reduction wave with $E_{\frac{1}{2}} = -0.13$ V. Concentrations of DDT below 10^{-6}M could also be determined and a coefficient of variations be obtained (Davidek and Janicek, 1963).

The γ isomer of hexachlorocyclohexane or γ-BHC has been determined in water samples down to 0.5 μg l^{-1} (Kosmaty and Bublik, 1970).

The pulse polarographic technique has been used for the determination of hexachlorobutadiene (Solonar and Bael, 1967), hexachlorocyclopentadiene and octachlorocyclopentene (Lyalikov, 1967) in water. Benzene was the favoured solvent for extraction of hexachlorobutadiene, and pulse polarography could be carried out in a solution containing benzene-ethanol-water. The reduction potential in this medium occurred at -1.12 V and the sensitivity was 0.4 mg l^{-1}. However, when a supporting electrolyte consisting of 0.2M tetraethyl ammonium iodide in 80% acetone was used, a sensitivity of 0.2 mg l^{-1} was achieved. Hexachlorocyclopentadiene and octachlorocyclopentene showed a linear dependence between ip and concentration when a buffer solution of pH $\geqslant$ 8.5 was used. For these two compounds, a medium of 0.1N LiClO$_4$, KCl or NaNO$_3$ in 4:1 C$_2$H$_5$OH–H$_2$O was used and the E_p values were -0.61 V, -0.70 V and -0.66 V respectively. The sensitivity under the above conditions was found to be 5×10^{-6} M.

E. POLYMERIC MATERIALS

A large variety of materials are used in the manufacture of plastics and polymers, and both raw materials and finished products can cause contamination of the aqueous environment. This can occur by direct disposal in effluents, or through degradation of the parent polymer. The sewage and effluent from the production of these materials will contain a mixture of species, and a preliminary separation method is often advantageous or even necessary. In a method for the polarographic determination of caprolactam (Eremin and Kopylova, 1973), a solvent extraction step with benzene was introduced to remove interfering organics from the sample. This was followed by conversion to ε-aminocaproic acid (ACA) by acid hydrolysis with 1N sulphuric acid. A cation exchange resin was used to separate the reaction product from the acid solution and after elution with 1M sodium bicarbonate, the d.c. polarogram was recorded in a sodium bicarbonate–formalin supporting electrolyte. From the limiting current measured at -1.12 V, concentrations down to 0.1 mg l^{-1} of ACA could be determined.

Vinyl esters have been polarographically determined after esterification (Filov, 1959, 1960) to produce vinyl alcohol and acetaldehyde. The reactions involved are:

$$R-COOCH=CH_2 + LiOH \rightarrow CH_2=CHOH + RCOO-Li$$

$$CH_2=CHOH \xrightarrow{\text{Isomerization}} CH_3CHO$$

The resulting acetaldehyde was polarographically determined in the supporting electrolyte 0.1M LiOH. Using the above reaction and polarographic techniques, vinyl formate, acetate, propionate and butyrate have been analysed, linearity being obtained in the range 5–500 mg l^{-1}.

In contrast to these compounds, which have been determined by virtue of reduction processes, melamine and cyanuric acid have been monitored using their oxidation behaviour. The choice (Zhantalai and Slisarenko, 1973a, b) of supporting electrolyte for the determination of melamine was a mixture of 10M NaOH, 0.8M KCl and 4 drops of 1% gelatin, added to 10 ml of effluent. Linearity was found in the region 2×10^{-6}–10^{-4}M. Using this supporting electrolyte, effluents could be analysed for melamine in the presence of cyanuric acid. This, in turn, could be determined in the concentration range 5×10^{-6}M–10^{-5}M, in the presence of melamine, amelide and amelin by its anodic wave at -0.04 V in a supporting electrolyte of 0.1M $NaHCO_3$.

Zhantalai and Slisarenko (1975) have determined melem in waste water and in melamine. Melem gave rise to two waves in solutions of pH < 3, the first wave at -0.84 V being used for analytical purposes. In a 2N H_2SO_4

supporting electrolyte, melem could be determined in the presence of s-triazines in the range 5×10^{-6} to 5×10^{-4} M with a relative error of $< \pm 5\%$.

F. SURFACTANTS

Surfactants are present in a wide variety of household and industrial materials. These include not only household cleaning products such as detergent powders and liquid cleaners, but preparations such as insecticides, polishes and cosmetics. Consequently, contamination of the aqueous environment by surfactants is widespread. Surfactants are compounds which contain both hydrophobic and hydrophilic groups. They may broadly be classified according to the charge of the hydrophilic species as anionic, cationic or nonionic. Polarographic methods have been developed for all three types (Jehring, 1966) but, in general, they are not specific to surfactant type and an effective separation is a prerequisite to determination in real situations. Ion exchange is the most generally used surfactant separation method, either anionic or cationic type being separated from nonionic by suitable choice of resin. This has been extensively studied by Linhart (1972), who concluded that separation using ion-exchange resin as a batch method has advantages over the more usual column operation.

(i) *Depression of maxima methods*

The widespread use of surface-active agents to suppress unwanted maxima in conventional d.c. methods has been applied in reverse—to determine surfactants by their depression of maxima produced by a suitable electroactive substance. The material most often used is oxygen (Dolezil and Kopanica, 1963) dissolved in the supporting electrolyte, although cations such as Cu(II) have also been used (Stackelberg and Shutz, 1943).

To carry out these estimations (Fujinaga and Okazaki, 1965; Kozarac et al., 1976), the d.c. polarographic behaviour of a suitable supporting electrolyte solution such as potassium chloride was obtained, omitting the normal de-oxygenation procedure. The water sample was then added to the cell and the i–E curve repeated. The difference in height of the maxima gave a measure of surfactant concentration (in % terms) in solution. This method has been used by Linhart (1972), who found that linear calibration curves, over the range 0–100 p.p.m., were obtained for pure surfactant solutions. Successful determinations were made (incorporating ion exchange separations) of surfactant levels in specified samples of natural waters and effluents.

(ii) *Methods based on the measurement of non-Faradaic admittance (tensammetry)*

An important property of surfactants is their ability to adsorb onto suitable substrates. This gives rise to tensammetric peaks when surfactant solutions are subjected to a.c. polarography using the d.m.e. Shinozuka *et al.* (1972) have determined several anionic surfactants at the $\mu g \, ml^{-1}$ level, whilst Ishu (1973) has analysed solutions of alkylbenzene sulphonates (and many other model water pollutants) by their tensammetric behaviour.

In the authors' experience, however, the method has little to commend it over other polarographic methods: non-linear calibration curves are often obtained (Gorodetskii *et al.*, 1976), whilst peak potentials are a function of surfactant concentration. Also, since in solutions of mixtures of adsorbing materials the most strongly adsorbed substance predominates in tensammetric behaviour, separation prior to polarography is necessary. This is so even though the surfactants may exhibit tensammetric peaks differing widely in potential.

(iii) *Other methods*

Since the majority of surfactant types possess no electro-active groups, conventional direct polarography is of little value. One exception is surfactants of the amine oxide type, which, in addition to reduction to the corresponding amine (Chambers, 1964), also exhibit a well-defined oxidation wave at the glassy carbon electrode.

Turning now to indirect methods, the analysis of anionic type surfactants (in the presence of non-ionic) has been achieved by Kambara and Hasabe, (1965). They made use of the property of the cationic dye methylene blue to form a neutral, chloroform extractable complex with anionic surfactants. This, indeed, forms the basis of the usual colorimetric estimation of anionic surfactants. After extraction, the methylene blue content of the complex (and hence the surfactant level) was estimated by a.c. polarography. Briefly, a solution of methylene blue was added to the acidified waste water sample and the resultant complex extracted into chloroform. A.c. polarography was then performed on this extract, using a solvent consisting of chloroform–water–methyl cellusolve in the ratio 1:1:5, with an ammonium nitrate buffer as supporting electrolyte. The resulting peak potential at -0.1 V (vs Hg pool) gave anionic surfactant concentrations down to $\sim 0.1 \, \mu g \, ml^{-1}$.

The previous method is applicable to all anionic surfactants. The authors (Hart *et al.*, 1976) have developed two methods which are specific to the most commonly used anionic surfactant-type—sodium alkylbenzene sulphonate (LAS)—involving nitration of the aromatic ring.

In the first procedure, the LAS was heated at 100°C for 30 minutes with a mixture of fuming nitric and concentrated sulphuric acids. After dilution of

7. ELECTROANALYSIS OF TRACE FOREIGN ORGANIC MATERIALS

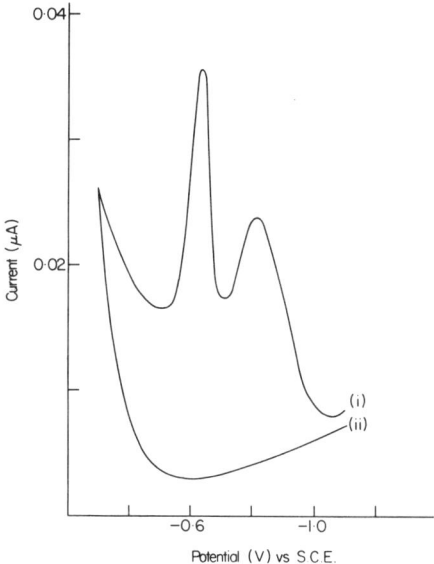

FIG. 6. Differential pulse polarograms of (i) 1.05 p.p.m. nitrated sodium pentadecyl benzene sulphonate, (ii) blank of nitration mixture in Britton–Robinson buffer pH 12.0, starting potential = −0.2 V.

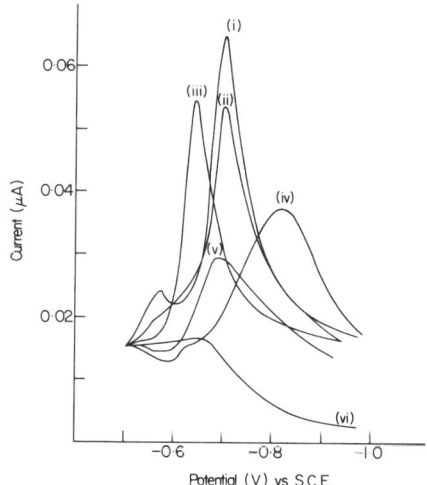

FIG. 7. Differential pulse polarograms of the nitro derivatives of LAS and two standard compounds recorded in Britton–Robinson buffer pH 12.0. (i) 20.2 µg of sodium ethyltridecyl benzene sulphonate; (ii) 17.8 µg of sodium hexylheptyl benzene sulphonate; (iii) 10.0 µg of sodium pentadecyl benzene sulphonate; (iv) 226.0 µg of phenol; (v) 20.0 µg benzoic acid; (vi) blank of nitration mixture.

the nitration mixture with distilled water, an aliquot was further diluted with Britton–Robinson buffer pH 12.0 which was subsequently subjected to polarography. Figure 6 shows the differential pulse polarogram of a typical LAS after treatment by the above procedure. The two peaks shown are consistent with the reduction of a dinitro species in two four-electron steps.

In the second procedure, the LAS was treated with fuming nitric acid for 15 minutes at 30°C. The excess fuming nitric acid was blown off with a stream of nitrogen, and the residue dissolved in Britton–Robinson buffer pH 12.0. The i–E curves of this solution were then obtained and Fig. 7 shows the differential pulse polarograms obtained when different LAS types, and two reference compounds, were nitrated by the latter method. One peak only was obtained which indicated the formation of a mono-nitro derivative reduced in one 4e step. One of the reference compounds, i.e. phenol, did not give a reduction peak which interfered with the LAS peaks. However, benzoic acid did give a peak with reduction potential very close to that of LAS, although smaller in size for a comparable concentration. Calibration plots for this second method were linear when 5–50 µg of LAS were nitrated, and 5 ml of buffer was used as supporting electrolyte. The coefficient of variation was 10% over the concentration range studied.

To determine LAS in sewage samples a 59 ml aliquot was evaporated to dryness in a 100 ml beaker. The cooled beaker was washed with two 5 ml portions of methanol and the washings added to a 10 ml conical centrifuge tube. A few boiling chips were added to the tube and the methanol carefully evaporated off. The centrifuge tube was allowed to cool to 30°C after which the second nitration method described was carried out. When this procedure was applied to sewage samples from three different sources, the differential

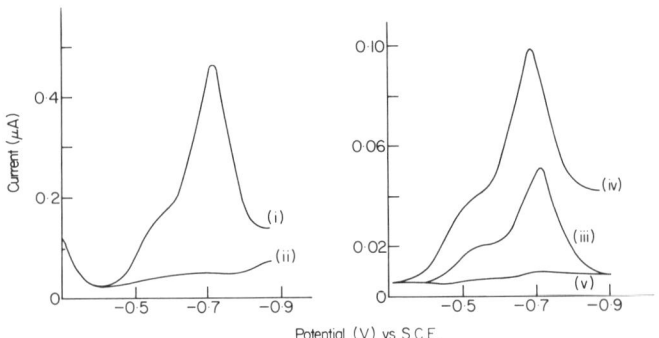

FIG. 8. Differential pulse polarograms of nitrated sewage residues and blanks of nitration mixture in Britton–Robinson buffer pH 12.0 starting potential = −0.3 V. (i) Residue from sewage stream a; (ii) blank; (iii) residue from sewage stream b; (iv) residue from sewage stream c; (v) blank.

pulse polarograms shown in Fig. 8 were obtained. The colorimetric methylene blue procedure of Fairing and Short (1956) was compared to the nitration procedure using samples of sewage from the same source as described in Fig. 8. The results obtained by the polarographic procedure were substantially lower than for the methylene blue method. These results are perhaps not surprising when one considers the following. Methylene blue complexes with all anionic surfactants, as well as some other anionic species. However, nitration can only take place in compounds that have a benzene ring, and therefore, it is possible to discriminate these compounds from other non-benzenoid organic compounds present in the sewage. Thus nitration followed by polarography appears to offer a more specific method for the determination of LAS in sewage, then the colorimetric methylene blue method.

G. MISCELLANEOUS MOLECULES

(i) *Explosives and plasticizers*

It has been shown by Whitnack (1975) that explosives such as nitroglycerine, 1,2-propylene glycol dinitrate, 2,4,6-trinitrotoluene, and 1,3,5-trinitro-1,3,5-hexahydrotriazine (RDX) can be determined directly in water without pretreatment. A continuous electrochemical analyser has also been described (Shaw et al., 1976).

The study by Whitnack showed that using a single sweep polarographic technique, a quick screening method was available to identify and determine

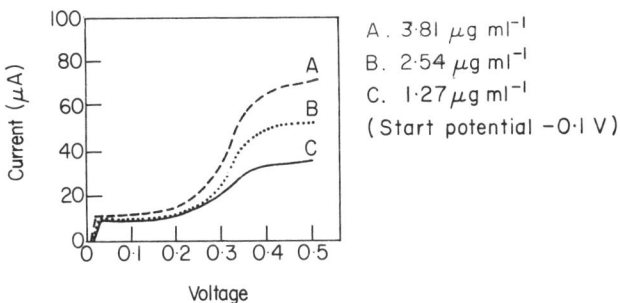

FIG. 9. 1,2-Propylene glycol dinitrate in sea water.

possible micropollutants. Figure 9 shows how the single sweep polarographic current increased with an increase in concentration of 1,2-propylene glycol dinitrate when analysis was carried out directly in sea water. This author also obtained similar polarograms for the other explosives when analysis was carried out in both lake and well water. Some determinations were possible in the sub-µg ml^{-1} range. In addition to explosive, the single-sweep technique

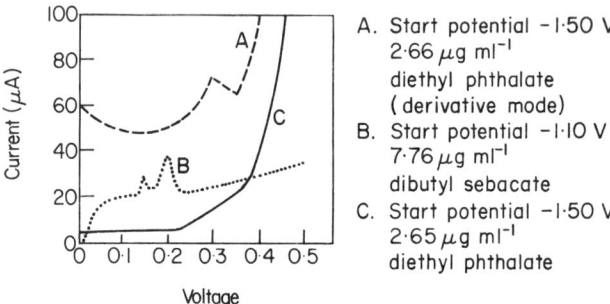

FIG. 10. Diethyl phthalate and dibutyl sebacate in sea water.

has been applied by Whitnack (1975) to the direct analysis of the plasticizers diethyl phthalate and dibutyl sebacate in sea water. For these compounds, concentrations of 2.66 and 7.76 mg l^{-1} were monitored directly in the water sample (Fig. 10).

(ii) *Lignin-sulphonic acids*

An indirect polarographic method for the determination of trace quantities of lignin-sulphonic acids using Nile blue has been developed by Erlebach et al. (1970). Nile blue gave two waves, in a supporting electrolyte containing

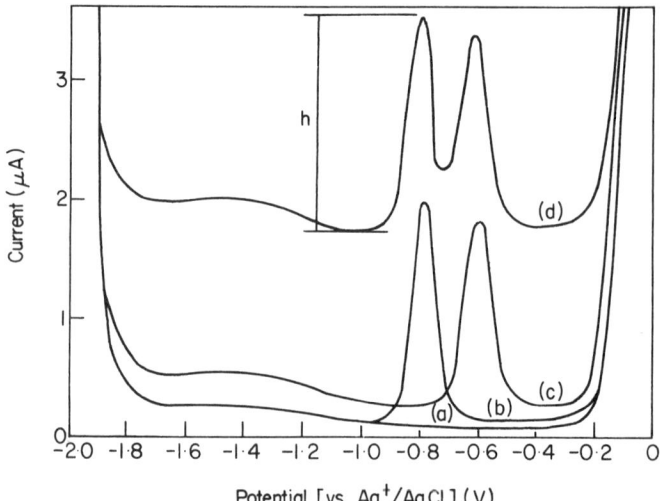

FIG. 11. Polarograms obtained from the analysis of solutions of sodium sulphide and methyl mercaptan in alkaline buffer. (a) Polarogram of the stock solution: 0.2M NaOH, 0.3M KCl, 750 mg Tomlinite/litre. (b) Polarogram of sodium sulphide: stock solution + 1.2 × 10^{-4}M Na$_2$S. (c) Polarogram of methyl mercaptan: stock solution + 19.4 × 10^{-4}M CH$_3$SH. (d) Polarogram of the mixture: stock solution + 1.2 × 10^{-4}M Na$_2$S + 19.4 × 10^{-4}M CH$_3$SH.

16% ethanol and buffered to a pH of 4.2. The second wave showed a decrease in adsorption characteristics in the presence of lignin-sulphonic acids and this has been used to determine 3–50 mg of the above per litre of paper mill solution. Polarographic methods for the determination of various lignin compounds have been reviewed by Zielinski and Surkiewicz (1974).

(iii) *Sulphur compounds*

A sensitive polarographic method, based on the formation of partially insoluble mercury salts has been developed by Renard *et al.* (1975), for the determination of thiols in pulping liquors. In an alkaline buffer medium, many inorganic and organic compounds present were found not to interfere with the determination of thiols (Fig. 11). This method compared favourably with other methods such as titrimetry with heavy metal cations, giving reproducible results to within 1%. Kostyleva and Evstifeen (1972) have determined tetraethylthiuram disulphide in sewage and natural waters with a sensitivity of 0.05 μg ml^{-1}, using an oscillopolarographic method.

(iv) *Azo compounds*

Azo dyes are used commercially for dying fabrics and can enter rivers via factory effluents. Franklin Smyth and Hassanzadeh (1976) has used d.p.p. to study the decay of several commercially used azo dyes under aquarium conditions in concentrations likely to be found in effluents. Differential pulse polarograms of CI direct blue 84, CI direct red 24, and CI acid red 73 were recorded with time by diluting small samples of the aquarium water

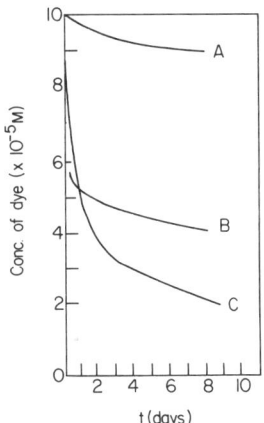

FIG. 12. Effect of time on the absorption of some diazo dyes by a mini aquarium (plants, 10 daphnae, sediment in 500 ml). Polarographic assay conducted in formate buffer pH 4.0 at 10^{-5}M. (A: CI Acid Red 73: B: CI Direct Blue 84: C: CI Direct Red 24.)

containing dye with formate buffer pH 4. This is illustrated in Fig. 12. The most toxic of the dyes studies was CI acid red 73 which apart from being reduced at the most positive potential (perhaps directly to aromatic amines) contained the α-azo-β-naphthol structure, well-known to be more toxic than the corresponding α-naphthol one.

A similar study was carried out by Hart et al. (1976) in which daphnia were present in aquarium water containing methyl red, methyl orange and p-nitrophenol and p-nitroaniline. Differential pulse polarograms were recorded directly in aquarium water and well-defined peaks were obtained. Polarographic reduction currents were measured over a period of several days and it was found that methyl orange was absorbed at a greater rate than either methyl red or the nitroaromatic compounds. An interesting observation was made while studying solutions containing daphnia in aquarium water containing p-nitrophenyl-azoresorcinol. Figure 13 shows

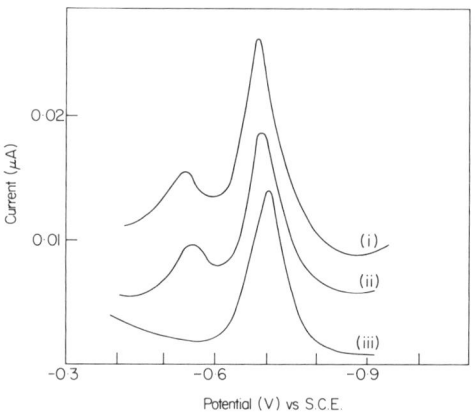

FIG. 13. Differential pulse polarograms of p-nitro-phenyl azo-resorcinol, 10^{-5}M in aquarium water. (i) Control solution, 1st day; (ii) control solution after 3 days; (iii) solution containing daphnia after 3 days.

the compound initially, and after 3 days, in the presence of daphnia and also a control. After 3 days the peak at -0.57 V had disappeared, and the peak at -0.70 V had shifted to a slightly more negative potential. This more negative peak coincided with that observed for the reduction of p-nitroaniline, which suggested that the daphnia had perhaps metabolized the azo compound to the corresponding amines. A subsequent experiment was carried out using identical analytical procedures but the daphnia used came from a different source. In this case, there was no difference in the polarograms of aquarium water containing daphnia initially and after three days. The reason for this was considered to be connected with the species, conditioning and age of daphnia used.

(v) *Total nitrogen*

Samples of effluent from municipal sources have been analysed by Moore (1969) and Albert et al. (1969), for total nitrogen, by microcoulometry. The method involved conversion of all organic and inorganic nitrogen compounds to ammonia by pyrolysis and catalytic processes. The ammonia was passed into a titration vessel containing 0.04% sodium sulphate and a bias of 100–112 mV was used to operate the vessel. A microcoulometer was incorporated which is generally operated at a constant sensitivity of 30 ohms. This allowed levels of nitrogen down to several parts per million to be determined. The basis of the nitrogen determination relied upon electrochemically generated hydrogen ions which reacted with the NH_3 produced by initial pyrolysis reduction processes. The precision and accuracy have been found to be $\pm 6\%$ and results obtained compared favourably to Nessler/Kjeldahl methods but were more rapid.

(vi) *Aliphatic acids*

A polarographic method for total monocarboxylic acids has been described by Shaova (1968). The method involved the steam distillation of a suitable volume of water (100–200 ml) to which had been added 10 ml 10% H_3PO_4. Distillation was carried out until 500 ml had been collected and to an aliquot of this was added tetramethyl ammonium iodide. The solution was transferred to the cell and after 20 minutes de-aeration with hydrogen, the polarogram was recorded from -1.00 V.

For the determination of formic, acetic, propionic and butyric acids, the polarographic wave due to formation of non-dissociated acids was used. The sensitivity of the method for total acids was 0.01 mg l^{-1} and an error of $\pm 10\%$ obtained in the range 10^{-4}–10^{-6} M l^{-4}.

III. CONCLUSIONS

In this chapter, we have endeavoured to show that electrochemical methods of analysis have a significant part to play in the determination of trace amounts of organic materials present in the aqueous environment. This is not to say that electroanalysis is necessarily the best method for all materials–any more than atomic absorption is the superior method for all heavy metal analyses in all matrices—but that each available method must be judged on its merits for the particular application.

Finally, the application of electroanalytical methods to the identification of a species/molecule by a "finger printing" technique is somewhat neglected and is drawn to the attention of the reader. Farwell et al. (1975) have used interrupted-sweep voltammetry to obtain fingerprint identification of many polychlorinated biphenyls and naphthalenes. This method, in common with others such as infra-red, could not be used for mixtures of compounds but required prior separation. However, applications of microprocessor techniques to this system would undoubtedly result in analysis of mixtures becoming possible. These results are discussed in greater detail in the chapter by Rowe and Smyth (Chapter 8).

REFERENCES

Adamovsky, M. (1966). *Vodni Hospodarstvi*, **16**, 102.
Afghan, B. K. and Goulden, P. D. (1971). *Envir. Sci. Tech.* **5**, 601.
Afghan, B. K., Goulden, P. D. and Ryan, J. F. (1972). *Anal. Chem.* **44**, 354.
Afghan, B. K., Kulkarni, A. V. and Ryan, J. F. (1975). *Anal. Chem.* **47**(3), 488.
Albert, D. K., Stoffer, R. L. and Oita, I. J. (1969). *Anal. Chem.* **41**(11), 1500.
Asplund, J. and Wanninen, E. (1971). *Anal. Letters*, **4**, 267.
Avedouard, Y., Suzanne, A., Vitorri, O. and Porthault, M. (1975). *Bull. Soc. Chem. Fr.* Nos 1–2, 130.
Ballinger, D. G. (1963). *J. Water Poll. Control Fed.* **35**(1), 116.
Bodyu, V. I. and Feldman, Ya. S. (1963). *Gidrolizn I. Lesokhim. Prom.* **16**(7), 11.
Bronstad, J. O. and Friestad, H. O. (1976). *Analyst*, **101**, 820.
Chambers, L. M. (1964). *Anal. Chem.* **36**(13), 2431.
Davidek, J. and Janicek, G. (1961). *Experientia*, **17**, 473.
Davidek, J. and Janicek, G. (1963). *Z. Anal. Chem.* **194**, 431.
Dolezil, M. and Kopanica, M. (1963). *Chemist-Analyst*, **52**, 76.
Dyatlovitskaya, F. G. and Berezouskii, F. I. (1962). *Gig. I. Sanit.* **27**, 50.
Dyatlovitskaya, F. G., Berezouskii, F. I. and Potemkina, S. K. (1963). *Gig. I. Sanit.* **28**, 38.
Eremin, Yu, G. and Kopylova, G. A. (1973). *Zavod. Lat.* **39**(9), 1065.
Erlebach, J., Lischke, P. and Kucera, Z. (1970). *Chem. Listy*, **64**(9), 984.
Fairing, J. D. and Short, F. R. (1956). *Anal. Chem.* **28**, 1827.
Farwell, S. O., Beland, F. A. and Geer, R. D. (1975). *Anal. Chem.* **47**, 895.
Filimonova, M. M. and Gorbunova, V. E. (1973). *Zh. Anal. Khim.* **28**(6), 1184.
Filimonova, M. M., Gorbunova, V. E. and Filimonova, B. H. (1975). *Zh. Anal. Khim.* **30**(2), 358.
Filov, V. A. (1959). *Gig. Truda i Prof. Posvy. Itogam Raboty*, 231–234.
Filov, V. A. (1960). *Gig. Truda i Prof. Zabol.* **4**(7), 54.
Fleszar, B. (1964). *Chem. Anal.* (Warsaw), **9**, 1075.
Franklin Smyth, W. and Hassanzadeh, H. (1976). *Z. Anal. Chem.* **280**, 299.
Fujinaga, T. and Okazaki, S. (1965). *Japan Analyst*, **14**(9), 832.
Ginsburg, V. I. and Frishman, T. A. (1959). *Zh. Anal. Khim.* **14**, 336.
Gorodetskii, Yu, S., Drondina, R. V. and Mamkov, A. A. (1976). *Izv. Akad. Nauk. Mold. SSR, SER, Fiz-Terh. Mat. Nauk*, **3**, 49.
Haberman, J. P. (1971). *Anal Chem.* **43**, 63.
Hart, J. P., Smyth, W. F. and Birch, B. J. (1976). *Proc. Anal. Div. Chem. Soc.* **13**(11), 336.

7. ELECTROANALYSIS OF TRACE FOREIGN ORGANIC MATERIALS

Henderson, C., Pickering, O. H. and Tarzwell, C. H. (1960). Taft. Sanit. Engineering Centre, Tech. Report W60-3:76.
Hoover, T. B. (1973). *In* U.S. National Technical Information Service, P.B. Report No. 222940/9. N.B.S., Washington.
Ishu, I. (1972). *Kogai Taisaku*, **8**, 847.
Ishu, I. (1973). *Mizu. Shori. Gijutsu*, **14**, 1019.
Ivanchenko, V. V. (1973). *Methody Opred. Pestits. Vode.* **1**, 145.
Jehring, H. (1966) *Tenside*, **3**(6), 187.
Kambara, T. and Hasabe, K. (1965) *Japan Analyst* **14**(6), 491.
Kosmatyi, E. S. and Bublik, L. I. (1970). *Khim. Prom. Ukr.* **4**, 43.
Kostyleva, V. S. and Evstifeen, M. M. (1972). *Gidrokhim. Issled.* 55.
Kozarac, Z., Zutic, V. and Vosovic, B. (1976). *Tenside*, **13**(5), 260.
Kuchumova, N. A., Bepuzo, C. H. and Mamomova (1963). *T. Vses. Nauchn. -Issled. Inst. Po Pererobotre I Ispol'z. Topliva.* **12**, 237.
Linhart, K. (1972). *Tenside*, **9**, 241.
Lyalikov, Yu. S. (1967). *Zh. Anal. Khim.* **22**, 1579.
Maienthal, E. and Taylor, J. K. (1973). "The Water Pollution Handbook", Vol. 4, pp. 1751–1880.
McKone, C. E., Byast, T. H. and Hance, R. J. (1972). *Analyst*, **97**, 653.
Melcer, I. and Melcerova, A. (1971). *Drev. Vyst.* **16**(1–2), 59.
Middleton, F. M. and Rosen, A. A. (1956). *Pub. Health Dept.* 71.
Moore, R. T. (1969). *Evniron, Sci. and Technol.* **3**(8), 741.
Novik, R. M. (1971). *Fiz-Khim. Methody. Anal.* 84.
Novik, R. M. and Pavlovskii, M. V. (1976). *Nov. Polyarogr., Tezisydokl. Vses. Soveshch., Polyarogr*, 6th, 188.
Ponomarev, Yu. P., Glazyrina, O. I., Kassai, T. V. (1974). *Fiz. Khim. Metody. Ochistki. Anal. Stochnykh Vod Prom. Predpr.* 91.
Renard, J. J., Kybes, G. and Bolker, H. I. (1975). *Anal. Chem.* **47**(8), 1347.
Ryckman, D. W. (1966). *J. Water Poll. Control Fed.* **38**, 458.
Shaova, L. G. (1968). *Ghidrokhim. Mater.* **46**, 171.
Shaw, D. A., David, D. J. and Tucker, H. C. (1976). *ISA Trans.* **15**(3), 227.
Shinozuka, N., Suzuki, H. and Hayano, S. (1972). *Japan Analyst*, **21**(4), 517.
Siegerman, H., O'Dom, G. and Flato, J. (1972). *Electrochem. Contrib. Environ. Prot.* 76.
Sobina, N. A., Kheifets, L. Ya., Bondarenko, L. M. and Glyadyaeva, L. A. (1976). *Zh. Anal. Khim.* **31**, 941.
Solonar, A. S. and Bael, N. G. (1967). *Gig. Sanit.* **32**, 66.
Stackelberg, M. V. and Shutz, M. (1943). *Kotl. Z.* **105**, 20.
Strafelda, F. (1970). *Chem. Tech. Prage., Anal. Chem.* **6**, 31.
Supin, G. S., Vaintraub, F. F. and Makarova, C. V. (1971). *Gig. Sanit.* **5**, 61.
Taylor, J. K., Zielinski, W. L. J. R., Maienthal, E. J., Durst, R. A. and Burke, R. W. (1972). U.S. National Technical Information Service, P.B. Report No. 21903513. N.B.S., Washington.
Vovik, Z. M. and Kozlova, I. V. (1972). *Prob. Anal. Khim.* **2**, 90.
Wernet, J. and Wahl, K. (1970). *Z. Anal. Chem.* **251**, 373.
Whitnack, G. C. (1961). *J. Electroanal. Chem.* **2**, 110.
Whitnack, G. C. (1975). *Anal. Chem.* **47**(4), 618.
Zaitsev, P. M. and Dichenskii, V. I. (1966). *Zavod. Lab.* **32**(7), 800.
Zhantalai, B. P. and Slisarenko, V. P. (1973a). *Zavod. Lab.* **39**(1), 6.
Zhantalai, B. P. and Slisarenko, V. P. (1973b). *Zavod. Lab.* **39**(2), 143.
Zhantalai, B. P. and Slisarenko, V. P. (1975). *Zavod. Lab.* **41**(3), 285.
Zhantalai, B. P., Sergeeva, A. S. and Kalichuk, L. R. (1976). *Zh. Khim. Abst.* **22I**, 298.
Zielinski, J. and Surkiewicz, S. (1974). *Przegl. Pap.* **30**(5), 182.

APPLICATIONS

In Agriculture

Chapter 8

ELECTROANALYSIS OF AGROCHEMICALS

R. R. ROWE

Murphy Chemical Ltd., Wheathampstead, Herts., England

and

MALCOLM R. SMYTH

*Department of Microbiology, Colorado State University,
Fort Collins, Colorado, U.S.A.**

I. INTRODUCTION

The application of voltammetric methods to the determination of agrochemicals in formulations, foodstuffs, crop residues, natural waters and body fluids has for long been recognized and practised in Eastern European countries. This is particularly true in Russia which is highly dependent on its agricultural industry and where polarography is a recognized analytical technique. Hence, the Russian analytical literature contains many references to the voltammetric determination of pesticides (taken to include insecticides, fungicides and rodenticides) and feed additives and the review by Supin (1972) should be consulted for references to this early work.

In Western countries, most interest in the application of voltammetry to agrochemistry has come from the laboratories of Martens and Nangniot in Gembloux, Belgium and from that of Gajan in Washington D.C., U.S.A. These workers published reviews on some of the early work carried out using polarography (Martens and Nangniot, 1963; Gajan, 1964a,b, 1965). More recently, Nangniot (1970) has published a comprehensive account of the applications of polarography in agriculture and biology.

With the rapid development of gas chromatographic methodology in the 1960's and the greater sensitivity offered by electron capture, flame photometric and alkali flame detectors over conventional d.c. polarography it is

*Present address: Institut für Chemie der Kernforschungsanlage, Jülich, West Germany.

not surprising that polarographic methods have been somewhat overlooked in this important area of analysis. This is reflected in the scant number of applications of the technique cited in such official manuscripts as the Pesticide Analytical Manual of the U.S. Food and Drug Administration (1970) and the Pesticide Manual of the British Crop Protection Council (1977).

It is the aim of this chapter, therefore, to review those organic agrochemicals that have been analysed by voltammetric methods in a variety of biological materials and formulations and to emphasize those areas where voltammetry could be put to greater use in situations of agricultural importance.

II. ORGANOCHLORINE COMPOUNDS

Organochlorine pesticides are usually determined using gas–liquid chromatography where the electron capture and coulometric detectors offer sensitivity and specificity not usually obtained with other methods. This is particularly true for residue determinations where samples containing 10 pg can be determined without too much difficulty.

The polarography of DDT, 1,1,1-trichloro-2,2-di(4-chlorophenyl) ethane was first reported by Keller et al. (1946) who showed that the alkyl chlorine atoms of DDT were reduced more easily than its aromatic ones in a supporting electrolyte of 0.1M TMAB/70% ethanol. Feher and Monien (1964) determined DDT in commercial formulations in a TMAB/acetone supporting electrolyte. Only one reduction wave was observed for both p,p'-DDT and o,p'-DDT at -1.07 V with a limit of detection of 5 μg ml^{-1}. Brezina and Romazanovich (1961) have described a method for determining DDT in the presence of DDD (dichlorodiphenyl dichloroethane). In a supporting electrolyte of 0.1M LiCl/70% ethanol, DDT gave one wave at -1.4 V while DDD was not reduced; in 0.05M TEAB/70% ethanol, however, DDT gave rise to two waves at -0.85 V and -2.3 V, while DDD gave one wave at -2.3 V.

Kemula and Kreminska (1960) have combined column chromatography and polarography to quantitatively separate the p,p'- and o,p'-isomers of DDT. They used a column packed with powdered rubber swollen with heptane. The DDT isomers were eluted using a solution of 0.05M TEAI/87% DMF. The eluate was collected from the column and the isomers determined at a reduction potential of -1.2 V. Using this system, p,p'-DDT was eluted before o,p'-DDT. Davidek and Janicek (1961) have tried to determine DDT in biological material using a direct method, but found that interferences in the matrix caused a deformation of the DDT peak. To overcome this problem they developed an indirect technique. DDT was nitrated at 90–95°C for 10 minutes using a mixture of concentrated sulphuric acid and fuming nitric acid. On cooling, the solution was diluted with water, followed by methanol to give a 50:50 water–methanol mixture. In this strongly acidic medium the tetra-nitro derivative was reduced in a single wave with a half wave potential of -0.13 V (vs s.c.e.). The limit of sensitivity of the method was quoted to be 0.4 μg ml^{-1}.

Gajan and Link (1964) have investigated DDT and various analogues of DDT using oscillopolarography. They found that the peak potential of DDT shifted with concentration from -0.70 V for a 20 μg ml^{-1} sample to -0.90 V for a 500 μg ml^{-1} sample, when using a silver wire reference electrode. The shift was from -0.58 V to -0.76 V for similar sample sizes when using a mercury pool electrode. For analytical purposes, a standard curve was prepared by diluting aliquots of stock DDT solution in acetone to 2.0 ml, adding 3.0 ml of ethanol and 5.0 ml of a 0.2M TMAB solution to give final concentrations between 5 and 50 μg ml^{-1}. Formulations and technical grade DDT were analysed by preparing solutions in acetone, adding ethanol and electrolyte as previously described, running the polarographic curve and comparing the peak height with the calibration curve. For use as a residue method, the DDT was eluted from a Florisil column with 100 ml of 6% eluent, the eluent concentrated to about 3 ml in a Kuderna–Danish concentrator and then evaporated to incipient dryness under a stream of nitrogen (Johnson, 1962). The residue was then dissolved in 3.0 ml of acetone and 3.0 ml of 0.2M TMAB, and the height of the i–E curve compared against standard samples run under the same conditions. Recoveries ranged from 75% from green beans to 90% from sprouts in the 1 to 5 μg ml^{-1} range. Of the compounds studied, only those with a trichloroethane group were reduced. o,p'-DDT, p,p'-DDT, methoxychlor and kelthane gave well defined waves at -0.78, -0.73, -0.78 and -0.74 V respectively (vs a silver wire electrode) but the method was unable to differentiate between these compounds in mixtures.

The γ-isomer of hexachlorocyclohexane ((I); also known as benzenehexachloride, γ-BHC and "Lindane") is employed as an insecticide and is reduced according to:

$$C_6H_6Cl_6 + 6e^- \rightarrow C_6H_6 + 6\,Cl^-$$

in solutions of pH >6. The resulting wave/peak has been used in several formulation assays (Ingram and Southern, 1948; Dragt, 1948; Streuli and Cooke, 1954; Fukami and Nakajima, 1954; Richardson and Miller, 1960; Supin and Budnikov, 1973) and this method found to compare favorably with bioassay.

γ-BHC has also been determined in aerosols (Koneva and Kutsenogii, 1972) and in soil (Kosmatyi and Bublik, 1974). The latter method involved extraction of the soil with hexane, filtering through Na_2SO_4, concentration of the extract and separation on silica gel. The method could determine γ-BHC in the presence of heptachlor (II) and heptachlor epoxide and the polarographic analysis was carried out in 0.05M TMAB containing either 10% (for γ-BHC or heptachlor) or 40% (for heptachlor epoxide) ethanol.

(I) — hexachlorocyclohexane structure

(II) — heptachlor structure

In a recent paper, Supin and Budnikov (1973) have investigated the adsorption properties of γ-BHC at a slowly dropping mercury electrode. By careful choice of accumulation time (7 sec) and applied potential (-0.4 V vs s.c.e.), they were able to concentrate γ-BHC at the Hg surface followed by measurement of the "stripping" peak obtained using fast scan pulsed oscillographic polarography. This method was found to be 60× more sensitive than the d.c. procedure and could determine down to 0.06 μg ml^{-1} γ-BHC in pure solution. Supin and Budnikov (1973) have also applied this technique to the determination of mucochloric acid (III: used as a starting material in the synthesis of certain pesticides) and quoted a detection limit of 0.2 μg ml^{-1}.

(III) — mucochloric acid structure: HO–C(=O)–C(Cl)=C(Cl)–C(=O)H

Berck (1962) has developed an indirect method for the determination of the grain fumigant chlorpicrin (IV). This involved liberation of Cl$^-$ ions by heating the substance in 60% propanol/3.5% monoethanolamine for 4 hours at 60°C. The mixture (8.5 ml) was then cooled, 1 ml 8N HNO$_3$ and 0.5 ml 1% gelatin added and the Cl$^-$ released determined by anodic polarography at $+0.22$ V vs s.c.e. He was also able to determine chlorpicrin in the presence of other fumigants, e.g. methyl bromide, ethylene dibromide, acrylonitrile and carbon tetrachloride and the method was successfully applied to the determination of these compounds in air samples taken from fumigated grain, flour and soil.

The polarographic behaviour of endosulfan (V; thiodan) has been studied by Colas et al. (1964). They used a silica gel column to separate a mixture of the two isomers of endosulfan followed by polarography in an aqueous acetone mixture (50/50) containing LiCl as the supporting electrolyte. This compound has also been isolated in the stomach contents of poisoned chickens using polarographic and gas–liquid chromatographic techniques (Panetsos and Kilikidis, 1973). The presence of parathion was also detected in this investigation.

Recently, Farwell et al. (1973; 1975) have reported on the use of interrupted-sweep voltammetric analysis for the identification of polychlorinated insecticides and other polychlorinated aromatics. The apparatus consisted of a three-electrode potentiostatically controlled circuit with a logic controlled interrupted linear voltage sweep mechanism. Using DMSO/0.1M TEAB as the supporting electrolyte, voltammograms for many polychlorinated aromatics have been obtained. Table I quotes the reduction potentials of several

TABLE I. *Reduction potentials of some organochlorine insecticides using the method of Farwell et al. (1973, 1975)*

Compound	Interrupt Potential, E_{2d}* (V vs s.c.e.)
Heptachlor	−1.656, −2.087, −2.319
Dieldrin	−1.675, −2.038, −2.233, −2.450
DDE	−1.814, −1.935, −2.177, −2.506
Chlordane	−1.753, −2.106, −2.333

* The analogue second derivative of the current was used to control the voltages at which the scans were interrupted.

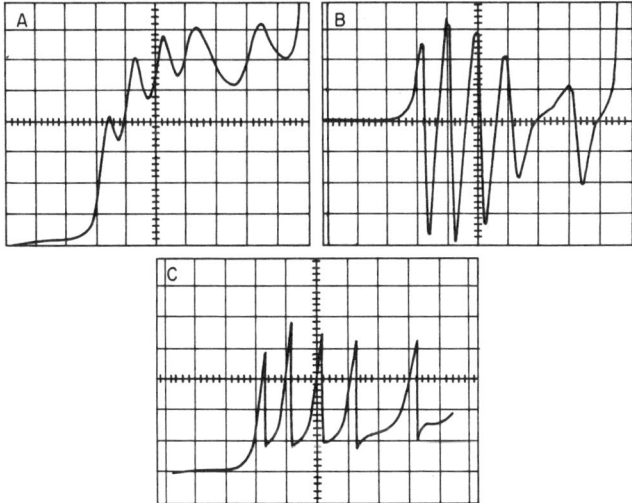

FIG. 1. Reduction voltammograms of 1,2,3,4-tetrachloronaphthalene (voltage range −0.7 to −2.6 V (vs s.c.e.). A. Normal scan at 0.5 V/division and 1000 μA full scale. B. Second derivative scan at 1.0 V/division and 1000 μA full scale. C. Interrupted sweep fingerprint at 0.5 V/division and 1000 μA full scale. (Farwell et al., 1975).

important agrochemicals. This method suffers, however, from the fact that one requires at least 9 µg of relatively pure compound in order to obtain a positive identification. It is therefore of little use in the analysis of pesticide residues in the environment but could find more application in the identification of isomeric products formed during the synthesis of polychlorinated agrochemicals. Reduction voltammograms of 1,2,3,4-tetrachloronaphthalene are shown in Fig. 1. The peak with the most negative reduction potential corresponds to reduction of naphthalene and the remaining reduction peaks represent the stepwise removal of the chlorine atoms.

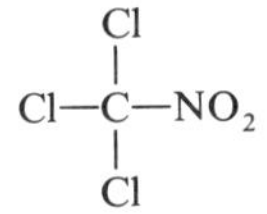

$$Cl-\underset{\underset{Cl}{|}}{\overset{\overset{Cl}{|}}{C}}-NO_2$$

(IV)

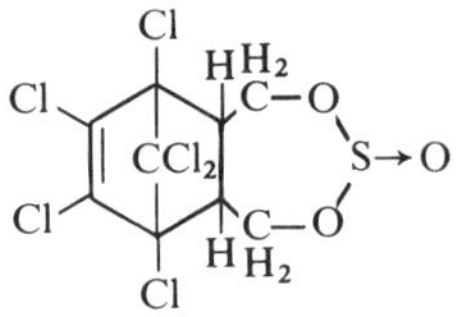

(V)

III. ORGANOPHOSPHORUS COMPOUNDS

With the large scale development of organophosphorus insecticides in the early 1950's, methods of analysis were based either on colorimetric procedures or on the determination of total phosphorus. These methods are inherently non-specific, however, and were slowly superseded by the development of g.l.c. procedures employing flame photometric or thermionic detection. Since many organophosphorus insecticides are unstable under the conditions required for g.l.c. procedures, increasing importance has been placed on the development of "cold" methods of analysis (e.g., t.l.c., h.p.l.c. with u.v. and fluoresence detectors). The polarographic behaviour of organophosphorus insecticides was not investigated in any depth until the early 1960s (Sohr, 1962; Nangniot, 1964) although several isolated methods had appeared before then. Since several organophosphorus pesticides also contain the $-NO_2$ group (e.g. parathion), the discussion here will be limited to methods which have been specifically developed around the phosphorus moiety. The polarographic behaviour of compounds such as parathion will be dealt with in the next section.

Jura (1955) has determined malathion (VI) and its main impurity deithyl fumarate (VII) in technical material. The method involved first measuring the wave due to diethyl fumarate in 0.3N HCl/30% C_2H_5OH (a medium in which malathion is polarographically inactive.) On treatment with alkali, both diethyl fumarate and malathion are quantitatively converted to fumaric acid which can subsequently be determined following acidification. The concentration of malathion can thus be determined by difference. The $E_{\frac{1}{2}}$ values of diethyl fumarate and fumaric acid were quoted to be -0.67 and -0.77 V (vs s.c.e.) in 0.3N HCl/30% C_2H_5OH. A modified version of this procedure has now been accepted as an official method of the Association of Official Analytical Chemists (AOAC) (Gajan, 1969: Pesticide Analytical Manual, 1970). This modified procedure has been used to determine malathion in crop residues with a limit of sensitivity of 0.3 µg ml^{-1}, based on 1 g crop in 1.0 ml cell solution.

$$\begin{array}{c} CH_3O \\ \diagdown \\ P \\ \diagup \\ CH_3O \end{array} \begin{array}{c} S \\ \diagup \\ \\ \diagdown \\ S-CH-CO_2C_2H_5 \\ | \\ CH_2-CO_2C_2H_5 \end{array}$$

(VI)

$$\begin{array}{c} O \\ \diagdown \\ C-O-C_2H_5 \\ | \\ CH \\ \| \\ CH \\ | \\ C-O-C_2H_5 \\ \diagup \\ O \end{array}$$

(VII)

In particular, Nangniot (1964; 1970) has studied the polarographic behaviour of a wide range of organophosphorus pesticides. Although phosphoric acid esters cannot be reduced at the d.m.e., compounds containing either the

$$\begin{array}{c} \diagdown \\ P- \\ \diagup \| \\ S \end{array} \quad \text{or} \quad \begin{array}{c} \diagdown \\ P-S- \\ \diagup \| \end{array}$$

moieties in their molecular structures can produce sharp adsorption peaks using a fast scan linear sweep voltammetric technique. The peaks obtained for 10 important organophosphorus compounds using a scan rate of 250 mV s^{-1} are shown in Fig. 2. This method could determine concentrations of the organophosphorus compounds at the µM level.

Sohr (1962), on the other hand, used oscillopolarography to monitor the adsorption properties of a series of trialkylphosphates and trialkylphosphites, triphenylphosphate, trimethylthiophosphate and benzyldimethylthiophosphate at the d.m.e. This investigation was limited, however, to the 10^{-2}–10^{-4}M level and Nangniot (1964) has suggested that linear sweep voltammetry could provide greater sensitivity for their determination in environmental situations. Apart from giving rise to cathodic incisions on oscillopolarography, thiophosphates also produce anodic waves at the d.m.e. This has been demonstrated by Supin et al. (1976) for some dialkyl (or alkylamine) thiophosphates and dialkyldithiophosphates. Supin and Budnikov (1973) have used a c.s.v. method of analysis for the determination of the insecticides phthalophos and benzophosphate in apples. This procedure involved

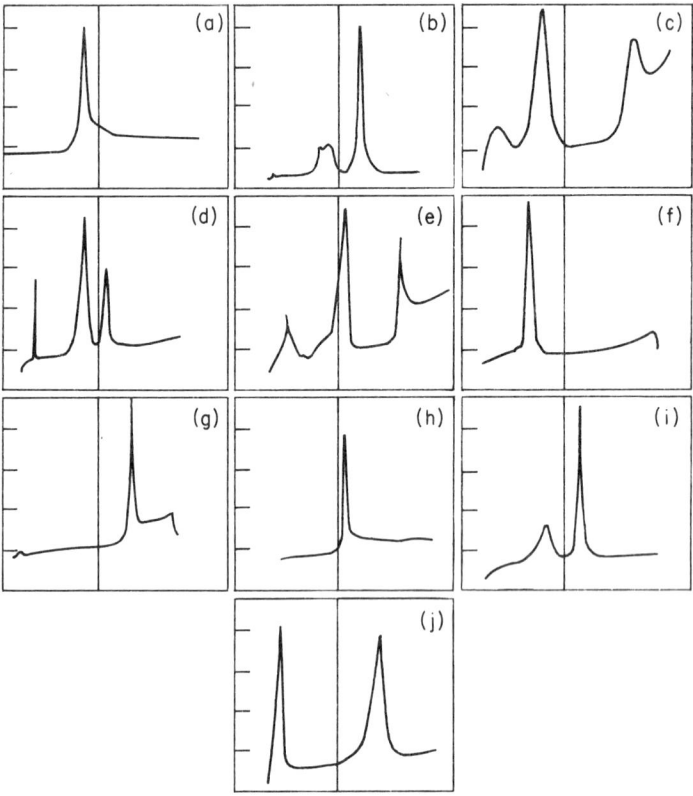

FIG. 2. Linear sweep voltammograms of several important organophosporus pesticides (Nangniot, 1964) (current ranges given in parentheses). (a) 3.30×10^{-4}M chlorthion in 0.2M KCl (0.5 µA); (b) 2.30×10^{-4}M prothoate in 0.2M NaOH (0.2 µA); (c) 1.29×10^{-4}M methyl azinphos in 0.2M NaOH (0.02 µA); (d) 2.82×10^{-4}M demeton-O-sulphoxide in 0.2M NaOH (0.03 µA); (e) 2.21×10^{-4}M malathion in 0.2M KCl (0.02 µA); (f) 1.22×10^{-4}M menazon in 0.2M NaOH (0.08 µA); (g) 1.70×10^{-4}M phenkapton in 0.2M KCl (0.03 µA); (h) 2.16×10^{-4}M parathion in 0.2M NaOH (0.5 µA); (i) 1.74×10^{-4}M dimethoate in 0.2M NaOH (0.2 µA); (j) 1.93×10^{-4}M demeton in 0.2M NaOH (0.02 µA).

extracting 200 g of apple with acetone, evaporation and treatment of the final extract with alkali. Under these conditions, the insecticides are hydrolysed to dimethyl- and diethylthiophosphate respectively. These compounds can then form insoluble salts with Hg and can be determined by c.s.v. using a scan rate of 1 V s^{-1}. The method could determine down to 0.2 µg kg^{-1} of the two insecticides in apple. Cox and Cheng (1974) have also developed a c.s.v. method of analysis for the determination of inorganic phosphate. This procedure involved the plating of a sparingly soluble Fe(II)-phosphate salt on a glassy carbon electrode. The method could determine down to 0.04 mg ml^{-1} under optimum conditions with no interference from Cl$^-$ or SO$_4^{2-}$ anions.

Gajan (1962) has applied polarography to the determination of systox-thiol, systox-thiono, di-syston and thimet and of their oxidation products. Analytically usable waves were obtained in a 2% aqueous solution of TEAOH. Residue determinations were carried out following separation by paper chromatography and recoveries in the region of 83.5% (for thimet) to 107.0% (for systox-thiol) were reported. Graham–Bryce (1970) has further investigated the polarographic behaviour of demeton (a 65:35 mixture of systox-thiol and systox-thiono), disulphoton and phorate (thimet) and has shown that the method outlined by Gajan (1962) is greatly dependent on the time that elapses between preparation of the sample and running the polarogram. This is because these compounds hydrolyse in alkaline media according to:

$$(RO)_2P(=X)-SR' \xrightarrow{OH^-} (RO)_2P(=X)-O^- + R'SH$$
(where X = O or S)

and it is the sulphydryl-containing product, i.e. R^{1}SH, that gives rise to adsorption waves at the d.m.e. This hypothesis was tested by relating the disappearance of the original insecticide (measured by g.l.c.) to the appearance of polarographic activity and by investigating the behaviour of hydrolysis products synthesized by an independent route. Graham-Bryce further emphasized the need to carefully standardize the conditions for the analysis of these compounds.

Gajan (1969) has reported a method for the determination of diazinon in apples and lettuce. Diazinon gives rise to a peak on linear sweep voltammetry at -0.90 V (vs Ag/AgCl) in a TMAB/CH$_3$COOH supporting electrolyte. The overall recovery of diazinon was quoted to be 104.2%. Gajan also recommended that polarographic methods for the analysis of diazinon, parathion, methylparathion and malathion be adopted as official methods for the AOAC. Diazinon has also been determined in technical products down to 2 μg ml^{-1} (Kosmatyi and Kavetskii 1975). Bates (1962) has determined guthion (methyl azinphos) in apples, pears, cucumbers and tomatoes using a supporting electrolyte of 0.05M KCl/0.1N CH$_3$COOH/60% acetone, pH 3.8. In this medium, guthion gave rise to a wave at -0.83 V (vs s.c.e.) due to the reduction of the carbonyl group in the molecule. The limit of detection was quoted to be 0.1 μg ml^{-1}.

Giang and Caswell (1957) have developed a method for the determination of dipterex (O,O-dimethyl-2,2,2-trichloro-1-hydroxyethylphosphonate) in various formulations. In a supporting electrolyte of 0.02N KCl (containing 0.002% gelatin as maximum suppressor) dipterex gave rise to a wave at

−0.68 V (vs s.c.e.). The two main impurities in the formulation, namely chloral hydrate and DDVP (O,O-dimethyl-2,2-dichlorovinylphosphate) were not found to interfere since they gave rise to waves at much more negative potentials (−1.61 and −1.53 V respectively).

Recently Davidek and Seifert (1971a) have developed an enzymopolarographic method for the determination of the organophosphorus pesticide intration. The method was based on the inhibition of anticholinesterase activity by the organophosphorus moiety (a reaction which parallels the *in vivo* biological activity of these compounds). Unreacted enzyme is then incubated with β-naphthyl acetate and the β-naphthol liberated measured by polarography following nitrosation. The method was applied to the analysis of intration in lettuce, cabbage, cherries and tomatoes. Naturally occurring enzymes which would be capable of hydrolysing β-naphthyl acetate were removed by precipitation with C_2H_5OH and subsequent centrifugation. No interference was observed in the presence of carotenes, xanthophyll, chlorophyll or anthocyanidines and the method was found to be relatively simple and rapid to perform. These authors have also compared enzymatic and polarographic methods for the determination of metation following separation of this organophosphorus pesticide and its oxygen and S-methyl analogues on silica gel layers (Davidek and Siefert, 1971b). The enzymatic method, which involved visual detection, was quoted to have an error of ±20% whereas the polarographic method was accurate to within ±6%.

Polarographic methods have also been described for the determination of organophosphorus pesticides based on the derivatization of other functional groups in the molecule. For example, dichlorfos (VIII) and trichlorfon (IX) have been determined following the liberation of glyoxal by alkaline hydrolysis and subsequent condensation with o-phenylenediamine (Seifert and Davidek, 1974). A similar method has also been developed for the determination of dichlorfos in milk following separation on silica gel (Davidek et al., 1976). In this case, however, the glyoxal liberated following alkaline hydrolysis was subsequently determined as quinoxaline. The recovery was quoted to be 85 to 88% at the 0.38–1.15 µg ml^{-1} level and 98% at the 111 µg ml^{-1} level.

IV. NITRO COMPOUNDS

Perhaps most applications of polarography for the determination of nitro-containing agrochemicals have come from investigations of nitrophenyl esters, e.g. parathion, fenitrothion, etc. The polarographic behaviour of parathion (X) was first reported by Bowen and Edwards (1950). These authors determined parathion in various formulations using a 0.5N KCl/0.1N acetic acid/50% acetone supporting electrolyte. In this medium parathion was found to have an $E_{\frac{1}{2}}$ value of -0.39 V (vs s.c.e.) whereas its two main metabolites, paraoxon (XI) and p-nitrophenol (also sources of contamination in formulations), were found to reduce at -0.47 and -0.69 V respectively. The method could determine parathion down to 20 µg mg^{-1} with an error of $\pm 1\%$. Various other workers have also investigated the polarographic behaviour of parathion, methyl parathion and EPN (XII) and reported small differences in the $E_{\frac{1}{2}}$ values of these compounds across the pH range (Shapiro, 1961; Yamano et al., 1962; Hasegawa, 1962). These differences cannot, however, be used for the quantitative determination of these compounds in mixtures.

$$\underset{(X)}{\begin{array}{c}C_2H_5O\\C_2H_5O\end{array}\!\!\!\!\overset{S}{\underset{\|}{P}}\!\!-O-\!\!\bigcirc\!\!-NO_2} \qquad \underset{(XI)}{\begin{array}{c}C_2H_5O\\C_2H_5O\end{array}\!\!\!\!\overset{O}{\underset{\|}{P}}\!\!-O-\!\!\bigcirc\!\!-NO_2} \qquad \underset{(XII)}{\begin{array}{c}C_2H_5O\\Ph\end{array}\!\!\!\!\overset{S}{\underset{\|}{P}}\!\!-O-\!\!\bigcirc\!\!-NO_2}$$

Smyth and Osteryoung (1978a) have made a detailed pulse polarographic study of parathion, methyl parathion, paraoxon, p-nitrophenol, EPN and PCNB. These authors reported little difference in the E_p values of the structurally related nitrophenyl esters over the pH range 2 to 12. They were also unable to determine parathion in the presence of paraoxon in the medium used by Bowen and Edwards (1950). They have studied the adsorption characteristics of parathion and paraoxon in TBAB solutions but were unable to use the differences observed in their behaviour for quantitative purposes. Structurally related nitrophenyl esters should therefore be separated by a chromatographic procedure prior to analysis. This is exemplified by the work of Koen and Huber (1970) who determined parathion and methyl parathion in lettuce following a liquid chromatographic separation.

The determination of p-nitrophenol in the presence of parathion has been reported by several authors (Bowen and Edwards, 1950; Ichim, 1972; Zietek, 1975, 1976; Smyth and Osteryoung, 1978a). In particular, d.p.p. can determine these compounds in mixtures in the low ng ml^{-1} range (Fig. 3). Zietek (1976) has used polarography to determine these compounds directly in blood but this method suffers from much higher limits of detection and

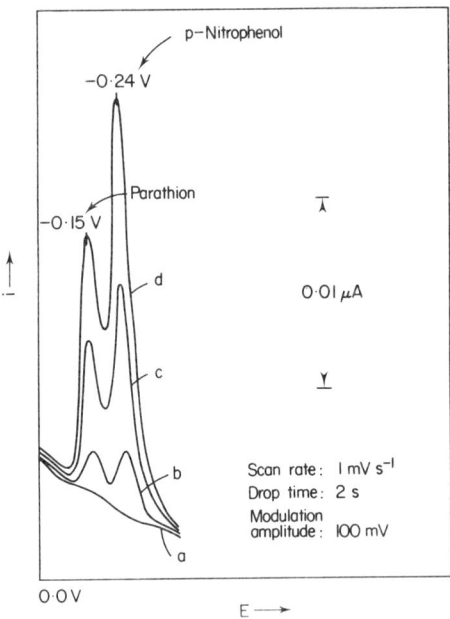

FIG. 3. D.p.p. determination of parathion and *p*-nitrophenol in BR buffer pH 3: (a) BR buffer pH 3; (b) 7.5 ng ml^{-1} each parathion and *p*-nitrophenol; (c) 22.5 ng ml^{-1}; (d) 37.5 ng ml^{-1}. (Smyth and Osteryoung, 1978a).

from the intrinsic problems associated with direct analysis in body fluids (Franklin Smyth and Smyth, Chapter 1).

Several authors have dealt with the determination of fenitrothion (XIII). Gruca and Mosinska (1970) applied polarography to the determination of fenitrothion in the presence of its major degradation products, 4-nitro-*m*-cresol and 4-nitroso-*m*-cresol in formulations. In a supporting electrolyte containing 0.04M LiCl, 0.04M triethanolamine and 0.5M acetic acid in 40% methanol, pH 5.4, fenitrothion had an $E_{\frac{1}{2}}$ value of -0.55 V. Interfering substances such as 4-nitro-*m*-cresol and the diphenyl derivative of fenitrothion had $E_{\frac{1}{2}}$ values of -0.67 and -0.65 V respectively. Other interfering substances were removed by t.l.c. Davidek and Seifert (1971b) have described an enzymo-polarographic method for the determination of fenitrothion and its *O*-analogue. The method involved separation on silica gel (3:1 light petroleum–acetone), incubation with an enzyme solution obtained from a beef liver homogenate and then spraying with 2-naphthyl acetate and Fast Blue B. Following elution of the spots with acetone, the compounds were determinated in borate buffer pH 9.2/40% acetone (containing 0.01% gelatin). The method was used to determine between 1–4 μg ml^{-1} fenitrothion.

$$\begin{array}{c} CH_3O \\ \diagdown \\ CH_3O \end{array} \! \! P \! - \! O \! - \! \underset{CH_3}{\underset{|}{\bigcirc}} \! - \! NO_2 \qquad \underset{N}{\bigcirc} \! - \! CH_2 \! - \! NH \! - \! CO \! - \! NH \! - \! \bigcirc \! - \! NO_2$$

(XIII)　　　　　　　　　　　　　　(XIV)

Recently Whittaker and Osteryoung (1976) have investigated the polarographic behaviour of "Vacor" rodenticide (XIV) and its main metabolite p-nitroaniline. These compounds were found to reduce at similar potentials in a range of supporting electrolytes and must therefore be separated by t.l.c. prior to analysis. Osteryoung et al. (1976) have used this procedure to determine p-nitroaniline in the liver of an accidental poisoning case. The method involved homogenization of 5 g tissue with 25 ml water, salting out 12 ml aliquots with 4 g NaCl and extracting with 2 ml THF. After centrifugation and drying the extract over Na_2SO_4, 5 or 10 µl of the THF layer was spotted on the t.l.c. plate and elution carried out with acetone–benzene (65:35). In this system, p-nitroaniline had an R_f of 0.75. Quantitation of the spot was achieved using d.p.p. while confirmatory evidence of the presence of p-nitroaniline was obtained using g.l.c. on a 1.5% OV-17/1.5% OV-210 column at 200°C.

Several nitro-containing compounds, e.g. furazolidone, dimetridazole and ronidazole are also employed in agrochemistry as feed additives. Hocquellet and Pevendrant (1972) have used superimposed a.c. polarography to determine five such compounds in animal feed. The method involved extraction with DMF/CCl_4 (1:3) at 70°C for 30–120 min. The clear extract (25 ml) was then shaken with 45 ml 0.2N HCl saturated with $Na_2B_4O_7$. Polarography was then carried out between -0.3 to -1.7 V (vs s.c.e.) and the method could simultaneously determine furazolidone in the presence of nitrofurazone (based on measurement of the peaks due to the nitro- and hydrazone reductions—the polarographic behaviour of these compounds is more fully dealt with by Browne in Chapter 4). The method was quoted to have an error of $\pm 5\%$ at 50 to 250 mg kg^{-1} level. More recently Slamnik (1974) has used d.c. polarography to determine dimetridazole and furazolidone in feed premixes. Several authors have also used polarographic methods to determine these feed additives in animal tissue. Parnell (1973), for instance, used a polarographic method to determine dimetridazole in pig tissue down to 0.1 µg g^{-1} in a 5 g sample. Cala et al. (1976) have developed a method for the determination of ronidazole (used in the treatment of dysentery) in pig tissue. They reported a limit of sensitivity of 2 ng g^{-1} using d.p.p. and extraction with DMF/CCl_4 (1:3) at 70°C for 30–120 min. The clear extract tissue. Michielli and Downing (1974) have also used d.p.p. to determine nicarbazin (equimolar complex of 4,4′-dinitrocarbanilide and 2-OH-4,6-dimethylpyrimidine) in chicken tissue down to 0.2 µg ml^{-1}.

TABLE II. *Half wave potentials of some dinitrophenol pesticides in various supporting electrolytes*

Compound	Supporting electrolytes			
	0.1N HNO$_3$[a]	0.1N acetic acid/[a] sodium acetate	0.1N NH$_4$OH/[a] NH$_4$Cl	B.R. buffer pH[b] 9.0
(i) dinitrophenols				
2,4-DNP	−0.06, −0.57, −0.96	−0.50, −0.88	−0.78, −1.04	
DNOC	−0.06, −0.42, −0.92	−0.45, −0.84	−0.74, −1.02	−0.50, −0.76
DNBP	−0.08, −0.40	−0.32, −0.76	−0.75, −1.04	−0.51, −0.75
Dinocap	−0.12, −0.31	−0.39, −0.67	−0.52, −0.78, −1.02	
Dinoterb				−0.50, −0.77
Mecinoterb				−0.625, −0.78
Dinobuton				−0.44, −0.61
(ii) dinitrophenol esters				
DNBP acetate	−0.08, −0.34	−0.41, −0.74	−0.69, −0.98	
Dinoterb acetate				−0.30, −0.85
Mecinoterb acetate				−0.49, −0.705, −0.85

[a] After Martens et al. (1961): $E_{\frac{1}{2}}$ values in d.c. polarography (vs Ag/AgCl). [b] After Rowe and Franklin Smyth (1978): E_p values in d.p.p. (vs s.c.e.).

TABLE III. *Half-wave potentials of some dinitroaniline herbicides (Southwick et al., 1976)*

Compound	$E_{\frac{1}{2}}$ values (V vs s.c.e.) at various pH's			
	1.5	5.1	7.4	9.2
Trifluralin	−0.19	−0.43	−0.54, −0.70	−0.64, −0.81
Benefin	−0.19	−0.43	−0.54, −0.70	−0.64, −0.81
Isopropalin	−0.17	−0.36	−0.56, −0.72	−0.65, −0.81
Dinitramine	−0.23	−0.51	−0.72	−0.81, −1.01
Nitralin	−0.16	−0.37, −0.53	−0.54, −0.71	−0.65, −0.79
Oryzalin	−0.16	−0.37, −0.53	−0.54, −0.68	−0.68, −0.81

The polarographic determination of dinitro compounds commonly used as agrochemicals has been reported by several authors. Martens *et al.* (1961) and Rowe and Franklin Smyth (1979) have investigated the polarographic behaviour of several dinitrophenols and their esters in a variety of supporting electrolytes (Table II). Southwick *et al.* (1976), on the other hand, have studied the polarographic behaviour of several dinitroaniline herbicides and attempted to correlate their mechanism of reduction to their relative toxicity in the environment. These compounds give rise to a single 8e$^-$ wave in acid solution (pH < 3) which then splits into two 4e$^-$ waves on increasing the pH (Table III). This is due to:

$$\underset{\underset{CF_3}{\underset{|}{}}}{O_2N\text{-}Ar\text{-}N(Pr)_2\text{-}NO_2} \xrightarrow[-H_2O]{4e^-, 4H^+} \underset{\underset{CF_3}{\underset{|}{}}}{HOHN\text{-}Ar\text{-}N(Pr)_2\text{-}NO_2} \xrightarrow[-H_2O]{4e^-, 4H^+} \underset{\underset{CF_3}{\underset{|}{}}}{HOHN\text{-}Ar\text{-}N(Pr)_2\text{-}NHOH}$$

It is surprising, therefore, taking into account the high sensitivity offered by polarographic methods for the analysis of these compounds, that so few working methods have appeared in the literature. Cimbura and Gupta (1964) have determined some dinitrophenol pesticides in biological materials. The method involved extraction of acidified samples with CCl_4 followed by a back extraction into 0.1N NaOH. The aqueous phase was then taken, the pH adjusted to 7.8, the solution centrifuged and polarography carried out on the clear supernatant between −0.2 to −1.2 V. Recoveries of dinitrophenol pesticides from blood were quoted to be of the order of 75%. Several authors have dealt with the determination of karathane (dinocap; dinitromethylheptylphenyl crotonate) in crop residues. Kikuchi and Morley (1964) determined karathane residues on peaches, apples, cucumbers, peas and apples after extraction with hexane and determination in McIlvaine buffer pH

3/50% ethanol. The recovery of karathane using this procedure was quoted to be about 85% and d.c. polarography could determine down to 0.5 µg ml^{-1} under optimum conditions. Kosmatyi et al. (1968), on the other hand, used a CCl_4 extraction followed by a t.l.c. separation to determine karathane in plant residues. The method could determine down to 0.5 µg kg^{-1}. Copin (1968a, b) has also used polarography to determine karathane residues on green beans. The method involved extraction with 0.2M acetic acid/sodium acetate buffer followed by recording the i–E curve directly in the extract. The limit of detection for this method was quoted to be 0.6 µg ml^{-1}.

Dinitro-o-cresol (DNOC) has been determined directly in blood and blood coagula at concentrations between 50 and 250 µg ml^{-1} (Mikolajek, 1969). Mosinska and Kotarski (1972) have used a chromatopolarographic method to determine 2-isopropyl-4,6-dinitrophenol in the presence of its main impurity 2,4-dinitrophenol in formulations. The method involved separation by t.l.c. (97:3 $C_6H_6:CH_3OH$) followed by polarography (in the presence of the t.l.c. adsorbent) in acetate buffer pH 5.4/50% ethanol. Sestakova and Skarka (1976) have used pulse polarography to determine nitrovin (a growth stimulator) in biofactor supplements of premixes and in feed between 1 to 80 mg kg^{-1}. The method involved a preliminary extraction with DMF, and furazolidone was found to interfere with the determination. Recently Filimonova and Gorbunova (1975) have described a method for the determination of N,N-diethyl-2,6-dinitro-4-trifluoromethylaniline in water and vegetables. The method involved extracting 1 litre water or 100 g vegetable tissue with hexane, purifying the extract on alumina (for tomatoes) or a cation exchange resin (for cabbage) and determination in 0.05N TEAOH/60% MeOH. The method was used to determine this compound in these matrices between 0.01 to 0.1 µg ml^{-1}.

V. SULPHUR-CONTAINING COMPOUNDS

Although many pesticides contain a sulphur atom within their molecular structure, polarographic methods of analysis for these compounds are often based on the reduction of another functional group in the molecule. In the case of parathion (X) for instance, the reduction of the $-NO_2$ group is usually employed for analytical purposes, although it has been shown that the S atom in this molecule plays an important role in the electrode reaction (Nangniot, 1964; Smyth and Osteryoung, 1978a). This is due to the adsorptive properties of S—Hg species which have been used for the determination of parathion in the presence of its O-analogue, paraoxon (Nangniot, 1964).

Compounds containing sulphur in the —SH form are particularly amenable to c.s.v. analysis due to their ability to form partially insoluble complexes with mercury (Franklin Smyth and Smyth, Chapter 1; Davidson, Chapter 5). The application of c.s.v. to the determination of some thiourea-containing agrochemicals has been investigated by Smyth and Osteryoung (1977). Thiourea (TU, XV) itself has been used as a fungicide, as an accelerator of sprouting in dormant tubers and to decrease the content of nitrifying bacteria in the soil. Phenylthiourea (PTU; XVI) exhibits both herbicidal and rodenticidal activity whereas α-naphthylthiourea (ANTU; XVII) has for long been used as a rodenticide. These compounds all exhibit anodic waves at the d.m.e.

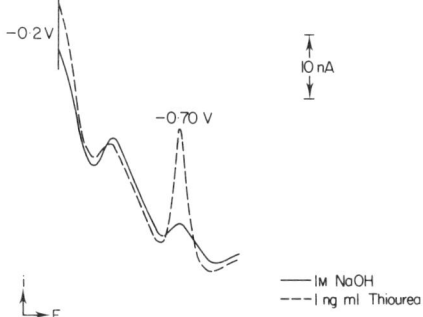

FIG. 4. D.p.c.s.v. of 1 ng ml^{-1} thiourea in 1M NaOH (Smyth and Osteryoung, 1977). ——1M NaOH; ---- 1 ng ml^{-1} TU.

in 1 N NaOH and can be determined using d.p.p. down to 1×10^{-7} M for TU and PTU and down to 2×10^{-7} M for ANTU. Since the anodic process is due to the formation of an insoluble mercury salt, c.s.v. offers much lower limits of detection than the d.p.p. method. This is illustrated in Fig. 4 for the determination of 1 ng ml^{-1} TU in 1N NaOH using d.p.c.s.v. Although this method is very sensitive, it does not offer the selectivity provided by d.p.p. which can be used to measure TU in the presence of PTU or ANTU. It could, however, provide a rapid screening method for patients suffering from defects in sulphur metabolism or for those who have come in contact with sulphur-containing pesticides in the environment.

Thiourea can also be determined voltammetrically following complexation with Cu(II) ions (Sohr and Wienhold, 1976) or by liberation of the S atom and subsequent determination of H_2S (Kosmatyi and Kavetskii, 1973). This latter procedure has also been used for the determination of other S-containing pesticides, e.g. diazinon, rogor and phenkapton and involved reduction of the pesticide by Al in HCl solutions (in the presence of Ni). The H_2S evolved is then determined by monitoring the decrease of Pb(II) concentration in the Pb(OAc)$_2$ trapping solution. The method could determine down to 0.25 µg ml^{-1} in pure solution.

Thiram (XVIII; tetramethylthiuram disulphide) is an active fungicitic agent. Its polarographic behaviour was first studied by Gregg and Tyler (1950) who reported that it exhibited two waves in d.c. polarography. They suggested that thiram formed a redox couple with the corresponding dithiocarbamate derivative, i.e.

$$\begin{array}{c}H_3C\\ \diagdown\\ N-\overset{\overset{\displaystyle S}{\|}}{C}-S-S-\overset{\overset{\displaystyle S}{\|}}{C}-N\diagup\\ H_3C\diagup \diagdown CH_3\end{array} \quad\rightleftharpoons\quad 2\quad \begin{array}{c}H_3C\diagdown\\ N-\overset{\overset{\displaystyle S}{\|}}{C}-SH\\ H_3C\diagup\end{array}$$

(XVIII)

and that the system obeyed Nernstian reversibility. Later work by Zuman et al. (1953), however, showed that since mercury salt formation was involved in the process, then the system could not be considered as truly reversible. Both Nangniot (1960a) and Brand and Fleet (1970a) have studied the effect of concentration on the i_{lim} values of the two waves exhibited by thiram on d.c. polarography. They have shown that whereas the height of the first wave i_1 was linearly dependent on concentration, that of a second wave i_2 was independent of concentration in the range 10^{-3} to 10^{-4} M. Wave i_1 was shown to be diffusion controlled whereas i_2 was attributed to the reduction of thiram while in an adsorbed state. The total wave height, i.e. $i_1 + i_2$ corresponds to a 2e$^-$ process and it would appear that the mechanism involves mercury adding across the disulphide linkage in the following manner:

$$(CH_3)_2N-\overset{\overset{S}{\|}}{C}-S-S-\overset{\overset{S}{\|}}{C}-N(CH_3)_2 \;\underset{}{\overset{+Hg}{\rightleftharpoons}}\; (CH_3)_2N-\overset{\overset{S}{\|}}{C}-S-Hg-S-\overset{\overset{S}{\|}}{C}-N(CH_3)_2$$

$$(CH_3)_2N-\overset{\overset{S}{\|}}{C}-S-Hg-S-\overset{\overset{S}{\|}}{C}-N(CH_3)_2 \;\underset{}{\overset{+2e^-,\,2H^+}{\rightleftharpoons}}\; 2\;(CH_3)_2N-\overset{\overset{S}{\|}}{C}-SH + Hg$$

Brand and Fleet (1970a) have also investigated the application of c.s.v. to the determination of thiram in aqueous solutions. They reported that the best results were obtained using a mercury plated platinum electrode in a solution of thiram containing an excess of ascorbic acid. This addition had the effect of chemically reducing the disulphide moiety in thiram to form free —SH groups which were then amenable to c.s.v. analysis. Pre-electrolysis was carried out at -0.1 V (vs s.c.e.) to remove the interference caused by the oxidation of "unreacted" ascorbic acid and of residual Cl^- ions in the buffer system. Using this method, they were able to determine thiram down to 10^{-8}M in pure solution. This method offers a much greater sensitivity over the d.c., linear sweep or a.c. techniques which have limits of detection for this compound of 6×10^{-6}, 3.5×10^{-6} and 1×10^{-6}M respectively.

Thiram exhibits the greatest stability in acidic solution. In alkaline media it is rapidly converted to the corresponding dithiocarbamate derivative. This parallels its preliminary metabolic fate although *in vivo* studies have shown that the dithiocarbamate derivative further breaks down to give CS_2 as the main metabolite.

$$\text{(XVIII)} \quad \xrightarrow[\substack{\text{pH 5} \\ CS_2}]{2GSH} \quad 2 \underset{CH_3}{\overset{CH_3}{\diagdown}} N-\overset{\overset{S}{\|}}{C}-SH + G-S-S-G$$

where G = glutathione

The polarographic behaviour of the dithiocarbamate anion has been extensively studied by several authors. Stricks and Chakravarti (1962) showed that the anodic process observed for these compounds involved:

$$\underbrace{R\,N-\overset{\overset{S}{\|}}{C}-S^{\ominus}}_{H^+} \xrightleftharpoons{pK_1} R\,N-\overset{\overset{S}{\|}}{C}-S^{\ominus} + H^+$$
$$\Updownarrow Hg^0$$
$$R\,N-\overset{\overset{S}{\|}}{C}-SHg + e^-$$

$$2\,RN-\overset{\overset{S}{\|}}{C}-SHg \rightarrow (R\,N-\overset{\overset{S}{\|}}{C}-S)_2\,Hg + Hg^0$$

The adsorption processes which accompanied this reaction were thought to be due to the orientation of the adsorbed species at the mercury surface. Halls *et al.* (1968a) have studied this process in greater detail and have also reported a method to determine mono- in the presence of dialkyl substituted dithiocarbamates. This is based on the fact that the monosubstituted compound is relatively stable in solutions of pH between 3.5 and 5.0, whereas the disubstituted compound breaks down to give CS_2. They also showed that the addition of a surface active agent such as gelatin improved the wave shape for analytical purposes. Canterford and Buchanan (1973) have applied d.p.p. to the determination of sodium diethyldithiocarbamate (NaDtc) and have quoted a detection limit of 2×10^{-6}M. Brand and Fleet (1968) reported detection limits for NaDtc of 10^{-6}M using both a.c. and l.s.v. The a.c. response to NaDtc showed two waves in the concentration range 10^{-3} to 10^{-4}M at -0.4 and -0.65 V respectively. The first wave disappeared in solutions of concentration less than 10^{-4}M. These authors also reported a c.s.v. method of analysis for this compound at a mercury plated platinum electrode. The curve relating peak current to concentration proved linear in the range 10^{-6} to 10^{-7}M and the use of a mercury plated platinum electrode was preferred over the h.m.d.e. since it was not affected by the rate of stirring or by spurious vibrations.

The polarographic behaviour of the ethylene-1,2-bisdithiocarbamate anion, i.e. EDC (XIX), has been studied by Brand and Fleet (1970b) using d.c., a.c., and c.s.v. techniques.

$$\begin{array}{c} \text{S} \\ \parallel \\ \text{CH}_2\text{—NH—C—S}^\ominus \\ | \\ \text{CH}_2\text{—NH—C—S}^\ominus \\ \parallel \\ \text{S} \end{array}$$

(XIX)

This species exhibits four waves on d.c. polarography and three waves on a.c. polarography. C.s.v. at a mercury plated platinum electrode yielded three peaks at -0.4 V (corresponding to the EDC anion), -0.7 V (corresponding to ethylene thiuram monosulphide) and -1.05 V (due to sulphide ions). They obtained a linear relationship between peak current (at -0.4 V) and concentration in the range 1×10^{-6} to 1×10^{-7}M following 2 minutes preelectrolysis at 0.0 V (vs s.c.e.) and then scanning at $100\,\text{mV s}^{-1}$. Halls et al. (1968b) have also studied the polarographic behaviour of EDC and have shown that the wave shape (on d.c. polarography) can be improved by the addition of ethanol or DMF. Addition of DMSO, AN or of gelatin had little or no effect on the process.

Engst and Schnaak (1974) have used a pulse polarographic method to determine 1,3-ethylenethiourea (a breakdown product of many ethylenebisdithiocarbamate pesticides) in potatoes, tomatoes, liver and meat. The method involved extraction with $CHCl_3$/ethanol (9/1), clean up on an alumina column, separation by paper chromatography and nitrosation to give a polarographically active moiety. The recovery was found to be around 70% and the method was used to determine concentrations of 1,3-ethylenethiourea between 0.05–$2\,\mu\text{g ml}^{-1}$ in these foods.

VI. CARBAMATES AND UREAS

Carbamates and ureas are often classed as one group of compounds in pesticide analysis owing to their similar chemical structure. The ureas of agrochemical importance are often substituted phenylureas (XX) whereas carbamates are usually either substituted phenylcarbamates (XXI) or substituted alkylcarbamates.

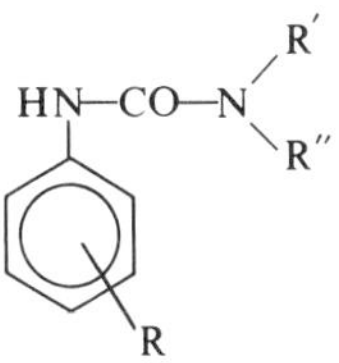

In general, these compounds either give rise to badly defined reduction waves or do not possess any inherent electroactivity. Nitrosation or nitration procedures are therefore employed to improve the sensitivity of the assay. Eberle and Gunther (1965) have investigated the use of nitrosation procedures for the determination of carbaryl, dimetilan, isolan, pyrolan and zectran. These compounds were first hydrolysed in a medium containing 0.5 ml acetic acid and 1.0 ml 0.1N H_2SO_4 for one hour under reflux. On cooling, the hydrolysed products were nitrosated in 5 ml 1N sodium nitrite at 5–10°C. The nitrosated products were then extracted into 2 × 10 ml ether,

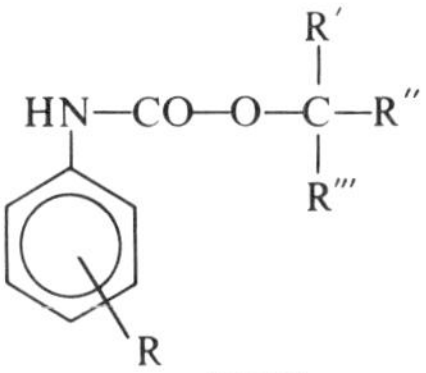

TABLE IV. *Reduction potentials (vs Ag/AgCl) of nitrosated carbamates*

Compound	$E_{\frac{1}{2}}$	Electrolyte	LOD
Carbaryl	−0.78 V	0.2M TBAB	0.8 μg ml^{-1}
Dimetilan	−0.69 V	0.5M NaOAc/NaCl	2.5 μg ml^{-1}
Isolan	−0.90 V	0.2M TBAB + 0.1 ml 0.5N NaOH	2.5 μg ml^{-1}
Pyrolan	−0.95 V	0.2M TBAB + 0.1 ml 0.5N NaOH	2.5 μg ml^{-1}

the combined ether layers dried over sodium sulphate and then concentrated to 0.1 ml. Following t.l.c. in benzene–methanol–acetone (80:20:20), the spots were scraped off, dissolved in 1 ml ethanol followed by 1 ml supporting electrolyte and the i–E curve recorded after deaeration. The reduction potentials of the nitrosated derivatives are given in Table IV.

Under these conditions zectran produced an N,N-dimethyl-N-nitroso compound which was found to decompose almost immediately. The nitrosation routes postulated by Eberle and Gunther (1965) for Pyrolan and Isolan are given below:

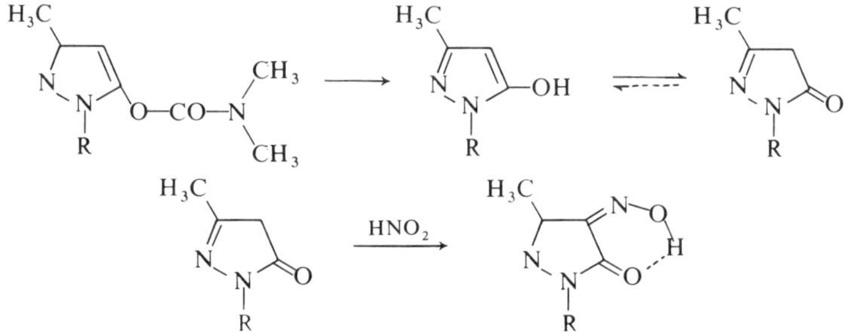

Gajan et al. (1965) have also determined carbaryl following nitrosation. In this instance, nitrosation was carried out in a medium containing 2.0 ml of glacial acetic acid and 2.0 ml of 1N $NaNO_2$ for 3 minutes. 6.0 ml of 50% aqueous KOH was then added and the mixture left to stand for 15 minutes.

TABLE V. *Half wave potentials of carbamates and ureas in various electrolytes (vs Hg pool)*

Compound	Electrolyte and $E_{\frac{1}{2}}$					LOD
	A	B	C	D	E	
EPTC	—	—	−1.00	−0.94	−0.83	25.4×10^{-6}M
Chlorpropham	—	—	−0.96	−0.13	−0.06	4.7×10^{-6}M
				−0.93		
				−1.45		
Barban	−1.07	−1.10	−0.88	−0.84	−0.95	15.5×10^{-6}M
	−1.23			−0.94	−1.20	
				−1.48		
Monuron	—	—	—	—	−0.54	
					−1.26	
Diuron	—	—	−0.98	−1.06	−0.65	8.6×10^{-6}M
					−0.90	
Linuron			−1.00	−1.04	−0.63	8.0×10^{-6}M
					−0.85	
Fluometron			−0.78	−0.86	−0.52	
Baturon			−1.03	−1.05	−0.87	
Methiuron		−0.43			−0.83	20.4×10^{-6}M
		−1.02				
Pebulate	−0.27	−1.36	−0.26	−1.05	−1.03	9.9×10^{-6}M
				−1.16		
Diallate			−1.20			
Triallate		−1.38		−0.54		16.5×10^{-6}M

Electrolytes: A. 0.2N Sulphuric acid; B. 0.2N Acetic acid/sodium acetate; C. 0.2N Tetramethylammonium chloride; D. 0.2N Ammonium hydroxide/ammonium chloride; E. 0.2N Sodium hydroxide.

5.0 ml of the solution was then transferred to a polarographic cell and the i-E curve recorded between −0.2 and −0.9 V. The method was applied to carbaryl residues on lettuce, apples, broccoli and potatoes at 10.0, 5.0 and 0.2 µg ml^{-1} levels. Recoveries greater than 95% were reported and other agrochemicals such as DDT etc. were found not to interfere. The nitrosated product of orthophenyl phenol did, however, give rise to a wave at the same reduction potential.

Engst et al. (1965) have developed a method for carbaryl based on the determination of a mixture of nitrated derivatives. The $E_{\frac{1}{2}}$ for the mixture of derivatives was found to be −0.83 V and limits of detections of 2 µg ml^{-1} with d.c. and 0.2 µg ml^{-1} with single sweep pulse polarography were reported. The determination of carbaryl in fruit and vegetables has also been reported by Porter et al. (1969). The method involved extraction with AN, and 0.2 to 10 µg ml^{-1} carbaryl could be determined with an extraction efficiency of 90 to 112%.

Hance (1970) has investigated a range of herbicides for polarographic activity, several of which were carbamates and ureas. The reduction potentials of these compounds in a variety of supporting electrolytes are given in Table V.

Watanabe (1972) has described a method for the determination of 1-naphthylmethylcarbamate and 2,3-dimethylphenylmethylcarbamate (XXII) in residues following nitrosation of the oxime group.

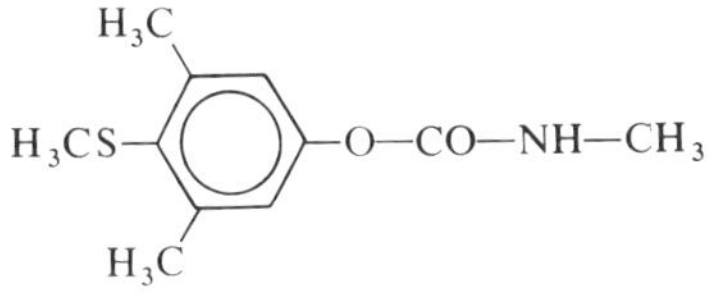

(XXII)

VII. HETEROCYCLIC NITROGEN COMPOUNDS

The polarographic activity of several s-triazine herbicides has been investigated by Hayes et al. (1967). They found that 4-amino-6-alkylamine-s-triazines (XXIII) were polarographically active in acid solution giving rise to a single

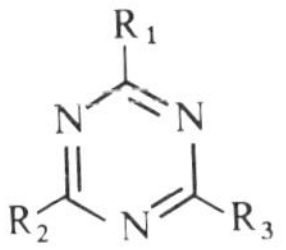

XXIII $R_1 = H; R_2 = NH_2; R_3 = NHEt$.
XXIV $R_1 = Cl$ or $Br; R_2 = NHR; R_3 = NHR$.
XXV $R_1 = N_3; R_2 = NHEt; R_3 = NHCMe_3$.
XXVI $R_1 = OMe; R_2 = NHPr; R_3 = NHPr$.

wave at about −1.0 V (vs s.c.e.). In solutions of pH 1, the limiting current was found to be linear with concentration up to 2×10^{-4} M. The limit of detection was 5×10^{-6}M with d.c. polarography and 2×10^{-7}M with single sweep polarography. The polarographic behaviour of prometryne, i.e. 2-methylthio-4,6-bis(isopropylamino)-s-triazine, was similar in all respects to that of triazines unsubstituted at the 2 position. The 2-chloro and 2-bromo-4,6-bis(alkylamino)-s-triazines (XXIV) were found to be polarographically active at pH values between 1.5 and 5.5 each giving rise to two waves. Detection limits were 3×10^{-6}M with d.c. and 2×10^{-7}M with single sweep polarography. 2-azido-4-ethylamino-6-t-butylamino-s-triazine (XXV) was active at all pH values, whereas the 2-methoxy-4,6-bis-(alkylamino)-2-triazines (XXVI) were polarographically inactive at all pH values.

The polarographic determination of picloram (XXVII) has been reported by several authors. Gilbert and Mann (1973) have described a pulse polarographic method for the determination of picloram in natural waters. This method could determine down to 0.02 µg ml^{-1} and the authors warned of the possible interference that could be caused by the presence of various metal ions in the water system. Filimonova and Gorbunova (1973) have determined picloram in water, soil and dried maize. The method (for dried maize) involved extraction with acetone/0.01N KOH followed by a liquid chromatographic separation on Al_2O_3. Picloram gave rise to a wave at -1.10 V (vs s.c.e.) in 0.1N HCl or 0.1N H_2SO_4 and the method could determine down to 0.2 µg ml^{-1} in water, 0.7 µg ml^{-1} in soil and 0.25 µg ml^{-1} in dried maize. Whittaker and Osteryoung (1979) have further investigated the polarographic behaviour of picloram using n.p.p. and d.p.p., and optimized conditions for its analysis using the latter technique.

Hance (1970) has investigated the polarographic behaviour of a range of herbicides (including various triazines and picloram) in a variety of supporting electrolytes. He has also reported that paraquat (XXVIII) and diquat (XXIX) gave rise to waves at -1.12 V and -0.99 V respectively in 0.2N H_2SO_4 and that limits of detection for these compounds using l.s.v. were of the order of 10^{-7}M. These compounds have also been investigated by other authors. Engelhardt and McKinley (1966) studied the polarographic behaviour of diquat in 0.1M KCl. In this medium, diquat gave rise to two waves at -0.56 and -1.06 V (vs s.c.e.) due to:

and the method could determine diquat down to 0.5 μg ml^{-1} in pure solution. Nangniot (1970) has further discussed how polarography can be used to determine diquat and paraquat in residues. Sheremet *et al.* (1973) have reported differences in the polarographic behaviour of the mono- and bisquaternary salts of paraquat which can be used in the analysis of formulations.

McLean and Daniels (1971) have described a method for the determination of piperazine in animal feed. The method involved dissolving 0.1 to 1 mg of material in 5 ml McIlvaine buffer pH 5 and 1 ml 37% aqueous HCHO. After 20 minutes, the solution was diluted to 25 ml with H_2O and the wave at -0.98 V (presumably due to an N-oxygenated derivative) measured.

VIII. ORGANOMETALLIC COMPOUNDS

The electrochemical behaviour of organometallic substances has been discussed by Watson (Chapter 10) and by Fleet and Fouzder (Chapter 9) and will not be dealt with in great detail here. In general, voltammetric methods of analysis for these compounds are based on the destruction of the complex followed by measurement either of the free metal ion concentration or that of the organic ligand.

Nangniot (1960b) has applied d.c. and l.s.v. to the determination of three metal-containing dithiocarbamate pesticides: ziram (XXX: zinc dimethyldithiocarbamate), zineb (XXXI: zinc ethylenebisdithiocarbamate) and ferbam (XXXII: ferric dimethyldithiocarbamate).

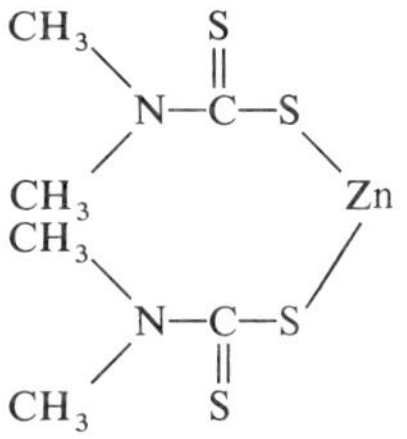

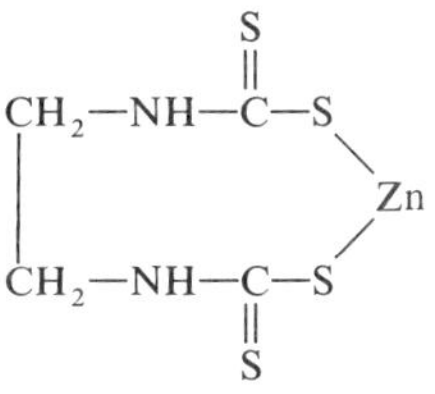

In 0.5M NH_4OH/0.5M NH_4Cl supporting electrolyte, ziram gave rise to an anodic wave at -0.41 V due to the oxidation of dimethyldithiocarbamic acid whereas in 0.2M NaOH it gave rise to a cathodic wave at -1.42 V due to the reduction of Zn(II) ions. Nangniot reported that in residue determinations, the oxidation wave due to dimethyldithiocarbamic acid was affected by interferences in the matrix and suggested that residue analyses be carried out in the latter supporting electrolyte. In this medium (0.2M NaOH), ziram could be determined down to 5 μg ml^{-1} using l.s.v. Nangniot also

reported detection limits of 1.0 and 0.2 µg ml^{-1} for zineb and ferbam in 0.1M NaOH and 0.5N KCl/0.5N sodium citrate/50% acetone respectively.

Budnikov et al. (1974) have investigated the polarographic behaviour of ferbam, marbam, ziram, zineb and maneb in organic solvents using d.c., a.c. and oscillopolarography. They found that ferbam and marbam were reversibly reduced in benzene/methanol (2/3) containing 0.2M LiClO$_4$ at −0.60 V (vs s.c.e.). These compounds could be determined between 0.1 to 3 µM using a.c. polarography. Ziram, zineb and maneb, however, were all found to be irreversibly reduced at −1.44, −1.30, and −1.30 V respectively.

An indirect pulse polarographic method has been employed for the determination of zineb on tobacco leaves (Lyalikov et al., 1965). Zineb was degraded by the addition of HCl/HNO$_3$, the mixture diluted with water and polarography carried out in 0.5M NaClO$_4$, pH 2. The wave used for analytical purposes was due to the reduction of Zn(II) ions. Halls et al. (1968c), on the other hand, have suggested a method for the determination of metal containing dithiocarbamates based on liberation of metal from the complex and subsequent determination of the anodic wave given by the dithiocarbamate anion. This method could determine concentrations of zineb down to 0.3 µg ml^{-1} wet weight vegetable tissue.

Budnikov et al. (1975) have determined the fungicide polycarbacin (a formulation containing 3 parts Zn ethylenebisdithiocarbamate + 1 part N,N′-ethylenethiuram disulphide) in apple peel. The method involved extraction of the peel with DMF and subsequent polarographic determination of polycarbacin in DMF/10% NaClO$_4$. The peak at −57 V (vs Hg pool), relating to the reduction of Zn(II) ions, was employed in the analytical investigation and the method could determine down to 0.8 µg ml^{-1}. No interference was observed due to decomposition products.

Because of the adsorption characteristics, Supin and Budnikov (1973) have been able to improve the detection limits obtained using oscillopolarography by employing a slow dropping mercury electrode and allowing the adsorbed species to accumulate at the mercury surface (at −0.4 V vs s.c.e.). In this way they were able to determine concentrations of zineb, maneb and ziram down to 0.1 µg ml in fruit and leaves.

Perhaps the most sensitive voltammetric method available for the analysis of these compounds is based on anodic stripping either of the complex or of the metal ions liberated following destruction of the complex. This is illustrated by the work of Sauberlich et al. (1969) who have been able to determine ferbam down to 0.5 ng ml^{-1} in pure solution.

Several other organometallic compounds of agrochemical importance have also been determined by voltammetric methods of analysis. For example, triphenyltin acetate can be determined following extraction into CH$_2$Cl$_2$, liberation of Sn following digestion with H$_2$SO$_4$ and subsequent determination as the tetrabromide (Gorbach and Bock, 1958). Vogel and Deshusses (1964) have determined triphenyltin acetate following extraction with $CHCl_3$ and digestion in 50% HCl and NH_4Cl. The $E_\frac{1}{2}$ of Sn(II) was quoted to be -0.47 V (vs s.c.e.) and the method was found to have a limit of detection of 2.5 µg ml^{-1}. Coussement (1972) has used polarography to determine triphenyltin acetate on potato leaves down to 0.32 µg cm^{-2} leaf tissue following extraction and calcination steps. Triphenyltin acetate has also been determined using anodic stripping voltammetry with a limit of detection for residues of 8 ng ml^{-1} (Nangniot and Martens, 1961).

The determination of organotin compounds and their degradation (and decomposition) products by anodic stripping voltammetry has been more fully investigated by Woggon (1972). He studied the influence of plating potential and time, volume of Hg drop and temperature on the anodic stripping of $RSnCl_3$ (where R = butyl or octyl), $Ph(CH_2)SnCl_2$, $SnCl_4$ and $PhCH_2Sn(O)OH$, and optimized conditions for their analysis in fungicide residues.

IX. MISCELLANEOUS COMPOUNDS

Hearth et al. (1966) have determined morestan in orange rind following a t.l.c. separation. The oscillopolarographic method could determine concentration of morestan down to 0.5 µg ml^{-1} and compared favourably with the colorimetric method.

Pyrethrums have been investigated by several workers: for example, Yamada et al. (1962) determined the alkyl and ethyl homologues of d,l-*cis-trans*-chrysanthemummonocarboxylate in 0.2M TEAB. The $E_\frac{1}{2}$ was quoted to be -1.25 V (vs s.c.e.). Ouwa et al. (1952) determined allethrin in the same supporting electrolyte ($E_\frac{1}{2} = -1.27$ V) and found allethrolene to interfere in some cases.

Westlake et al. (1969) have determined ciodrin (XXXIII) in fortified animal tissues by oscillopolarography of its hydrolysis product acetophenone. After extraction with AN,

$$CH_3-CH-O-\underset{\underset{(XXXIII)}{}}{\overset{O}{\underset{\|}{C}}}-\underset{\underset{}{H}}{\overset{}{C}}=\underset{\underset{}{CH_3}}{\overset{}{C}}-O-\overset{O}{\underset{\|}{P}}-(OCH_3)_2 \xrightarrow{H^+} \underset{Ph}{\overset{CH_3}{\diagdown}}C=O$$

liquid partition into hexane and coagulation clean-up, the extract was distilled from 2N H_2SO_4 and 10% dichromate, collected and subjected to polarography in 0.1M boric acid/0.1M NaOH pH 10.25. The $E_\frac{1}{2}$ of acetophenone was found to be -1.71 V (vs s.c.e.) and the limit of detection was quoted to be 0.1 µg ml^{-1}.

The determination of captan and phaltan is usually based on an alkaline hydrolysis reaction:

$$\text{(phthalimide)} + OH^- \longrightarrow \text{(o-COO}^-\text{, CO—NHR)}$$

The factors governing this reaction have been more fully dealt with by Nangniot (1970). Lyalikov et al. (1970) have further described a direct method for the analysis of captan based on its reduction in acetone/H_2O (3/7) containing citrate-phosphate buffer pH 3.4. Using pulse polarography, μM captan could be determined by this method. Recently Seifert et al. (1975) have described an indirect method for the determination of phaltan residues on strawberries and cotton seeds based on the reaction of phaltan with L-cysteine to form phthalimide. The $E_{\frac{1}{2}}$ of phthalimide was quoted to be -0.82 V in 0.1N H_2SO_4/10% acetone and the method had a limit of detection of 0.5 μg ml^{-1}. Recoveries from strawberries and cotton seeds were of the order of 80 to 90%.

N-phenyl-N-isopropylchloroacetamide has been determined in soil following extraction with benzene, clean-up on an alumina column and determination in TEAI/60% methanol. Recoveries were of the order of 80 ±5% and the method could determine down to 0.5 μg kg^{-1} in soil (Filimonova et al., 1977).

Zhantalai and Slisarenko (1973) have determined cyanuric acid in $NaHCO_3$, pH 8.5 between 5 μM–0.1 mM. In this medium, cyanuric acid gave rise to an anodic wave (due to oxidation of one of the hydroxyl groups in the molecule) at -0.04 V (vs s.c.e.). In this method Cl$^-$ was found to interfere up to 10 mM whereas other interferences such as ammelide, ameline and melamine were found to give rise to waves at much more negative potentials.

Murano et al. (1972) have determined pentachlorobenzaldoxime in formulations using a DMF–0.5N HCl/H_2O (3:1:1) supporting electrolyte. In this medium, the compound gave rise to a 4e$^-$ reduction at -0.595 V (vs s.c.e.) and the interfering influence of pentachlorobenzaldehyde was recovered by t.l.c. The method was used to determine pentachlorobenzaldoxime between 0.08 to 0.25 μg ml^{-1}.

Dulak (1975) has used polarography to determine 4,5-di Cl -2-ϕ-SH-pyridazin-2-one (an intermediate in the synthesis of the herbicide pyrazon) in formulations. This compound gave rise to a wave at -0.8 V (vs s.c.e.) in methanol–Sorenson buffer–gelatin (6.9:3:0.1) and no interference was shown by the starting compounds phenylhydrazine or 2,3-diCl-4-oxobut-2-enoic acid. Recently Smyth and Osteryoung (1978b) have reported the polarographic determination of the azomethine-containing pesticides cyolane, cytrolane, chlordimeform and drazoxolan.

Several authors have used polarography to study structural aspects of agrochemistry. Regauti (1973) has studied the reduction of 1,2-benzoisothiazole derivatives in DMF and reported that the sequence of reduction potentials was influenced by the same structural variations which depress their antimycotic activity. Sherman et al. (1976) have studied the polarography of some benzothiadiazoles and benzofurazan carbonitriles (known to possess herbicidal activity) and have shown how substitution affects their redox behaviour.

X. CONCLUSION

Polarography, although probably not matching the sensitivity of element-specific analytical techniques, does however have a useful role to play in the analysis of pesticides at both the formulation and residue level. It can provide a quick and accurate screening technique and be a useful tool in the elucidation of metabolite pathways. Polarographic methods should therefore be developed and used side by side with other techniques such as gas chromatography and high performance liquid chromatography for the analysis of agrochemicals in the environment.

REFERENCES

Bates, J. A. R. (1962). *Analyst*, **87**, 786.
Berck, B. (1962). *J. Agr. Food Chem.* **10**, 158.
Bowen, C. V. and Edwards, F. I. (1950). *Anal. Chem.* **22**, 706.
Brand, M. J. D. and Fleet, B. (1968). *Analyst*, **93**, 498.
Brand, M. J. D. and Fleet, B. (1970a). *Analyst*, **95**, 1023.
Brand, M. J. D. and Fleet, B. (1970b). *Analyst*, **95**, 905.
Brezina, M. and Romazanovich, N. P. (1961). *Zavod. Lab.* **27**, 1953.
Budnikov, G. K., Toropova, V. F., Ulakhovich, N. A. and Viter, I. P. (1974). *Zh. Anal. Khim.* **29**, 1204.
Budnikov, G. K., Supin, G. S., Ulakhovich, N. A. and Shakurova, N. K. (1975). *Zh. Anal. Khim.* **30**, 2275.
Cala, P. C., Downing, G. V., Michielli, R. F. and Wittick, J. J. (1976). *J. Agr. Food Chem.* **24**, 764.
Canterford, D. R. and Buchanan, A. S. (1973). *J. Electroanal. Chem.* **44**, 291.
Cimbura, G. and Gupta, R. C. (1964). *Proc. Can. Soc. Forensic Sci.* **2**, 350.
Colas, A. et al. (1964). *Phytopharmacie*, **13**, 3.
Copin, A. (1968a). *Bull. Rech. Agron. Gembloux*, **2**, 254.
Copin, A. (1968b). *Bull. Rech. Agron. Gembloux*, **2**, 261.
Coussement, S. (1972). *Ann. Gembloux*, **78**, 41.
Cox, J. A. and Cheng, K. H. (1974). *Anal. Letters*, **7**, 659.

Davidek, J. and Janicek, G. (1961). *Experientia*, **17**, 473.
Davidek, J. and Seifert, J. (1971a). *Die Nahrung*, **15**, 691.
Davidek, J. and Seifert, J. (1971b). *J. Chromat.* **59**, 446.
Davidek, J., Seifert, J. and Dolezalova, Z. (1976). *Milchwissenschaft*, **31**, 276.
Dragt, G. (1948). *Anal. Chem.* **20**, 737.
Dulak, K. (1975). *Z. Anal. Chem.* **274**, 123.
Eberle, D. O. and Gunther, F. A. (1965). *J. Assoc. Off. Anal. Chem.* **48**, 927.
Engelhardt, J. and McKinley, W. T. (1966). *J. Agric. Food Chem.* **14**, 377.
Engst, R. and Schnaak, W. (1974). *Die Nahrung* **18**, 597.
Engst, R., Schnaak, W. and Woggon, H. (1965). *Z. Anal. Chem.* **207**, 30.
Farwell, S. O., Beland, F. A. and Geer, R. D. (1973). *Bull. Environ. Contam. Tox.* **10**, 157.
Farwell, S. O., Beland, F. A. and Geer, R. D. (1975). *Anal. Chem.* **47**, 895.
Feher, F. and Monien, H. (1964). *Z. Anal. Chem.* **204**, 19.
Filimonova, M. M. and Gorbunova, V. E. (1973). *Zh. Anal. Khim.* **28**, 1184.
Filimonova, M. M. and Gorbunova, V. E. (1975). *Zh. Anal. Khim.* **30**, 358.
Filimonova, M. M., Gorbunova, V. E. and Filimonov, B. F. (1977). *Zh. Anal. Khim.* **32**, 140.
Fukami, H. and Nakajima, H. (1954). *Botya Kogaku*, **19**, 83.
Gajan, R. J. (1962). *J. Assoc. Off. Anal. Chem.* **45**, 401.
Gajan, R. J. (1964a). *Res. Reviews* **5**, 80.
Gajan, R. J. (1964b) *Res. Reviews* **6**, 75.
Gajan, R. J. (1965). *J. Assoc. Off. Anal. Chem.* **48**, 1028.
Gajan, R. J. (1969). *J. Assoc. Off. Anal. Chem.* **52**, 811.
Gajan, R. J. and Link, J. (1964). *J. Assoc. Off. Anal. Chem.* **47**, 1109.
Gajan, R. J., Benson, W. R. and Finnochiaro, J. M. (1965). *J. Assoc. Off. Anal. Chem.* **48**, 958.
Giang, P. A. and Caswell, R. L. (1957). *J. Agr. Food Chem.* **5**, 753.
Gilbert, D. D. and Mann, J. M. (1973). *Int. J. Environ. Anal. Chem.* **2**, 221.
Gorbach, S. and Bock, R. (1958). *Z. Anal. Chem.* **163**, 429.
Graham-Bryce, I. J. (1970). *Pestic. Sci.* **1**, 73.
Gregg, E. C. and Tyler, W. P. (1950). *J. Amer. Chem. Soc.* **72**, 4561.
Gruca, M. and Mosinska, K. (1970). *Pr. Inst. Przem. Org.* No. 2, 89.
Halls, D. J., Townsend, A. and Zuman, P. (1968a). *Anal. Chim. Acta*, **41**, 51.
Halls, D. J., Townsend, A. and Zuman, P. (1968b). *Anal. Chim. Acta*, **41**, 63.
Halls, D. J., Townsend, A. and Zuman, P. (1968c). *Analyst*, **93**, 219.
Hance, R. J. (1970). *Pestic. Sci.* **1**, 112.
Hasegawa, K. (1962). *Kagaku Keisatsu Kenkyusho Hokoku*, **15**, 62.
Hayes, M. H. B., Stacey, M. and Thompson, J. M. (1967). *Chem. and Ind.* 1222.
Hearth, F. E., Ott, D. A. and Gunther, F. A. (1966). *J. Assoc. Off. Anal. Chem.* **49**, 774.
Hocquellet, P. and Pevendrant, A. (1972). *Analusis*, **7**, 192.
Ichim, A. (1972). *Igiena*, **21**, 225.
Ingram, G. B. and Southern, H. K. (1948). *Nature, Lond.* **161**, 437.
Johnson, L. (1962). *J. Assoc. Off. Anal. Chem.* **45**, 363.
Jura, W. H. (1955). *Anal. Chem.* **27**, 525.
Keller, H., Hochweler, M. and Holban, H. (1946). *Helv. Chem. Acta*, **29**, 512.
Kemula, W. and Kreminska, K. (1960). *Chemia, Analit.* **5**, 611.
Khan, S. U. (1975). *Res. Reviews*, **59**, 21.
Kikuchi, T. and Morley, H. V. (1964). *J. Tokyo Univ. Fish*, **50**, 57.
Koen, J. G. and Huber, J. F. K. (1970). *Anal. Chim. Acta*, **51**, 303.
Koneva, T. N. and Kutsenogii, K. P. (1972). *Probl. Anal. Khim.* **2**, 58.
Kosmatyi, E. S. and Bublik, L. I. (1974). *Ukr. Khim. Zh.* **40**, 1316.
Kosmatyi, E. S. and Kavetskii, V. N. (1973). *Zh. Anal. Khim.* **28**, 1028.

8. ELECTROANALYSIS OF AGROCHEMICALS

Kosmatyi, E. S., Polonskaya, F. J. and Tverskaya, B. M. (1968). *Khim. Sel. Khoz.* **6**, 354.
Kosmatyi, E. S. and Kavetskii, V. N. (1975). *Zav. Lab.* **41**, 286.
Lyalikov, Y. S., Bodya, V. L. and Kozlova, L. V. (1965). *Zav. Lab.* **31**, 1190.
Lyalikov, Y. S., Solanar, A. S. and Kilar, P. F. (1970). *Izv. Akad. Nauk. Mold. S.S.R., Ser. Biol. Khim. Nauk.* 63.
Martens, P. H. and Nangniot, P. (1963). *Res. Reviews*, **2**, 26.
Martens, P. H., Nangniot, P. and Dardenne, G. (1961). *Meded. Landbouw. Opzoekst. (Gent)*, **26**, 1523.
McLean, J. O. and Daniels, O. L. (1971). *J. Assoc. Off. Anal. Chem.* **49**, 774.
Michielli, R. F. and Downing, G. V. (1974). *J. Agr. Food Chem.* **22**, 449.
Mikolajek, A. (1969). *Mikrochim. Acta*, No. **6**, 1229.
Mosinska, K. and Kotarski, A. (1972). *Chemia. Analit.* **17**, 327.
Murano, A., Umeda, I. and Oba, S. (1972). *Japan Analyst*, **21**, 644.
Nangniot, P. (1960a). *Meded. Landbouw. Opzoekst. (Gent)*, **25**, 1285.
Nangniot, P. (1960b). *Bull. Inst. Agron, Stat. Rech. Gembloux*, **28**, 365, 373, 381.
Nangniot, P. and Martens, P. H. (1961). *Anal. Chim. Acta*, **24**, 776.
Nangniot, P. (1964). *Anal. Chim. Acta*, **31**, 166.
Nangniot, P. (1970). *In* "La Polarographie en Agronomie et Biologie". Duculot, Gembloux.
Osteryoung, J. G., Whittaker, J. W., Tessari, J. and Boyes, V. (1976). Proceedings 172nd ACS meeting, "Pesticide Residues in Large Animals", San Fransicso.
Ouwa, I., Inove, Y., Ueda, J. and Ohno, M. (1952). *Botyu Kogaku*, **17**, 106.
Panetsos, A. and Kilikidis, K. (1973). *Hellen. Kteniatr.* 75
Parnell, M. J. (1973). *Pestic. Sci.* **4**, 643.
"Pesticide Analytical Manual" (1970). Vol. 1, Section 12F, p. 1. U.S. Food and Drug Administration.
Pesticide Manual of British Crop Protection (1977). Edition No. 5 (H. Martin and C. R. Worthing, eds). British Crop Protection Council.
Porter, M. L., Gajan, R. J. and Burke, J. A. (1969). *J. Assoc. Off. Anal. Chem.* **52**, 117.
Regauti, V. (1973). *Farmaco, Ed. Sci.* **28**, 243.
Richardson, L. T. and Miller, D. M. (1960). *Can. J. Bot.* **38**, 163.
Rowe, R. R. and Franklin Smyth, W. (1979). To be published.
Sauberlich, H., Woggon, H. and Jehring, H. (1969). *Z. Anal. Chem.* **246**, 185.
Seifert, J. and Davidek, J. (1974). *Z. Lebensm.-Unters. Forsch.* **55**, 266.
Seifert, J., Davidek, J. and Egertova, J. (1975). *Pestic. Sci.* **6**, 173.
Sestakova, I. and Skarka, P. (1976). *Biol. Chem. Vyz. Zvivat.* **12**, 321.
Shapiro, L. M. (1961). *Nancha Isslehe. Lab. Min. Zdrarookhr. Beloorussk U.S.S.R.* 95.
Sheremet, N. G., Supin, G. S., Kashkin, B. M. and Mel'nikov, N. W. (1973). *Zh. Anal. Chim.* **28**, 1269.
Sherman, E. O., Jr, Lambert, S. M. and Pilgram, K. (1976). *J. Heterocyclic. Chem.* **11**, 763.
Slamnik, M. (1974). *Talanta*, **21**, 960.
Smyth, M. R. and Osteryoung, J. G. (1977). *Anal. Chem.* **49**, 2310.
Smyth, M. R. and Osteryoung, J. G. (1978a). *Anal. Chim. Acta.* **96**, 335.
Smyth, M. R. and Osteryoung, J. G. (1978b). *Anal. Chem.* **50**, 1632.
Sohr, H. (1962), *Chem Zvesti* **16**, 316.
Sohr, H. and Wienhold, K. (1976). *Anal. Chim. Acta*, **83**, 415.
Southwick, L. M., Willis, G. H., das Gupta, P. K. and Keszthelyi, C. P. (1976). *Anal. Chim. Acta*, **82**, 29.
Streuli, C. A. and Cooke, W. D. (1954). *Anal. Chem.* **26**, 970.
Stricks, W. and Chakravarti, S. K. (1962). *Anal. Chem.* **34**, 508.
Supin, G. S. (1972). *Probl. Anal. Khim.* **2**, 145.

Supin, G. S. and Budnikov, G. K. (1973). *Zh. Anal. Khim.* **28**, 1459.
Supin, G. S., Kayushima, E. N., Itskova, A. L. and Mande'baum, Ya. A. (1976). *Zh. Obshch. Khim.* **46**, 2312.
Vogel, J. and Deshusses, J. (1966). *Helv. Chim. Acta.* **47**, 181.
Watanabe, T. (1972). *Noyaku Kensasho Hokoku*, 57.
Webster, J. G. and Dawson, J. A. (1952). *Analyst* **77**, 203.
Westlake, A., Gunther, F. A. and Westlake, W. E. (1969). *J. Agr. Food Chem.* **17**, 1157.
Whittaker, J. and Osteryoung, J. G. (1976). *Anal. Chem.* **48**, 1418.
Whittaker, J. and Osteryoung, J. G. (1979). *Anal. Chem.* (to be published).
Woggon, H. (1972). *Z. Anal. Chem.* **260**, 268.
Yamada, R., Sato, T. and Iwata, J. (1962). *Botyu Kogaku.* **17**, 31.
Yamano, H., Matsumara, T. and Fukado, K. (1962). *Kagaku Keisatsu Kenkyusho Hokoku*, **15**, 62.
Zhantali, B. P. and Slisarenko, V. B. (1973). *Zav. Lab.* **39**, 6.
Zietek, M. (1975). *Mikrochim. Acta*, 463.
Zietek, M. (1976). *Mikrochim. Acta*, 549.
Zuman, P., Zumanova, R. and Soucek, A. (1953). *Chem. Listy*, **47**, 1522.

Chapter 9

THE ELECTROCHEMICAL BEHAVIOUR AND ANALYSIS OF ORGANOMETALLIC COMPOUNDS OF MERCURY, TIN, LEAD AND GERMANIUM

NANI B. FOUZDER and BERNARD FLEET

Department of Chemistry, Imperial College of Science and Technology London, England

I. INTRODUCTION

During the last ten to twenty years there has been a noticeable expansion in the industrial production of organometallic compounds. These compounds play an increasingly vital role in modern technology, with the range of their applications including stabilizers, catalysts, pesticides, fungicides and medicaments. Organomercury compounds are the most widely produced, followed by organotin and organolead derivatives. The increasing, and in many cases uncontrolled, use of these compounds has been followed by considerable concern about the toxicological effects of organometallic compounds and their degradation and metabolic products, since such organic derivatives are in general more toxic than the parent metal species.

Analytical methods have therefore been developed for the determination of the parent organometallic compounds and their degradation and metabolic products. Electroanalytical methods have shown themselves very useful in this context.

The present review covers the salient features of the electrochemical behaviour of organometallic derivatives of mercury, tin, lead and germanium from which optimum methods of analysis both at the formulation and trace level can be derived.

II. MERCURY

A. INTRODUCTION

Organomercurials have been used as fungicides, pesticides and medicinals for most of this century. By the 1950s however the toxicological effects of mercury were becoming apparent. Numerous cases of wildlife poisoning caused by eating mercury treated seeds were reported and even more serious was the tragedy caused by methyl mercury production in Minamata Bay, Japan, which contaminated the fish and shellfish population. Over 100 people were severely poisoned by mercury. It has been observed that the alkyl mercury compounds are unique in that their toxic effects can cause irreversible damage to the central nervous system while the effects of aromatic and inorganic mercury are in most cases, reversible. While the extensive pollution from inorganic mercury was an accepted hazard of the chloralkali industry and other manufacturing processes, the work of Jensen and Jernelov (1969) showed that inorganic mercury compounds could be converted to mono- or dialkylmercury derivatives by various aquatic microorganisms including those present in the sludge on lake bottoms.

B. GENERAL POLAROGRAPHIC BEHAVIOUR

(i) *Monoorganomercury compounds*

The electrochemical behaviour of organomercury compounds has been extensively studied over the last twenty-five years both from the analytical and the mechanistic viewpoints. The earliest work on this topic was carried out by Kraus (1913) who electrolysed ammonia solutions of alkylmercury halides and found that a free radical, $RHg^{\cdot}$ was formed which disproportionated to R_2Hg at room temperature. Some thirty-five years later Costa (1948) observed two polarographic waves for the reduction of alkylmercury salts, and attributed these to the following irreversible processes.

$$RHg^+ + e \rightarrow RHg^{\cdot}$$
$$RHg^{\cdot} + H^+ + e \rightarrow RH + Hg$$

Vojir (1952, 1957) also observed two reduction waves for phenylmercury compounds and showed from a.c. oscillographic experiments that the first wave was reversible. Benesch and Benesch (1951, 1952) isolated diphenylmercury as the product of the first reduction step but unfortunately considered the first wave irreversible from a consideration of its shape and $E_{\frac{1}{2}}$–pH dependence. However, later investigations have shown conclusively that this process was reversible for most organomercury salts, and that the free radical formed in the first step was dimerized at the surface of the electrode (Butin et al. 1967; Heaton and Laitinen, 1974).

$$2\,RHg^{\cdot} \rightleftarrows (RHg)_2$$

The equilibrium constant of the above reaction, assuming dimerization was fast and reversible, has been found to be $10^{-5.7}$.

Controlled potential electrolysis at low temperature yielded an unstable electrically conducting material, the composition of which corresponded to $(RHg)_n$ (Dessy et al., 1966a; Webb et al., 1970). This compound gave no e.s.r. signal (Gowenlock et al., 1958; Gowenlock and Trotman, 1957). Controlled potential electrolysis at room temperature on the plateau of the first wave showed the electrolysis products to be dialkylmercury, R_2Hg and mercury metal. It was thought previously that the following reaction took place (Kashin et al., 1972)

$$2\,RHg^{\cdot} \rightarrow R_2Hg + Hg$$

but Butin et al. (1974) and Ershler et al. (1974) have shown that at the mercury surface, dialkylmercury, R_2Hg, was not in equilibrium with free-radicals, $RHg^{\cdot}$, but with the dimer, $RHg.HgR$, known as "organic calomel". Organo-mercury free radicals initially formed were very rapidly converted to "organic calomel", $(RHg)_2$, the lifetime of which was found to be very short.

The lifetimes of organomercury free radicals vary according to the solution conditions and nature of the radical species. Dessy et al. (1966a) did not observe any cyclic voltammetric peak corresponding to the oxidation of the phenylmercury free radical in dimethoxyethane (glyme) although chrono-potentiometry of phenylmercury salts indicated that the free radical under-

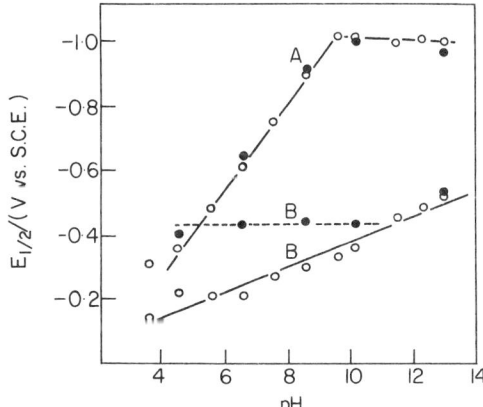

FIG. 1. pH-dependence of 4-methyl-1-pentene/mercury(II)acetate, conc. 2×10^{-4}M (A) second wave, (B) first wave: (O) 50% methanol, 0.002% Triton-X-100 plus buffer medium, (●) as above plus 0.1M potassium bromide. (Data from Fleet, B. and Jee, R. D., 1970).

goes oxidation. Fleet and Jee (1970) observed a 5 s half-life for some olefin-mercury free radicals in 50% methanol. The free radicals are stabilized by adsorption on mercury while unadsorbed alkylmercury free radicals dimerize rapidly. The lifetime of methyl mercury in 50% methanol, determined by chronocoulometric experiments, had been found to be ca 0.2 s, while for pyridyl-3-mercury free radical on mercury in aqueous buffer the lifetime has been estimated to be 10 ms (Degrand and Laviron, 1968b).

The anomalous pH dependence of the first wave observed by several workers has been shown by Fleet and Jee (1970) to be due to the effect of complexation with buffer components of the medium. Earlier workers observed that the addition of complexing anions (e.g. thiocyanate, chloride, bromide and iodide) caused negative shifts in $E_{\frac{1}{2}}$ proportional to their complexing affinity for mercury. These anions had no effect on the half wave potential of the second wave.

Using a supporting electrolyte of negligible complexing properties, i.e. sodium nitrate, the effect of acetate, bromide and hydroxide ions were investigated for 4 methyl-1-pentene mercury(II)acetate. From plots of $E_{\frac{1}{2}}$ vs pH (Fig. 1) the species in solution appeared to be RHgOAc, RHgBr and RHgOH. In a similar study Butin et al. (1967) rather optimistically calculated stability constants for the formation of RHgBr and $RHgBr_2^-$.

The second polarographic wave for monoorganomercury compounds has been found to be rather distorted and drawn-out. But the corresponding cyclic voltammetric peak was well-defined, indicating that the electron-transfer was relatively fast. The irreversible drawn-out appearance of the second wave was due to competition of the electrochemical process with the chemical dimerization of the free radical and to the stability of the hydrocarbon product and consequent slowness of the reverse reaction. Both the half-wave potential and the limiting current of this step have been found to depend on pH in acidic pH's but not at higher pH values. These observations were the basis of the mechanistic scheme for the free radical reduction in acidic and in basic media (Fleet and Jee, 1970).

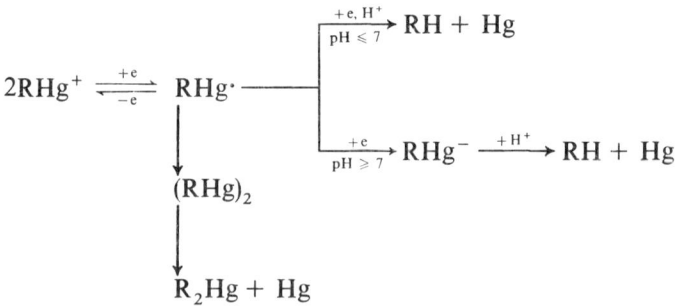

Substituent effects on the reduction of RHgX type compounds. Studies on the influence of the substituent R on the reduction of RHgX type compounds have shown that the half-wave potential values increase in the order allyl < aryl < alkylmercury halides (Denisovich and Gubin, 1973a). However, for *p*-substituted phenylmercury chlorides in 90% dioxane it has been found that the half-wave potential of the first wave was little affected by the change of the substituent, but that of the second wave changed to a considerable extent. In general, introduction of an electron-donating substituent shifted the half-wave potential towards positive values, while the introduction of electrophillic groups led to negative potential shifts. Linear relationships have been found to exist between the electron affinity of the R· radicals and the half-wave potential of the second wave for many hydrocarbon radicals, from which it may be concluded that the electronic change during electrochemical reduction was mainly localized in the partially-filled orbital of R·. For the RHgX compounds in which the carbon atom is joined to the Hg atom through sp^3 hybridization, the half-wave potential was correlated with the Taft induction constant σ^* for the respective radicals (Denisovich and Gubin, 1973a).

(ii) *Diorganomercury compounds*

Disubstituted organomercury compounds have been found to be reduced via the cleavage of the C—Hg bond in a two-electron irreversible process to produce the carbanion, i.e.

$$R_2Hg + 2e \rightarrow Hg + 2R^- \xrightarrow{+2H^+} 2RH$$

Diferrocenylmercury was only reduced with difficulty in dimethylformamide and the reduction wave was observed at very high negative potentials (Denisovich and Gubin, 1973b).

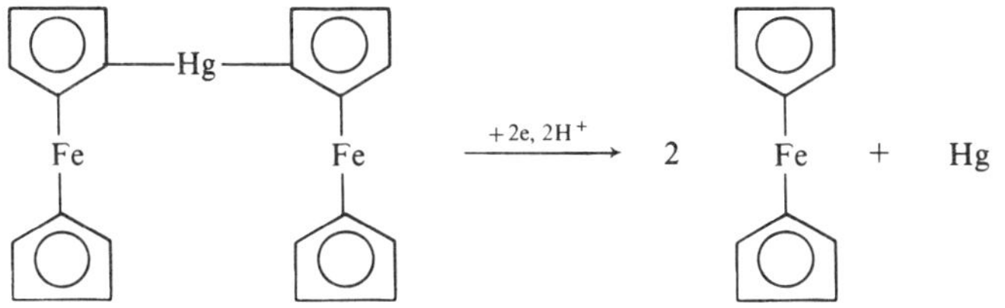

An anodic oxidation wave was also observed for the ferrocene ion at a positive potential.

For carboranylmercury compounds, in addition to the wave corresponding to the reduction of the C—Hg bond, an additional wave appeared at higher negative potentials corresponding to the reduction of the carborane nucleus (Denisovich and Gubin, 1973b).

Substituent effects on the reduction of R_2Hg type of compounds. The thermodynamic stability of the outgoing carbanion R^- has been found to affect the magnitude of the half-wave potential (Dessy *et al.*, 1966a; Hush and Oldham, 1967; Butin *et al.*, 1967). A Hammett type of correlation between the half-wave potential and the *structure* of R of the following form

$$E_{\frac{1}{2}} = 2\sigma\rho$$

has been observed (Butin *et al.*, 1967) for a series of dibenzylmercury compounds with various substituents in the benzene ring $(X-C_6H_4-CH_2)_2Hg$.

For structurally similar compounds of the R_2Hg type having identical transfer-coefficient values a good linear relationship has been found to exist between the half-wave potentials and the pK_a values of the respective hydrocarbon acids (Denisovich and Gubin, 1973a). This relationship provides a basis for the determination of the pK_a values for the hydrocarbon acids by the polarographic method.

C. POLAROGRAPHIC ANALYSIS OF ORGANOMERCURIALS

Due to their toxic effects organomercurials are rarely used as pharmaceuticals; they were previously quite widely used as diuretics. They still find fairly wide use, however, as pesticides and fungicides, seed dressings and antifouling additives in paints.

Brezina and Zuman (1958) have summarized the earlier work on organomercury pharmaceuticals and their conclusions on the optimum analytical conditions are still valid for other applications. Phenylmercury derivatives such as ethylmercurithiosalicylate and p-chloromercuribenzoate are reduced in two waves. The first wave, corresponding to the formation of the organomercury radical, shows some complications from adsorption which can be overcome by the addition of a surface active agent such as gelatine or Triton-X-100 (0.01%). A variety of pH values have been used for analysis of this type of compound but a pH 9 Britton–Robinson buffer would seem to be the best (Wuggatzer and Cross, 1952).

Hopes (1966) has described a method for the polarographic analysis of phenylmercury fungicides. The sample was digested with hydrochloric acid and the organomercurial then extracted with chloroform. The chloroform extract was carefully evaporated to dryness and the residue dissolved in an aqueous pH 9 buffer before polarographic analysis.

Detailed electrochemical studies of methylmercury (Heaton and Laitinen, 1974) and of mercury(II) olefin compounds (Fleet and Jee, 1970) show overall similarity of the reduction processes and the authors have defined the optimum analytical conditions. Both of these studies have shown that the half-wave potential for the reduction of the organomercurial is influenced by the components of the buffer system due to complexation with mercury. Heaton and Laitinen showed that a linear dependence of differential pulse peak current on concentration was obtained for methylmercury over the range 10^{-4} to 10^{-7}M.

Fleet and Jee (1969) determined Hg(II) olefin addition compounds by a.c. polarography in 9:1 $CH_3OH:H_2O$ with 0.1M sodium hydroxide or nitrate as supporting electrolyte. The method has been proposed for the analysis of a wide range of olefins and offers the further possibility, in view of the variation in rates of formation of the olefins, of employing a differential reaction rate technique for analysis of mixtures. The sample of olefin (0.1–2meq) dissolved in methanol was placed in a 10 ml volumetric flask. An excess of methanolic mercury(II) acetate solution (stabilized with 0.1 ml

TABLE I. *Alternating current polarographic data for a selected range of olefins*

Compound	E_s, V	i_p/c	Half-peak width, mV	Time of formation (room temperature) min
4-Methyl-1-pentene	−0.51	1.84	125	60
Cyclohexene	−0.46	2.13	137	90
2,5-Dimethyl-1,5-hexadiene	−0.44	12.3	80	60
Styrene	−0.47	1.85	121	60
α-Methylstyrene	−0.45	2.03	121[b]	40
Allyl alcohol	−0.57	2.09	129	60[a]
Allyl acetate	−0.55	2.05	125	720[a]
Allylacetone	−0.51	1.94	113	2
Vinyl acetate	−0.55	2.29	121	2
N-vinyl-2-pyrrolidone	−0.49	2.03	165[c]	7
N-vinyl carbazole	−0.41	4.54	80	2
Vinyl n-butyl ether	−0.51	1.85	129	2
Vinyl isobutyl ether	−0.51	1.81	121	2
Vinyl 2-chloroethyl ether	−0.51	2.11	121	2
Crotyl alcohol	−0.51	1.80	177[d]	720[a]

E_s = summit potential w.r.t. s.c.e. i_p/c = current in μA for a 1×10^{-4}M solution. Taken from slope of concentration plot. Instrument sensitivity: a deflection of 2.36 μA was observed for the "calibration" setting of the Univector. [a] 2 ml of 0.2M sodium nitrate solution 10 ml added as catalyst. [b] Decreases to 109 mV at 1×10^{-4}M. [c] Decreases to 121 mV at 1×10^{-4}M. [d] Decreases to 129 mV at 1×10^{-4}M. (Data from Fleet and Jee, 1969).

of glacial acetic acid to 100 ml of 0.1M solution) was added so that the final concentration would be approximately $2-3 \times 10^{-4}$M. After a suitable time had elapsed for the formation of the organomercury compound (see Table I) 1 ml of 1.0M aqueous sodium hydroxide was added as supporting electrolyte, and the solution made up to 10 ml with methanol. For those compounds which required a catalyst, 2 ml of 0.2M sodium nitrate (methanolic solution) was added at the same time as the mercury(II) acetate. This solution was deaerated by passage of nitrogen for 5 min, and the a.c. polarogram recorded. It was found to be very important that the 10 ml volumetric flask or other vessels were free from traces of water, otherwise precipitation of mercury(II) oxide occurred as soon as the supporting electrolyte was added and the reaction between olefin and mercury(II) acetate ceased.

Typical a.c. polarographic data from a range of organomercurials are shown in Table I.

III. TIN

A. INTRODUCTION

Organotin derivatives represent the most rapidly expanding group of organometallic compounds. They have found extensive use as wood preservatives, as stabilizers for PVC plastics, as fungicides and as antifouling paint additives. Dialkyltin compounds, particularly the carboxylates and the mercaptocarboxylates and some monoalkyltin compounds, are extensively used for stabilization of polyvinyl chloride plastics and as polymerization catalysts. Without organotin stabilizers PVC turns black and brittle due to its partial degradation on exposure to ultra-violet light or heat. Dibutyltin dimaleate and dibutyltin dilaurates are especially used for the control of helminthic and protozoal infections in poultry. Dialkyltin compounds are also used as low temprature catalysts and curing agents. The toxicity of organotin compounds is well documented (Thayer, 1974). In 1954 "Stalinon" a medical preparation containing diethyltin diiodide marketed in France as a treatment for staphylococcal infections, caused the deaths of over 100 persons and the permanent disablement of a similar number. While all organotin compounds are toxic, the alkyl derivatives are more toxic than the aryl ones and the tri-substituted derivatives more toxic than the di- or tetra-substituted derivatives. Environmentally, organotin compounds have an advantage over alternative organometallics in that the inorganic tin formed on degradation is nontoxic. In view of the importance of these compounds their electrochemical behaviour has been extensively investigated by the authors' group with the primary aims of developing electroanalytical procedures.

B. GENERAL POLAROGRAPHIC BEHAVIOUR

(i) *Triorganotin compounds*

The polarographic behaviour of trialkyl and triaryltin compounds has been extensively studied. In aprotic solvents, trialkyl and triaryltin compounds have been found to give two waves corresponding to successive reduction to form the anion species and subsequently the organotin hydride

(Dessy et al., 1966b). In mixed aqueous–organic solvents the first wave has been found to be accompanied by an adsorption prewave (Vanachayangkul and Morris, 1968). The overall mechanism of reduction is represented by the following scheme (Booth and Fleet, 1970).

$$Ph_3Sn^+ \underset{-e}{\overset{+e}{\rightleftharpoons}} Ph_3Sn\cdot \xrightarrow{+e} Ph_3Sn^- \xrightarrow{+H^+} Ph_3SnH$$
$$\downarrow$$
$$Ph_3Sn\text{-}SnPh_3$$

The reduction of several triphenyltin compounds has been found to show similar polarographic behaviour, irrespective of the nature of their anions (Booth and Fleet, 1970). All the waves have been found to be time independent in aqueous buffer (pH 7.0) containing 50% ethanol. Under these conditions the second wave was accompanied by a large maximum which was easily suppressed by the addition of Triton X-100. The first two waves were independent of pH. At pH 7.3, the first wave was linearly dependent on concentration up to ca 10^{-4} M, above which it was independent of concentration. The second wave appeared only at concentrations above 10^{-4} M and was linearly dependent on concentration up to ca 10^{-3} M. Both the second and the third waves have been found to be diffusion controlled. The third wave initially observed at pH 4.75 reached a maximum at pH 7.3 and then decreased in the form of a dissociation curve, being accompanied beyond pH 9.6 by the appearance of a pronounced minimum at about -1.65 V vs s.c.e. Addition of tetrabutylammonium io has been found to cause the disappearance of the third wave, but lithium, barium and lanthanum ions showed no significant effect on the polarographic waves. As the tetrabutylammonium ion is known to be strongly adsorbed at the electrode, this effect could be due to a competition between tetrabutylammonium ion and triphenyltin free radicals for the available electrode surface. Triphenyltin compounds also showed two a.c. polarographic peaks. The first peak was tensammetric in origin due to the adsorption step while the second peak was due to the first reduction process. The concentration dependence of the two peaks was in accordance with their d.c. polarographic behaviour. The second peak appeared at approximately 10^{-4} M and was linearly dependent on concentration while the first adsorption peak was independent of concentration above 10^{-4} M but linearly dependent on concentration below 10^{-4} M down to a limit of detection of 5×10^{-6} M. The phenomenon of strong adsorption of the triphenyltin free radicals on the electrode surface has been successfully used by Botho and Fleet (1970) to determine this compound by anodic stripping voltammetry at concentrations down to 10^{-8} M.

A detailed study of the polarographic behaviour of trialkyltin compounds showed an overall similarity to the triaryltin compounds. The behaviour of

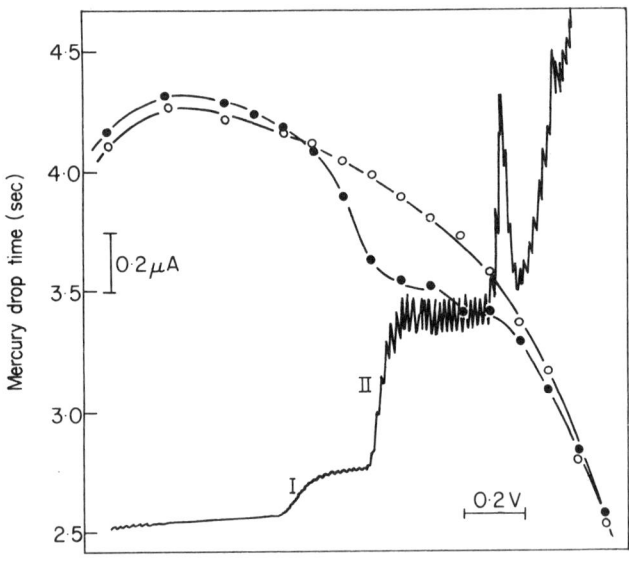

FIG. 2. Polarographic current-potential curves for a 2.76×10^{-4} M TBTO solution in 0.1 M acetic acid/0.1 M ammonia buffer (pH 7) in 50% (v/v) ethanol. Start potential, -0.2 V. Droptime curve for the same solution (●) and that of the buffer (○). (Data from Fleet, B. and Fouzder, N. B., 1975a).

bis(tributyltin) oxide (TBTO), shown in Fig. 2 has, however, been found to be somewhat different (Fleet and Fouzder, 1975a). In acetate buffer (pH 7) containing 50% ethanol TBTO showed four waves. The height of the first wave depended linearly on concentration over the range 10^{-7} to 10^{-4} M. The second wave appeared only above 10^{-4} M and was also linearly dependent on concentration up to approximately 10^{-3} M. The first and third waves were found to be adsorption waves while the second wave was kinetic in nature. The second wave was the normal reduction wave and corresponded to the reduction of the tributyltin cation produced by the hydrolysis of bis-(tri-n-butyltin) oxide to the tributyltin free radical, which was strongly adsorbed onto the electrode giving rise to the first adsorption prewave. The kinetic nature of the second wave was explained by postulating that bis-(tri-n-butyltin) oxide was not reduced at the potential corresponding to the plateau of the first and second wave which corresponded to the reduction of tributyltin cations. The hydrolysis of bis-(tri-n-butylin) oxide which produced the reducible species tributyltin hydroxide was the slow step controlling the

9. ELECTROCHEMISTRY OF ORGANOMETALLIC COMPOUNDS 271

kinetics of the electrode process within this potential region. The third wave of bis-(tri-n-butyltin) oxide corresponded to the adsorption of the products of the second reduction process which involved simultaneous reduction of tributyltin free radicals and bis-tri-n-butyltin oxide molecules to tributyltin hydride. The following side reactions could take place concurrentely.

$$Bu_3Sn\cdot + Bu_3Sn\cdot \rightarrow Bu_6Sn_2$$

$$Bu_3Sn^- + Bu_3Sn^+ \rightarrow Bu_6Sn_2$$

$$Bu_3Sn^- + H^+ \rightarrow Bu_3SnH$$

$$Bu_3SnH + Bu_3SnOH \rightarrow Bu_6Sn_2 + H_2O$$

The formation of hexaalkyl- and hexaaryl-ditin compounds at potentials corresponding to the plateau of the second wave has been confirmed (Devaud, 1966; Fleet and Fouzder 1975a). Exhaustive electrolysis of trialkyltin halides and of triphenyltin acetate at potentials corresponding to the respective third and fourth waves and subsequent isolation and characterization of the products confirmed the formation of tributyltin hydride (Devaud, 1966). The mechanism of reduction of bis-tri-n-butyltin oxide could therefore schematically be represented as follows (Fleet and Fouzder, 1975a).

$$(Bu_3Sn)_2O \underset{-H_2O}{\overset{+H_2O}{\rightleftarrows}} 2\,Bu_3SnOH \rightleftarrows 2\,Bu_3Sn^+_{aq} + 2OH^-$$

$$Bu_3Sn^- \underset{-e}{\overset{+e}{\rightleftarrows}} Bu_3Sn\cdot \underset{-e}{\overset{+e}{\rightleftarrows}} Bu_3Sn^-$$
$$\quad\quad\quad\quad |\quad\quad\quad\quad\quad\quad\quad |$$
$$\quad\quad\quad (Bu_3Sn)_2\quad\quad\quad Bu_3SnH$$

and

$$(Bu_3Sn)_2O \xrightarrow{+4e,\,H_2O} 2Bu_3Sn^- + 2OH^-$$

The reduction of tri-t-butyltin compounds at the dropping mercury electrode was shown to undergo an unusual process which did not lead to hexa-t-butylditin but to 1,1,2,2-tetra-t-butyl-1,2-di(2-hydroxy-2-methyl-ethyl)ditin This deviation from the normal reduction pattern has been attributed to steric hindrance from the bulky tertiary butyl groups (Lerouz and Devaud, 1974).

(ii) *Diorganotin compounds*

The polarographic behaviour of diethyltin dichloride was first investigated by Riccoboni and Popoff (1949), who postulated a two-electron process for this compound. This view was further supported by other investigations. Morris (1967) observed that the two-electron step was greatly distorted by formation of an insoluble polymer product $(R_2Sn)_n$ on the electrode surface, and the wave limited by adsorption of the polymer. At higher concentrations the wave has been found to split into three—an adsorption pre-wave, a reduction wave and a desorption wave. More recently Zezula and Markusova (1972) claimed to have observed four reduction waves for the reduction of dimethyltin dichloride. These authors have ascribed the first wave to an adsorption process and the second wave to a reversible two-electron step:

$$R_2Sn^{2+} \underset{-2e}{\overset{+2e}{\rightleftharpoons}} R_2Sn: \rightarrow \text{Products (unidentified)}$$
$$\downarrow$$
$$(R_2Sn:)_n$$

The third wave was suggested to be due to a two-electron reduction of the dialkyltin hydroxide species formed by hydrolysis of the dialkyltin salt and the fourth wave to reduction of the diradical. Although the formation of polymer was confirmed by its isolation and characterization (Devaud, 1967, several authors have disagreed as to the overall reduction mechanism (Flerov et al., 1969; Flerov and Tyurin, 1970; Leroux and Devaud 1973).

The following mechanism of reduction of dialkyltin compounds has been proposed following a detailed study of their electrochemical behaviour (Fleet and Fouzder, 1975b)

$$Bu_2Sn^{2+} \underset{-e}{\overset{+e}{\rightleftharpoons}} Bu_2Sn\cdot^+ \longrightarrow Bu_2Sn: \xrightarrow{+2e, 2H^+, 2H_2O} Bu_2SnH_2$$
$$\qquad\qquad\qquad \downarrow \qquad\qquad \downarrow$$
$$\qquad\qquad\quad (Bu_2Sn)_2^{2+} \quad (Bu_2Sn:)_n$$

The first wave was found to correspond to the one-electron reversible reduction of dibutyltin dication to dibutyltin ion radical. The dibutyltin ion radical was sufficiently stable in 80% (v/v) ethanol solution for the second wave to be observed but in the presence of higher proportions of water the radical was dimerized more rapidly and the second wave was not observed.

The kinetic nature of the second step was interpreted as follows: during electroreduction dibutyltin diradical species were produced which polymerized to form $(Bu_2Sn)_n$. This polymer formed an insulating film on the mercury electrode surface and further reduction of the dibutyltin ion radicals was then only possible after penetration through the insulating film. This diffusion was the slowest and rate-controlling step in the electrode process. The overall mechanism of reduction has been further confirmed by cyclic voltammetry. A steady state cyclic voltamogram with a depolarizer concentration of 1.55×10^{-4} moles l^{-1} and a scan rate of 200 mV s^{-1} showed three cathodic and three anodic peaks (Fig. 3). Peaks Ia and Ib had a peak separation of 60 mV which corresponded to a one-electron reversible process. Peak IIa corresponded to the irreversible reduction of dibutyltin ion-radical to dibutyltin diradical and Peak IIb to the oxidation of dibutyltin diradical. Peaks IIIa and IIIb were due to the two-electron reduction of

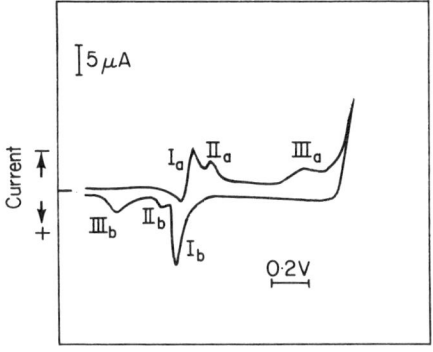

FIG. 3. Cyclic voltammogram of a 1.55×10^{-4} moles l^{-1} dibutyltin dilaureate solution in 80% (v/v) ethanol at pH 7.0; Start potential, -0.1 V; Scan rate, 200 mV s^{-1}. (Data from Fleet, B. and Fouzder, N. B., 1975b.)

dibutyltin diradical to dibutyltin dihydride during the cathodic sweep and the subsequent oxidation of dibutyltin dihydride during the reverse sweep respectively.

A differential pulse polarogram of dibutyltin dilaureate in acetate buffer (pH 7.0) containing 80% (v/v) ethanol showed one well-defined peak corresponding to the first reversible step and two other ill-defined peaks (Fig. 4). This first peak was linearly dependent on concentration over the range 5×10^{-9} to 3×10^{-4} M and hence was of analytical value.

The electrochemical behaviour of diorganotin mercaptides was found to be very similar to other diorganotin compounds with the exception that two

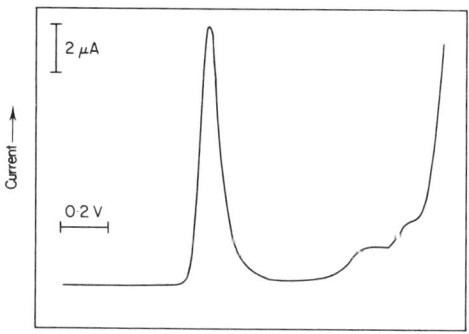

FIG. 4. Differential pulse polarogram of a 1.55×10^{-4} M dibutyltin dilaureate in 80% (v/v) ethanol at pH 7.0. Start potential -0.2 V (Data from Fleet, B. and Fouzder, N. B., 1975b.)

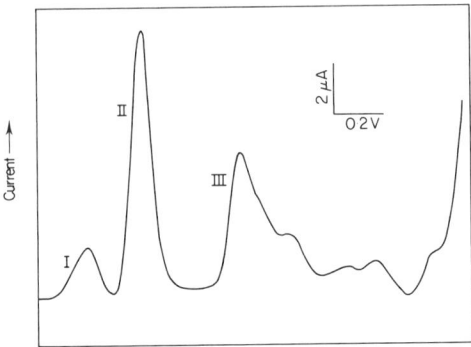

FIG. 5. Differential pulse polarogram of a 1.85×10^{-4}M di-n-octyltin dithioglycolic acid isooctyl ester solution in 80% (v/v) ethanol at pH 7.0 (acetate buffer). Start potential, 0.0 V. (Data from Fleet, B. and Fouzder, N. B., 1975c.)

new anodic peaks appeared at potentials higher than the main reduction peak (Fig. 5). One of these anodic peaks was an adsorption peak while the other corresponded to the oxidation of the mercaptide (Fleet and Fouzder, 1975c).

$$R_2Sn(SR)_2 \xrightarrow[Hg]{-2e} R_2Sn^{2+} + Hg(SR)_2$$

Both the oxidation peak and the first reduction peak could be used for analytical measurement over the concentration range from 10^{-4} to 5×10^{-7} M.

(iii) *Monoorganotin compounds*

The electrochemical behaviour of monoorganotin compounds has not so far been studied in any detail and the mechanism of electrochemical reduction for this class of compound is not clear. Methyl, ethyl, or butyltintrichloride have given several reduction waves in aqueous ethanol medium over a wide range of pH. The waves were both time and pH-dependent and greatly distorted by adsorption of the depolarizer as well as the reduction products. Slow hydrolytic reactions of the compound to stannoic acid was most probably the major cause of the complicated time and pH-dependence of the waves. However, the ultimate product of reduction has been found to be an insoluble polymer, which has been isolated and characterized as $(RSN)_n$ (Devaud *et al.*, 1967). The normal reduction has been postulated to be a single irreversible three-electron step: for example, in acidic medium (Mehner *et al.*, 1968)

$$RSnOOH + 3H^+ + 3e \longrightarrow RSn\cdot + 2H_2O$$
$$nRSn\cdot \longrightarrow (RSn)_n$$

or, in alkaline medium,

$$RSnOO^- + 2H_2O + 3e \longrightarrow RSn\cdot + 4OH^-$$

$$nRSn\cdot \rightarrow (RSn)_n$$

The normal reduction was inhibited by the polymeric film formed at the electrode. In neutral or slightly basic solution (pH 7–9.5) the wave has been found to be kinetic in nature and this was attributed to slow conversion of an electroinactive species, present in this pH region, into the electroactive species.

The behaviour of phenyltintrichloride has been found to be somewhat different from that of the alkyltintrichlorides (Devaud and Souchay, 1967; Devaud and Laviron, 1968). In acidic medium, three distinct waves were observed which were attributed to the reduction of the species, $C_6H_5SnCl_3$, $C_6H_5Sn^{3+}$ and partly to the hydrolysis and condensation products respectively. The reduction was irreversible and autoinhibited due to adsorption of the reduction product. The overall reduction process corresponded to a three-electron process.

It has recently been shown (Fleet and Fouzder, 1978b) that differential pulse polarography of phenyltintrichloride gave a peak in acetate buffer (pH 7.0) containing 80% (v/v) ethanol corresponding to the process

$$PhSnCl_3 \cdot \xrightarrow{+3e} PhSn\cdot + 3Cl^-$$

The phenyltin free radicals formed at the electrode polymerized rapidly according to the following reaction:

$$nPhSn\cdot \rightarrow (PhSn)_n$$

The polymer was isolated after exhaustive controlled potential electrolysis and characterized. The reduction process was complicated by the hydrolysis of organotin compound to benzene stannoic acid.

$$PhSnCl_3 + H_2O \rightarrow PhSnOOH + 3Cl$$

This hydrolysis was unusually slow but was accelerated in alkaline solution. This behaviour was reflected in the pH-dependence and time-dependence of the limiting current. It has been postulated that the product of hydrolysis, benzene stannoic acid, was not reduced at the same potential as phenyltin trichloride so that with increasing time the extent of hydrolysis increased and consequently the limiting current decreased. The rate and extent of hydrolysis also increased with increasing pH of the medium so that the limiting current decreased rapidly with increasing pH in alkaline medium.

Phenyltintrichloride also gave an anodic wave having $E_{\frac{1}{2}}$ of -0.1 V vs s.c.e. Devaud and Souchay (1967) ascribed the anodic wave to the oxidation of stannoate, $PhSnOO^-$. Fleet and Fouzder (1978a) however have considered it to be due to oxidation of the organotin according to the following reactions

$$PhSnCl_3 + 2H_2O - 3e \xrightarrow{Hg} PhSnOOH + 3H^+ + \tfrac{3}{2}Hg_2Cl_2$$

and

$$PhSn^{3+} - 2e \xrightarrow{Hg} Sn^{4+} + PhHg^+$$

In alkaline solution the following reactions could take place.

$$PhSnCl_3 + 4OH^- - 3e \rightarrow PhSnOO^- + 2H_2O + \tfrac{3}{2}Hg_2Cl_2$$

and

$$3Cl^- - 3e \xrightarrow{Hg} \tfrac{3}{2}Hg_2Cl_2$$

It was assumed that stannoate was not oxidized at the potential under consideration.

C. POLAROGRAPHIC ANALYSIS

For trace analysis using d.p.p. it is recommended that the first adsorption peak corresponding to the formation of the organotin free radical be used. At higher concentrations corresponding to formation of more than a monolayer the first main reduction peak should be used (Booth and Fleet, 1970). Utilizing the first adsorption peak Fleet and Fouzder (1975a) have carried out the analytical determination of fentin residues in pesticide formulations by the differential pulse polarographic technique (Table II).

TABLE II. *Organotin content of pesticide formulations**

Pesticide formulation	Organotin content (%)
Fentin acetate tech.	76.9
Brestan	45.4
Duter	45.4

* From Fleet and Fouzder (1952a).

Booth and Fleet (1970) have shown that fentin residues can be determined in the region 10^{-7}–10^{-8} M by anodic stripping voltammetry using the adsorption of the triphenyltin radical as the pre-electrolysis step. While there was some slight loss of radical during the pre-electrolysis step due to dimerization or deactivation the method was found to be reasonably precise. It was necessary to control the length of the pre-electrolysis step so that less than a monolayer of free radical was deposited on the electrode surface. When more than a monolayer was deposited, two distinct stripping peaks were obtained corresponding to the removal of the bulk and surface monolayer, respectively. Using this procedure levels of fentin in potatoes were determined following extraction of the macerated sample with acetonitrile and direct measurement in the aqueous organic extract (Table III).

TABLE III. *Samples of potatoes analysed for fentin* *

Sample	Skin		$\frac{1}{4}$-inch layer		Bulk		
	µg ml^{-1}	Potato, %	µg ml^{-1}	Potato, %	µg ml^{-1}	Potato, %	Total µg ml^{-1}
1	0.03	8	0.01	70	0.001	22	0.008
2	0.05	12	0.001	47	0.001	41	0.007
3	ND^a	8	0.01	60	ND^a	32	0.005
4	ND^a	8	0.01	51	ND^a	41	0.005
5	0.001	8	0.002	64	ND^a	28	0.001

^{a} ND, None detected. * Booth and Fleet (1970).

The procedure was as follows: triphenyltin acetate was extracted from a 100 g sample previously dried for 30 minutes at 90°C, with 100 ml of acetonitrile. The extract was passed down a 2 cm diameter column containing 5 cm layer of anhydrous sodium sulphate to remove any remaining water and 2.5 cm of alumina to remove oils and starchy material. The volume of the eluent was then reduced to less than 10 ml using a Kuderna evaporator. The extract was then mixed with acetate buffer (pH 7.5) and Triton-X 100 (0.002%) and the final solution was diluted so that it contained 50% (v/v) ethanol. For the anodic stripping process pre-electrolysis was carried out at -1.0 V vs s.c.e. for two minutes using a rotating mercury coated platinum wire electrode (1.0 × 0.1 cm, although recent advances in the technique would now favour the use of glassy carbon or wax impregnated graphite as the substrate). The anodic peak corresponding to the oxidation of the adsorbed triphenyltin free radical was measured. Standard additions of fentin to the potato samples prior to the extraction process indicated that the procedure was within 10% of the quantitative recovery for 0.4 µg ml^{-1} of fentin, and the method is approximately one hundred times more sensitive than conventional spectrophotometric procedures.

Polarographic determination of dialkyltin compounds can be carried out over the concentration range 10^{-4} to 10^{-9} M by differential pulse polarography using acetate buffer (pH 7.0) containing 80% (v/v) ethanol as the background electrolyte and utilizing the first reduction peak. By using differential pulse polarography under the same solution conditions as for diorganotin compounds, monoorganotin compounds could be determined down to 5×10^{-7} M. However as the peak height of this class of compounds decreased with time, polarographic measurements should be carried out as quickly as possible. The decrease in peak current over a 15–20 minutes period was of the order of a few percent (Fleet and Fouzder, 1978a) for these compounds.

Polarographic analysis of mixtures of organotin species could also be carried out. Figure 6 shows the differential pulse polarogram of pure phenyltintrichloride alone and also after the addition of di- and triphenyltin compounds in acetate buffer (pH 7.0) containing 80% (v/v) ethanol. It can be

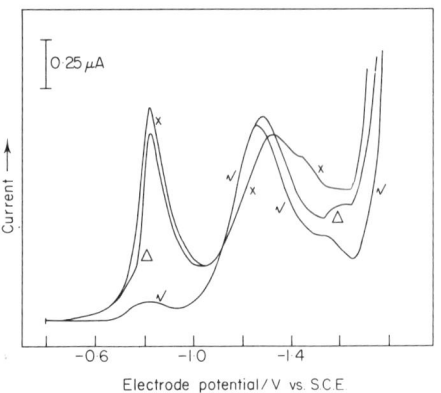

FIG. 6. Differential pulse polarogram of a 2.23×10^{-5} moles 1^{-1} phenyltin tricholride solution in acetate buffer (pH 7.0) containing 80% (v/v) ethanol.
($\sqrt{}$) pure phenyltin trichloride solution.
($\triangle$) in the presence of diphenyltin dichloride (2.91×10^{-5}M).
($\times$) in the presence of di- and tri-phenyltin compounds (2.91×10^{-5}M and 1.43×10^{-5}M respectively). (Data from Fleet, B. and Fouzder, N. B., 1978b.)

seen from this figure that the major peaks corresponding to the primary reduction of mono-, di- and triphenyltin chlorides are at -1.28, -0.84 and -0.83 V vs s.c.e. respectively. A linear calibration curve for phenyltintrichloride in the presence of di- and tri-phenyltin compounds was observed.

Jehring and Mehner (1963a,b, 1964) determined traces of tributyltin chloride in dibutyltin compounds using a potassium chloride supporting electrolyte solution at pH 10 containing 30% isopropanol as co-solvent.

Oscillographic polarography has also been used to measure as little as 0.005% tributyltin chloride in the presence of other butyltin chlorides and tetrabutyltin (Geyer and Rotermund, 1969).

Bork and Selivokhin (1969a, 1969b) have also carried out analytical determinations of organotins in mixtures using 2–5 M hydrochloric acid in 30 or 40% (v/v) alcohol.

Fleet and Fouzder (1978b) separated dibutyltin compounds from their mixture with tributyltin oxide by high pressure liquid chromatography using a silica gel column, 1:1 methanol—1M potassium nitrate being the eluent. The separated organotins were monitored amperometrically using a glassy carbon wall jet electrode detector system. Determinations of di-n-butyltin dichloride and tri-n-butyltin chloride in the presence of mono-n-butyltintrichloride with sodium diphenyldithiocarbamate has been described by Issleib et al. (1968). The determination was carried out in solutions containing 5% of sodium acetate trihydrate in methanol, which dissolved the dithiocarbamates of the di-n-butyltin and tri-n-butyltin compounds and shifted the reduction wave of the mono-n-butyltin compound to more negative potentials. Mixtures of di and trialkyltins could also be analysed potentiometrically by first titrating the total organotin potentiometrically, using standard 8-hydroxyquinoline solutions at an applied potential of -1.4 V vs s.c.e. in ammonia–ammonium nitrate buffer (0.2 M). The latter measurement gave the dialkyltin concentration. The procedure was accurate to within 1% for titrations of 2×10^{-3}–2×10^{-4} M solutions of dialkyltin compounds (Plazzogna and Pilloni, 1967).

Hassova and Pribyl (1970) have devised an amperometric procedure for the analysis of dialkyltins based on the reaction of these compounds with oxalic acid in slightly acidic medium to form the corresponding tin oxalate. Dibutyl and diethyltin compounds were determined in 5 M acetic acid–ethanol (1:1) using a dropping mercury electrode at a potential of -0.9 V vs Ag/AgCl. Trialkyltin impurities did not interfere in the determination while the interference by monoalkyltin impurities could be easily suppressed by EDTA.

Tagliavini and Plazzogna (1962) determined hexamethylditin in tetramethyltin coulometrically with amperometric end-point indication using 0.5M alcoholic ammonium bromide as the background electrolyte. Tagliavini (1966) also used electrolytically generated iodine, bromine or silver ion for the amperometric titration of several hexaorganoditin compounds such as hexamethylditin, hexaphenylditin, etc. The method was based on the following reaction

$$R_3Sn - SnR_3 + X_2 \rightarrow 2R_3Sn$$

where X_2 is bromine or iodine generated electrolytically in the system.

IV. LEAD

A. INTRODUCTION

Organic derivatives of lead also have widespread industrial application, primarily as antiknock additives in petrol, but also, although to a much smaller extent, as polymerization catalysts, antifouling agents and as alkylating agents for the production of organomercurial fungicides. Like the tetraalkyltin compounds the tetraalkyllead ones are toxic due to the formation of the trialkyllead cation which like the trialkyltin cation acts by inhibition of oxidative phosphorylation while the dialkyllead compounds are toxic due to their ability to complex enzyme sulphydryl groups. The study of the toxicology of lead compounds has assumed considerable importance due to the increasing concern over environmental lead pollution.

B. GENERAL POLAROGRAPHIC BEHAVIOUR

(i) *Triorganolead compounds*

Despite the extensive research work carried out in other areas of organolead compounds, there has been scant attention to the electrochemical behaviour of organolead compounds until recently. Costa (1948) and later Korshunov and Malyugina (1961) investigated the polarography of trialkyllead compounds and observed two reduction waves. The first wave has been attributed to the one-electron reduction of the trialkyllead species to a trialkyllead free radical, while the second step was attributed to the reduction of hexaethyldilead produced at the electrode in the preceding step. Two waves were also observed in non-aqueous solution. Dessy et al. (1966b) have proposed that in dimethoxyethane medium triphenyllead compounds are reduced in two steps as follows:

$$Ph_3Pb^+ \begin{cases} \xrightarrow{+e} Ph_3Pb\cdot \xrightarrow{+Hg} Ph_2Hg + Pb \\ \xrightarrow{+2e} Ph_3Pb \end{cases}$$

The product of the first step was found to be the triphenyllead free radical which readily dissolved into the electrode to form diphenylmercury, while the second step involved the two-electron reduction of the triphenyllead cation to form, initially, the triphenyllead anion.

Colliard and Devaud (1972), however, have since shown that the polarographic behaviour of organolead compounds in aqueous ethanol solution was complicated due to the initial reaction of this compound with the mercury of the d.m.e. to produce metallic lead which was subsequently oxidized at the

9. ELECTROCHEMISTRY OF ORGANOMETALLIC COMPOUNDS

electrode resulting in the complex appearance of the i–E curves. For the reduction of triphenyllead hydroxide in 0.1N perchloric acid/20% methanol these same authors observed two major waves and two other smaller complex waves. These authors attributed each reduction wave to the direct reduction of the triphenyllead cation. The reduction processes may be summarized in the following scheme:

$$Ph_3Pb^+ \begin{cases} \xrightarrow[-0.1\text{ V}]{+e} Ph_3Pb\cdot \xrightarrow{+\frac{3}{2}Hg} \frac{3}{2}Ph_2Hg + Pb \\ \xrightarrow[-0.5\text{ V}]{+e,\ \frac{3}{2}Hg} \frac{3}{2}Ph_2Hg + Pb \\ \xrightarrow[-1.10\text{ V}]{+4e,\ 3H^+} 3C_6H_6 + Pb \\ \xrightarrow{+2e,\ H^+,\ Hg} C_6H_6 + Ph_2Hg + Pb \end{cases}$$

and

$$Pb \xrightarrow[-0.4\text{ V}]{-2e} Pb^{2+}$$

More recent work (Fleet and Fouzder, 1978a) has shown that triphenyllead acetate also gave two well-defined waves in acetate buffer (pH, 7.0) containing 50% ethanol (Fig. 7). The first wave was preceded by two rather complex small waves, one anodic and one cathodic. Both the normal waves

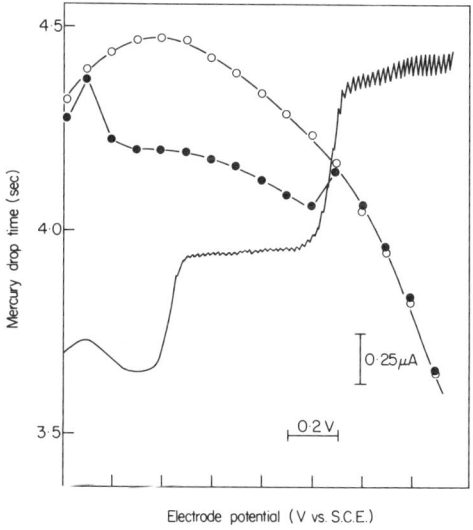

FIG. 7. Polarographic current-potential curve of a 1.68×10^{-4}M solution of triphenyllead acetate in acetate buffer (pH 7.0) containing 50% (v/v) ethanol. Start potential, -0.1 V. Drop-time curve of the same solution (●), and that of the background solution (○). (Data from Fleet, B. and Fouzder, N. B., 1978a.)

were found to be diffusion-controlled and irreversible, and linearly dependent on concentration. Controlled potential coulometry at potentials corresponding to the plateau of the first and second normal wave gave n-values of one and four respectively. But under identical conditions controlled potential coulometry at -0.325 V showed that the electrolysis current initially increased cathodically and then decreased rapidly through zero to an anodic value that corresponds to a final n-value of about 0.5. This unusual behaviour was due to the oxidation, in this potential region, of metallic lead produced in a process subsequent to the primary electrochemical step.

The following mechanism of reduction of triphenyllead acetate has been postulated by Fleet and Fouzder (1978a):

$$Ph_3Pb^+ \xrightarrow{+e} Ph_3Pb\cdot_{ads} \begin{cases} \xrightarrow{+e} Ph_3Pn^- \xrightarrow{+H^+} Ph_3PbH \\ \xrightarrow[-Pb]{+Hg} 3PhHg\cdot \xrightarrow{+3e} 3PhHg^- \end{cases}$$

$$\downarrow$$
$$Ph_6Pb_2$$
$$\downarrow$$
$$Ph_4Pb + Pb \qquad\qquad 3C_6H_6 + Hg$$

(with $+3H^+$ on the right branch)

$$Ph_3Pb^+ + Hg \rightarrow Ph_2Hg + PhHg^+ + Pb$$

$$Pb - 2e \rightleftharpoons Pb^{2+}$$

According to this scheme the first major wave corresponded to the one-electron reduction of triphenyllead cation to triphenyllead free radical which was strongly adsorbed at the electrode surface. The reduction was irreversible and accompanied by a typical Brdicka adsorption pre-wave which was greatly distorted by the opposing oxidation wave of metallic lead produced from the chemical breakdown of the adsorbed triphenyllead free radicals. This caused the formation of the S-shaped wave which preceded the normal reduction wave I. The second step was found to be rather complicated: the triphenyllead free radical produced in step I could undergo further reduction to a triphenyllead anion or alternatively could dissolve into the dropping mercury electrode to produce phenylmercury free radical which would be immediately reduced further to the phenylmercury anion and free metallic mercury. Both of these processes could take place simultaneously, together with the possible dimerization and disproportionation processes. This overall scheme has been supported by cyclic voltammetry experiments and by voltammetry using the glassy carbon electrode.

(ii) *Diorganolead compounds*

The electrochemical behaviour of diphenyllead compounds in aqueous alcoholic medium has been found to be very similar to that of the corresponding triphenyllead derivatives. Two main waves were observed, with the first wave preceded by an apparent adsorption wave.

Colliard and Devaud (1972) attributed the two main waves to the following processes

$$Ph_2Pb^{2+} \xrightarrow{+2e} Ph_2Pb:$$

and

$$Ph_2Pb^{2+} \xrightarrow{+2H^+ + 4e} 2 C_6H_6 + Pb$$

The first wave was greatly distorted by the adsorption process involving oxidation of the metallic lead produced by dismutation of the diradical species $Ph_2Pg^{..}$

$$3 Ph_2Pb^{..} \rightarrow Ph_6Pb_2 + Pb$$

or by reaction with the mercury of the electrode according to

$$Ph_2Pb^{..} + Hg \rightarrow Ph_2Hg + Pb.$$

The reduction of dialkyllead compounds in aqueous–alcoholic medium has been studied by Morris (1969) who claimed to have observed a two-electron reduction wave. The product of this process was assumed to be a dialkyllead compound which was rapidly decomposed by disproportionation to tetraalkyllead and metallic lead and by transmetallation to diethyl mercury. The relative amount of tetraalkyllead was found to increase with increasing concentration of the diethyllead cation.

Recent experiments (Fleet and Fouzder, 1978a) have shown, however, that dialkyllead compounds gave two well-defined one-electron reduction steps in aqueous ethanolic medium, corresponding to the scheme

$$R_2Pb^{2+} \xrightarrow{+e} R_2Pb^{.+} \xrightarrow{+e} R_2Pb \xrightarrow{Hg} R_2Hg + Pb$$

The waves were diffusion controlled and irreversible. Two more rather ill-defined polarographic waves were also observed and the behaviour was quite complicated because of the reaction with mercury. All the waves have been found to be time independent. The height of wave I was little influenced at pH 7.0 but in highly acidic or alkaline medium it split into two waves. These two waves were due to the reduction of hydrolysis products of the dibutyllead compound. Dibutyllead has also been found to react with borate ion so that results observed in borate buffer had little significance. It is of interest that the dibutyllead radical ion showed no evidence of strong absorption on mercury. There was a marked difference between aryl and alkyllead compounds in this respect.

(iii) *Monoorganolead compounds*

Colliard and Devaud (1973) studied the polarographic behaviour of monoorganolead compounds. For phenylleadtriacetate in aqueous ethanol they observed three successive one-electron steps. The product of the first step, phenyllead dication was further hydrolysed to phenol or alternatively it decomposed to form diphenylmercury. The metallic lead formed during decomposition was oxidized to the lead ion. The product of the second reduction step was phenyllead cation which also reacted with mercury to form diphenylmercury or decomposed to phenol and lead. The situation was further complicated by reduction of the lead ion which is superimposed on the primary process. The reduction product of the third step was phenyllead free radical which was slowly decomposed into diphenylmercury and lead.

The overall mechanism of the reduction process may be represented as follows:

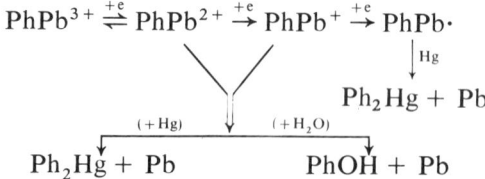

In fresh solutions however, the products of the first and the second steps were stabilized by complex formation with acetate ions.

C. POLAROGRAPHIC ANALYSIS

In acetate buffer medium (pH 7.0) containing 50% ethanol triphenyllead compounds showed two well-defined differential pulse polarographic peaks which were diffusion controlled and linearly dependent on concentration over a wide concentration range (Fig. 8). Both peaks were suitable for analytical measurement although in order to avoid problems of chemical side reactions, it was essential that the initial potential be sufficiently negative (-0.3 V vs s.c.e. is recommended). The first peak showed a linear concentration range of 2×10^{-4} to 5×10^{-7} M. A similar medium could be used for analysis of dialkyllead compounds over the conventional differential pulse polarographic range 10^{-4} to 10^{-7} M and the reaction of the compounds with metallic mercury did not interfere with the analytical measurement.

Since the half-wave potentials of the organolead compounds have been found to depend on the number and nature of the organic species attached to lead, simultaneous analysis of mixture of organolead compounds could be

9. ELECTROCHEMISTRY OF ORGANOMETALLIC COMPOUNDS

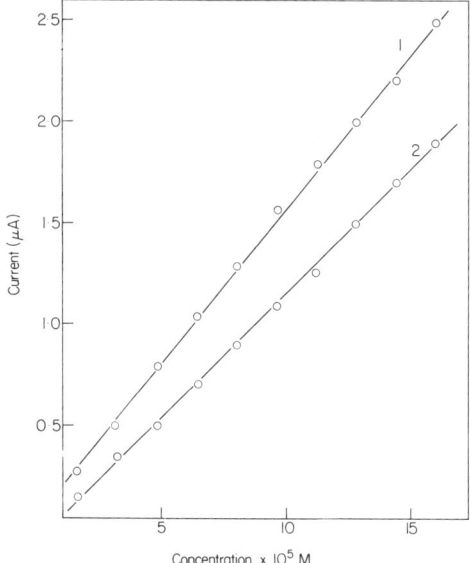

FIG. 8. Dependence of pulse polarographic limiting current and differential pulse polarographic peak current on concentration of triphenyllead acetate: (1) pulse polarographic wave I, (2) differential pulse polarographic peak I. (Data from Fleet, B. and Fouzder, N. B., 1978a.)

carried out. It should be noted that tetraalkyllead is reduced only at very high negative potentials, so determination of tetraalkyllead has generally to be carried out by indirect methods based on the decomposition of the organic compound into inorganic lead and its subsequent polarography. Thus Hubis and Clark (1955) described a rapid polarographic method for the determination of tetraethyllead in petroleum in which the petroleum sample is dissolved in 2-ethoxyethanol (cellosolve)–hydrogen chloride solution which simultaneously decomposed the organic lead to lead chloride and extracted the latter. De Vries *et al.* (1959) have also described a polarographic method for the determination of hexaalkyldilead and triethyllead chloride in tetraethyllead. The method involved a direct polarographic measurement at $E_{\frac{1}{2}} = -0.24$ V vs s.c.e. using 1:1 (v/v) benzene–methanol mixture with lithium chloride as the background electrolyte. A similar method for the determination of hexaethyldilead in tetraethyllead and in triethyllead has also been described by Vertyulina and Korschunov (1959). Tagliavini *et al.* (1961) carried out coulometric determination of hexaethyldilead with amperometric end-point indication. The method was based on the following reaction

$$R_3Pb\text{—}PbR_3 + X_2 \rightarrow 2R_3PbX$$

where X is iodine produced electrolytically in the system under constant current. Plazzogna and Pilloni (1967) and Pilloni (1968) carried out amperometric analysis of mixtures of diorgano- and triorganolead compounds at the 0.01–0.05 mM level with potassium ferrocyanide using 50% aqueous methanol/0.5M lithium nitrate background electrolyte at -0.55 V vs s.c.e. The triorganolead was determined by adding an excess of sodium tetraphenylborate and then titrating the unconsumed tetraphenylborate with thallous nitrate solution at -0.475 V vs s.c.e. in 0.01 M perchloric acid and -0.05 M sodium perchlorate solution as background electrolyte. A potentiometric method for the determination of inorganic lead (ca 0.4–1% range) in the presence of organolead compounds was also described (Plazzogna and Pilloni, 1967).

Pilloni and Plazzogna (1964) and Pilloni (1967) have described coulometric determination of tetraalkyllead compound using bromine or mercurous ion. The titration of the organolead with bromine was based on the following reaction

$$R_4Pb + Br_2 \rightarrow R_3PbBr + RBr.$$

An anodic stripping method for the determination of organolead in petroleum has been described by Roschig and Matschiner (1967). A solution of bromine in chloroform has been used to decompose the tetraalkyllead followed by extraction of lead ions into 0.1M nitric acid. In this method pre-electrolysis was carried out at -0.8 V vs s.c.e. for 1–5 mins and lead determined at -0.39 V vs s.c.e. The limit of detection was 8 parts of lead per 10^{13}, with a coefficient of variation of 7%.

V. GERMANIUM

A. INTRODUCTION

Interest in organogermanium chemistry has increased steadily in recent years. Its primary importance is as an intermediate in synthetic organic chemistry but it also has potential application in various other areas such as hydrolysis-proof coatings, in photoconductive materials and as a drug for the treatment of cancer. The toxicological effect of organogermanium compounds is far less marked than their corresponding tin and lead analogues but the di- and trialkyl derivatives do show some slight antibacterial activity.

B. GENERAL POLAROGRAPHIC BEHAVIOUR

(i) *Triorganogermanium compounds*

Allred and Bush (1968) and Curtis and Allred (1965) used electrochemical techniques including polarography and cyclic voltammetry to calculate the energy of the lowest unoccupied molecular orbital of 4-(trimethylgermyl)-biphenyl and 4,4-bis(trimethylgermyl)-biphenyl in N,N'-dimethylformamide and observed that the results were in good agreement with e.s.r. data. A study of organogermanium compounds in 1,2-dimethoxyethane was carried out by Dessy *et al.* (1966b) who proposed the following scheme of reduction for these compounds

$$Ph_3GeCl \xrightarrow{+e} [Ph_3Ge\cdot] \xrightarrow{HS} Ph_3GeH$$

$$Ph_2GeCl_2 \xrightarrow{+2e} Ph_2GeH_2$$

$$Ph_3GeGePh_3 \xrightarrow{+2e} 2\,Ph_3Ge^- \quad (HS = solvent)$$

The triphenylgermanium radical which can abstract a hydrogen atom from the solvent has been found to have a very short lifetime in this medium so that attempts at detection by cyclic voltammetry were unsuccessful. Boczkowski and Bottei (1973) carried out a detailed investigation of the electrochemical behaviour of triphenylgermanium halides in 1,2-dimethoxyethane. These authors showed that the reduction of triphenylgermanium chloride and bromide follows the scheme shown below:

$$Ph_3GeX \xrightarrow{+e} [Ph_3Ge\cdot] \xrightarrow{HS} Ph_3GeH$$
$$\downarrow$$
$$Ph_3GeGePh_3$$

The first step was an irreversible one-electron transfer which resulted in the formation of a triphenylgermyl free radical, which showed no corresponding oxidation peak by cyclic voltammetry. For Ph_3GeI, however, these authors observed two waves, the first a kinetic wave and the second an irreversible diffusion controlled wave.

In a recent investigation Fleet and Fouzder (1978a) have shown that triphenylgermanium bromide gives one rather small drawn-out reduction wave in nonaqueous solvents such as dimethylsulphoxide or acetonitrile. However, in aqueous ethanolic medium this compound showed two waves (Fig. 9). The first wave had the characteristics of an adsorption wave and showed a linear dependence on concentration below the 10^{-4}M level. The limiting current was dependent on pH and showed a significant decrease at pH 7.6. The limiting current was influenced by the solvent composition

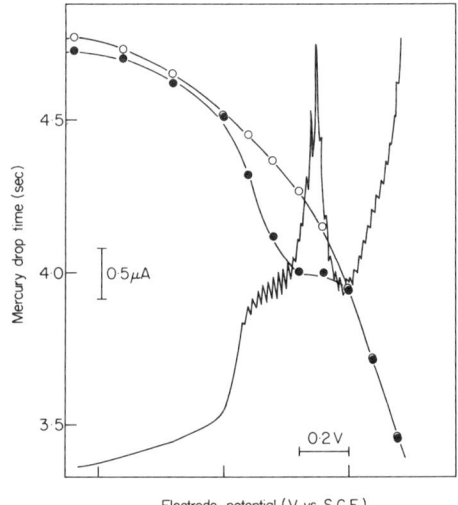

Fig. 9. Polarographic current-potential curve of a 4.90 × 10^{-4}M solution of triphenylgermanium bromide in acetate buffer (pH 7.0) containing 50% (v/v) ethanol solution. Start potential, −0.4 V. Droptime curve of the same solution (●) and of the background electrolyte (○). (Data from Fleet, B. and Fouzder, N. B., 1978a.)

showing a sharp decrease at ethanol concentration >60%(v/v). The second wave had a rather unusual appearance and decreased rapidly with increasing ethanol concentration. Cyclic voltammetric measurements have shown both waves to be irreversible. It has been posulated that the small wave in nonaqueous solvents corresponds to the irreversible one-electron reduction of triphenylgermanium bromide. The transient free radicals formed in this step rapidly abstract hydrogen from the solvent to give triphenylgermane.

$$Ph_3GeBr \xrightarrow{+e} [Ph_3Ge\cdot] \xrightarrow{HS} Ph_3GeH$$

In aqueous ethanol medium, particularly at pH ⩽ 7.0, the free radicals were strongly adsorbed and rapidly abstracted a proton according to the reaction

$$Ph_3Ge\cdot_{ads.} \xrightarrow{+H^+} Ph_3GeH^+$$

The triphenylgermanium hydride cation, Ph_3GeH^+, was found to be much more reactive than the triphenylgermanium bromide and could undergo further reduction to triphenylgermanium hydride. Strong adsorption of the products of electrolysis led to the appearance of a typical Brdicka adsorption pre-wave. A fraction of the transient triphenylgermanium free radicals could also dimerize in parallel reactions, particularly in neutral or in alkaline solution and at higher concentrations. Both the dimer and the hydride have been isolated and identified by IR spectroscopy after exhaustive controlled potential electrolysis (Boczkowski and Bottei, 1973). In addition to the cathodic waves an anodic wave was also observed in the polarogram, which corresponded to the mercury dissolution process. The reduction mechanism may be summarized in the following scheme (Fleet and Fouzder, 1978a)

$$Ph_3GeBr \xrightarrow{+e} [Ph_3Ge^\cdot] \begin{array}{c} \xrightarrow{pH \geq 7, HS} Ph_3GeH \\ \xrightarrow{+H^+, pH \leq 7} Ph_3GeH^+ \xrightarrow{+e} Ph_3GeH \end{array}$$
$$\downarrow$$
$$Ph_3GeGePh_3$$

(ii) *Di and monoorganogermanium compounds*

At the present time there is no information available on the electrochemical behaviour of these classes of compound.

C. POLAROGRAPHIC ANALYSIS

The first reduction wave for triphenylgermanium compound has been utilized for analytical purposes over the concentration range 5×10^{-4} down

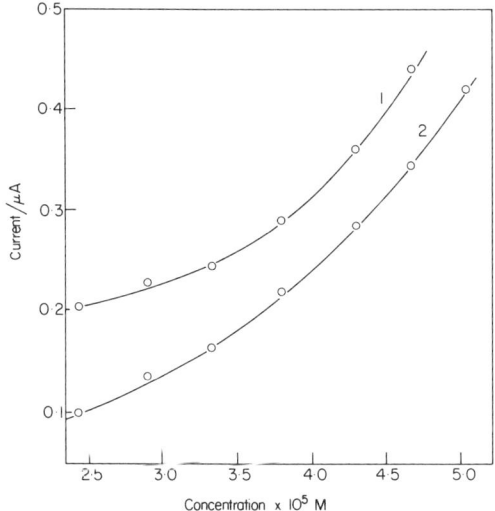

FIG. 10. Concentration dependence of the differential pulse polarographic peak current for triphenylgermanium bromide; (1) peak 1 (2) peak 2. (Data from Fleet, B. and Fouzder, N. B., 1978a.)

to ca 10^{-7}M. The second wave was unsuitable for analytical purposes, being less well-defined and showing a complex concentration dependence. A slight non-linearity was observed for the pulse and differential pulse calibration curves based on wave 1 (Fig. 10). As has been mentioned previously this is fairly characteristic for electrode reactions involving free radical products or intermediates where some degree of regeneration of depolarizer occurs. Empirical calibration over the concentration region of interest is recommended. The optimum solution conditions are aqueous ethanol/0.1M acetate buffer (pH 6.5–7.0) (1:1).

VI. CONCLUSIONS

The rapidly expanding literature on organometallics confirms the increasing commercial importance of these compounds. A clear understanding of the electrochemical behaviour of these compounds is essential, firstly for the development of analytical procedures and secondly for the definition of experimental conditions for electrosynthetic processes. From the analytical viewpoint electrochemical methods, in particular polarography, offer unique information in that they provide a highly sensitive method of analysis both for the primary material and the various intermediates in chemical or electrochemical reactions. This feature is of vital importance in environmental studies where the identification of metabolites or breakdown products, which can be far more toxic than the parent compound, is essential.

Although electrochemical studies of organometallic compounds have been in progress since the beginning of the century it is only in the last decade with the advance of modern electrochemical and spectroscopic analytical techniques that any serious understanding of the mechanism of electrode processes has emerged. From the preceding discussion the overall reduction mechanism for these compounds may be represented by a modified scheme, based on that proposed by Dessy and Bares (1972).

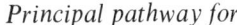

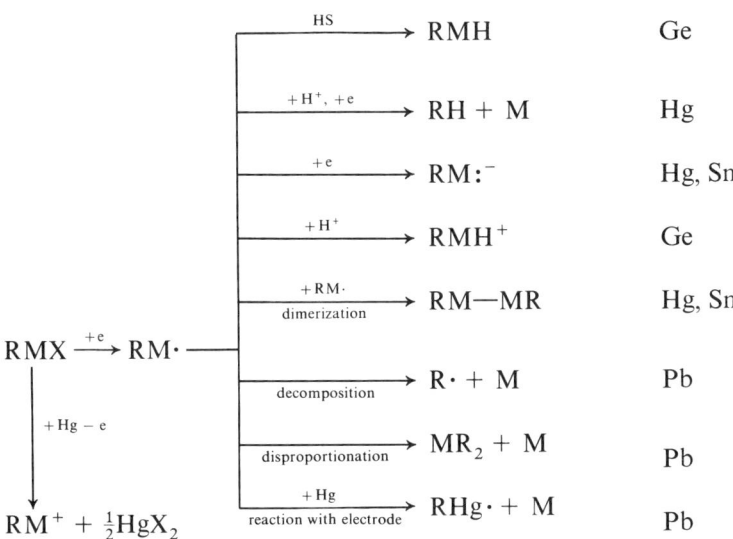

The primary electrochemical reduction step involving the formation of the free radical is most important for electroanalytical purposes. The second reduction step gives rise to some interesting metalloid anion species which may be synthetically very useful. However, the chemical reactivity and catalytic behaviour of electrochemically generated organometallic species of these metals has yet to be fully exploited.

REFERENCES

Allred, A. L. and Bush, L. W. (1968). *J. Amer. Chem. Soc.* **90**, 3352.
Benesch, R. E. and Benesch, R. (1951). *J. Amer. Chem. Soc.* **73**, 3391.
Benesch, R. and Benesch, R. E. (1952). *J. Phys. Chem.* **56**, 648.
Boczkowski, P. J. and Bottei, R. S. (1973). *J. Organometal. Chem.* **49**, 389.
Booth, M. D. and Fleet, B. (1970). *Anal. Chem.* **42**, 825.
Bork, V. A. and Selivokhin, P. I. (1969a) *Plast. Massy*, **10**, 60.
Bork, V. A. and Selivokhin, P. I. (1969b). *Tr. Mosk, Khim-Technol. Inst.* **62**, 249.
Brezina, M. and Zuman, P. (1958). "Polarography in Medicine, Biochemistry and Pharmacy", p. 527. Interscience, New York.
Butin, K. P., Beletskaya, I. P., Kashin, A. N. and Reutov, P. A. (1967). *J. Organometal. Chem.* **10**, 197.
Butin, K. P., Ershler, A. B., Strelets, V. V., Kashin, A. N., Beletskaya, I. P., Reutov, O. A. and Marcushova, K. (1974) *J. Organometal. Chem.* **64**, 171.
Colliard, J. P. and Devaud, M. (1972). *Bull. Soc. Chim. France*, 4068.

Colliard, J. P. and Devaud, M. (1973). *Bull. Soc. Chim. France*, 1541.
Costa, G. (1948). *Am. Chim. (Rome)*, **38**, 655.
Curtis, M. D. and Allred, A. L. (1965). *J. Amer. Chem. Soc.* **87**, 2554.
Degrand, C. and Laviron, E. (1968a). *Bull. Soc. Chim. France*, 2228.
Degrand, C. and Laviron, E. (1968b). *Bull. Soc. Chim. France*, 2233.
Denisovich, L. I. and Gubin, S. P. (1973a). *J. Organometal. chem.* **57**, 87.
Denisovich, L. I. and Gubin, S. P. (1973b). *J. Organometal. chem.* **57**, 99.
Dessy, R. E. and Bares, L. A. (1972). *Accounts chem. Res.* **5**, 415.
Dessy, R. E., Kitching, W., Psarras, T., Salinger, R., Chen, A. and Chivers, T. (1966a). *J. Amer. Chem. Soc.* **88**, 460.
Dessy, R. E., Kitching, W. and Chivers, T. (1966b) *J. Amer. Chem. Soc.* **88**, 453.
Devaud, M., (1966) *J. Chim. Phys.* **63**, 1335.
Devaud, M. (1967) *J. Chim. Phys.* **64**, 791.
Devaud, M. and Laviron, E. (1968). *Rev. Chim. Minerale*, **5**, 427.
Devaud, M. and Souchay, P. (1967). *J. Chim. Phys.* **64**, 1778.
Devaud, M., Souchay, P. and Person, M. (1967). *J. Chim. Phys.* **64**, 646.
De Vries, J. E., Lauw-Zecha, A. and Pellecer, A. (1959) *Anal. Chem.* **31**, 1995.
Ershler, A. B., Strelets, V. V., Butin, K. P. and Kashin, A. N. (1974). *J. Electroanal. Chem.* **54**, 75.
Fleet, B. and Jee, R. D. (1969). *Talanta*, **16**, 1561.
Fleet, B. and Jee, R. D. (1970). *J. Electroanal. Chem.* **25**, 397.
Fleet, B. and Fouzder, N. B. (1975a). *J. Electroanal. Chem.* **63**, 59.
Fleet, B. and Fouzder, N. B. (1975b). *J. Electroanal. Chem.* **63**, 69.
Fleet, B. and Fouzder, N. B. (1975c). *J. Electroanal. Chem.* **63**, 79.
Fleet, B. and Fouzder, N. B. (1978a). *J. Electroanal. Chem.* (in press).
Fleet, B. and Fouzder, N. B. (1978b) *J. Electroanal. Chem.* (communicated).
Flerov, V. N. and Tyurin, Y. M. (1970). *Soviet Electrochem.* **6**, 1357.
Flerov, V. N., Spiridonova, N. V. and Tyurin, Y. M. (1969). *Tr. Gork. Politekh. Inst.* **25**, 45.
Geyer, R. and Rotermund, U. (1969). *Acta. Chim. Acad. Sci. Hung.* **59**, 201.
Gowenlock, B. G. and Trotman, J. (1957). *J. Chem. Soc.* 2114
Gowenlock, B. G., Pritichard-Jones, P. and Ovenall, D. V. (1958). *J. Chem. Soc.* 535.
Haasova, L. and Pribyl, M. (1970). *Fresenius' Z. Anal. Chem.* **249**, 35.
Heaton, R. C. and Laitinen, H. A. (1974). *Anal. Chem.* **46**, 547.
Hopes, T. M. (1966). *J. Assoc. Off. Anal. Chem.* **49**, 840.
Hubis, W. and Clark, R. O. (1955). *Anal. Chem.* **27**, 1009.
Hush, N. S. and Oldham, K. B. (1967). *J. Electroanal. Chem.* **6**, 34.
Isslieb, K., Matschiner, H. and Naumann, S. (1968). *Talanta*, **15**, 379.
Jensen, S. and Jernelov, A. (1969), *Nature, Lond.* **223**, 753.
Jehring, H. and Mehner, H. (1963a). *Z. Chem.* **3**, 33.
Jehring, H. and Mehner, H. (1963b). *Z. Chem.* **3**, 473.
Jehring, H. and Mehner, H. (1964). *Z. Chem.* **4**, 273.
Kashin, A. N., Ershler, A. B., Strelets, V. V., Butin, K. P., Beltskaya, I. P. and Reutov, O. A. (1972). *J. Organometal. Chem.* **39**, 237.
Korshunov, I. A. and Malyugina, N. I. (1961). *J. Gen. Chem. USSR.* **31**, 982.
Kraus, C. A. (1913). *J. Amer. Chem. Soc.* **35**, 1732.
Leroux, P. and Devaud, M. (1973). *Bull. Soc. Chim. France*, 2254.
Leroux, P. and Devaud, M. (1974). *Bull. Soc. Chim. France*, 2763.
Mehner, H., Jehring, H. and Kriegsman, H. (1968). *J. Organometal. Chem.* **15**, 107.
Morris, M. D. (1967). *Anal. Chem.* **39**, 476.
Morris, M. D. (1969). *J. Electroanal. Chem.* **20**, 263.
Pilloni, G. (1967). *Farmaco Ed. Prat.* **22**, 666.

Pilloni, G. (1968). *Corsi Semin. Chim.* 98.
Pilloni, G. and Plazzogna, G. (1964). *Ric. Sci. R.C.*, *A*, **4**, 27.
Plazzogna, G. and Pilloni, G. (1967). *Anal. Chim. Acta.* **37**, 260.
Riccoboni, L. and Popoff, P. (1949). *Att. 1st. Veneto Sci., Venezio*, **107** (11), 123.
Roschig, M. and Matschiner, H. (1967) *Chem. Tech. Berlin*, **19**, 103.
Tagliavini, G. (1966). *Anal. Chim. Acta*, **34**, 24.
Tagliavini, G. and Plazzogna, G. (1962). *Ric. Sci. R.C.*, *A*, **2**, 356.
Tagliavini, G., Bulluco, V. and Ricoboni, L. (1961). *Ric. Sci. R.C.*, *A*, **31**, 338.
Thayer, J. S. (1974). *J. Organometal. Chem.* **76**, 265.
Vanachayangkul, A. and Morris, M. D. (1968). *Anal. Lett.* **1**, 885.
Vertyulina, L. N. and Korschunov, I. A. (1959). *Khim. Nauka Prom.* **4**, 136.
Vojir, V. (1952). *Chem. Listy*, **46**, 129.
Vojir, V. (1957). *Coll. Czech. Chem. Commun.* **16**, 488.
Webb, J. L., Mann, C. K. and Walborsky, H. M. (1970). *J. Amer. Chem. Soc.* **92**, 2043.
Wuggatzer, W. L. and Cross, J. M. (1952). *J. Amer. Pharm. Assoc., Sci. Ed.* **41**, 80.
Zezula, I. and Markusova, K. (1972). *Coll. Czech. Chem. Commun.* **37**, 1018.

Chapter 10

POLAROGRAPHIC BEHAVIOUR AND ANALYSIS OF THE ORGANIC COMPOUNDS OF ARSENIC

A. WATSON*

Max Planck Institut für Metallforschung, Institut für Werkstoffwissenschaften, Laboratorium für Reinstoffe, Schwäbisch Gmünd, W. Germany

I. INTRODUCTION

The organic compounds of arsenic have found a wide range of uses in the biological context. Today the phenyl arsonic acids find widespread application as additives to animal feeding stuffs. Others in former years were of considerable importance as agents against organisms causing infectious diseases and, although now largely superseded in many of these applications, some are still important in the treatment of African trypanosomiases. Some of these compounds are used as pesticides and defoliants and some have been or are of significance in chemical warfare.

In the region of eighteen papers have been published on the polarography of these compounds, covering most of the types of compound in more or less detail. This chapter consists of a review of the polarographic behaviour of organoarsenic compounds and the resulting analytical applications.

The polarography of these compounds offers a good example of the selectivity possible in polarographic analysis, which in this case allows differentiation of organo-arsenicals according to oxidation state and the number of organic substituents on the arsenic atom. Analysis of both the phenyl arsonic acids and the considerably more toxic lower oxidation state phenyl arsenoxide in mutual mixtures is possible by polarography in acidic solutions without prior separation. By a suitable choice of pH it is possible to determine diphenyl and triphenyl impurities often present in the mono-

---

* Present Address: Chemistry Department, Paisley College of Technology, Paisley, Scotland.

phenyl arsonic acid. These analytical applications are discussed against the background of their biological significance and in comparison with other existing analytical techniques.

These compounds show a wide variation of polarographic behaviour and give a good illustration of the various types of electrochemical processes displayed in polarography and the techniques for their investigation. These compounds display different types of limiting current, different current-potential, current-time and half wave potential concentration relationships, polarographic maxima and minima, and a wide variety of adsorption and inhibition effects. The effect of pH, the electrochemical reaction paths, macro- and micro- (microcoulometry) controlled potential electrolyses and polarographic Hammett relationships are also discussed. Other related techniques such as cathode ray and Heyrovsky–Forejt oscillographic polarography are briefly mentioned.

For experimental details and precise numerical data the original publications must be consulted.

The following compounds are dealt with in the text.

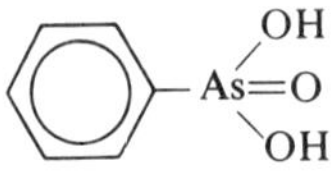

Phenyl arsonic acid

Phenyl arsenoxide

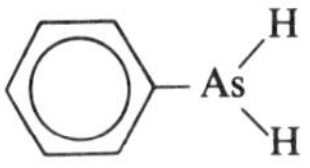

Phenyl arsine

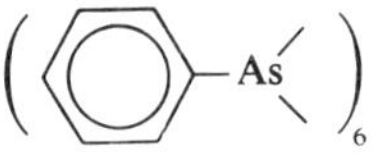

Arsenobenzene

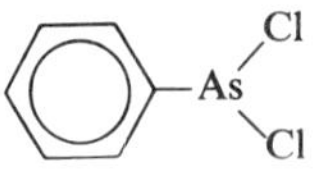

Phenyl dichloroarsine

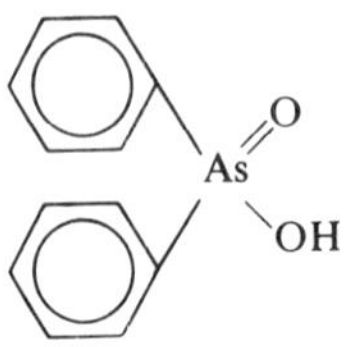

Diphenyl arsenic acid

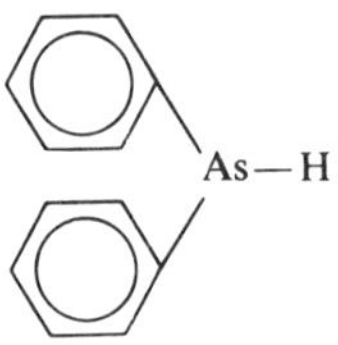

Diphenyl arsine

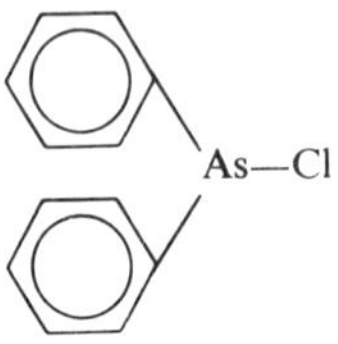

Diphenyl chloroarsine

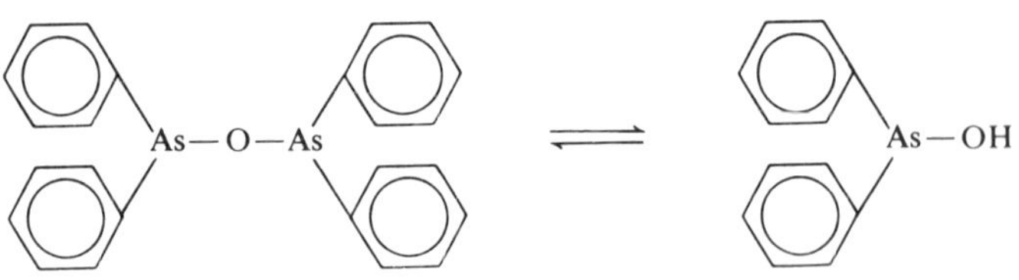

Diphenyl arsine oxide

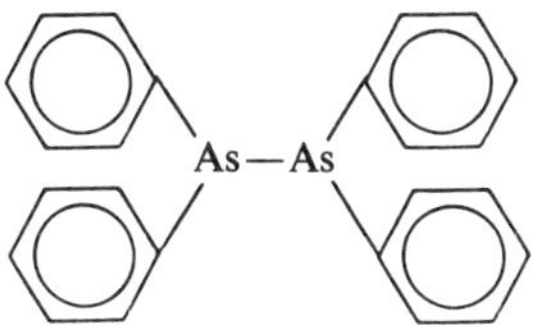

Tetraphenyl diarsine

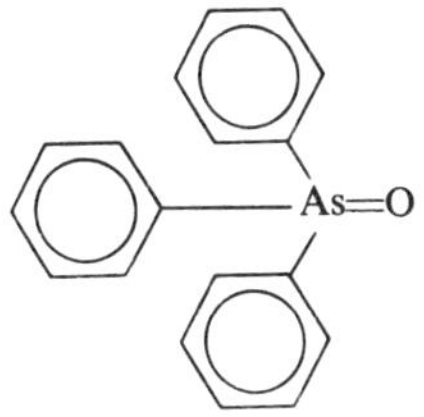

Triphenyl arsine oxide

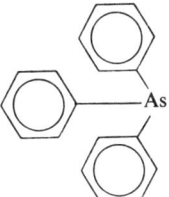

Triphenyl arsine

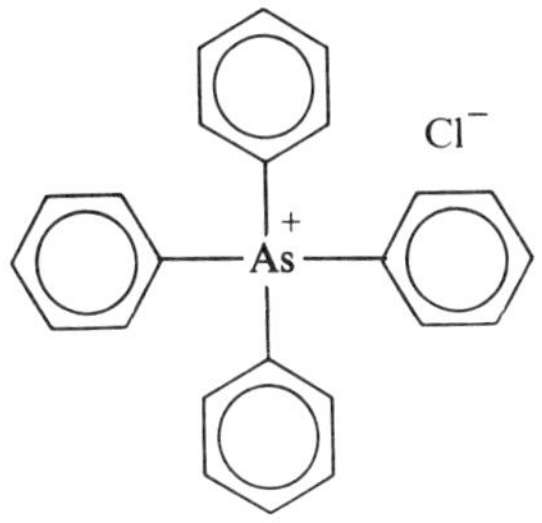

Tetraphenyl arsonium chloride

II. THE ARSONIC ACIDS

These are probably the most important of the organic compounds of arsenic. The 4 amino, the 4 nitro and the 4 hydroxy-3-nitro-phenyl arsonic acids in particular are widely used as additives in animal feeding stuffs in the 0.005–0.1% concentration range as quoted by Morrison (1968). Other uses are as analytical reagents for heavy metals, as reviewed by Welcher (1968a), and as an easily prepared starting point in the synthesis of other organic compounds of arsenic (Doak and Freedman, 1970a). A few medical applications remain today for example, in the treatment of African trypanosomiases, amoebic dysentery, etc. (Tech. Rep W.H.O. (1969)). They are among the least toxic of the organic compounds of arsenic being excluded from the body within twenty-four hours, unlike the arsenic(III) compounds. They are stable crystalline solids and moderately strong acids in aqueous solutions.

The earliest paper on these compounds of electrochemical interest was that of Fichter and Elkind (1916). This paper dealt with the electrolytic products obtained at large cathodes of several metals. Although in this work the potential was not controlled and occurred before the organic compounds of arsenic were completely characterized, the paper did indicate the general trends of the reactions. They found three phenyl arsonic acids to be reduced at a mercury electrode to the corresponding phenyl arsine in very strongly acid (8M HCl) solutions and to arsenobenzenes in dilute acid solutions (2M HCl). The next papers were those of Breyer (1938, 1939) which were an early example of the attempt to relate half-wave potentials to substitution effects and acid dissociation values. The work of Maruyama and Furuya (1957a,b) established that the waves due to amino phenyl arsonic acids were suitable for analytical purposes. Wallis (1959) examined the cathode ray polarography of 2-nitro phenyl arsonic and two complex azo dyestuffs containing the arsonic group. This work was primarily concerned with adsorption effects. Yoshidu (1970) also examined the d.c. and a.c. polaro-

TABLE I. *The phenyl arsonic acids studied and their half wave potentials under the conditions given in the references*

Derivative	Watson et al. (1970)	Watson and Svehla (1975a)	Breyer (1938)	Breyer (1939)	Maruyama and Furuya (1957b)
H	−0.82	−0.945	−1.274	−1.26	−0.955
4-OH	−0.89	−0.963	−1.366	−1.35	
2-OH	−0.81				
2,4-diOH	−0.88	−0.990			
4-OH, 3-CH_3CONH				−1.33	
4-OH, 3-NH_2		−0.975		−1.45	
4-NH_2		−0.980	−1.464	−1.44	−0.985
3-NH_2					−0.976
2-NH_2		−0.938			−0.960
4-NH_2, 3-NH_2				−1.47	
4-CH_3CONH			−1.300	−1.28	
4-CH_3COO				−1.33	
4-CH_3O			−1.354	−1.34	
4-CH_3		−0.955	−1.290	−1.28	
4-Cl		−0.905		−1.32	
2,4-diCl			−1.361	−1.34	
4-NO_2		−1.000			
3-NO_2		−0.950			
2-NO_2		−0.86			
3-NO_2, 4-OH		−1.015			
5-NO_2, 2,4-diOH	−0.87	−1.030			
n-propane arsonic acid		−1.11			

10. POLAROGRAPHY OF THE ORGANIC COMPOUNDS OF ARSENIC

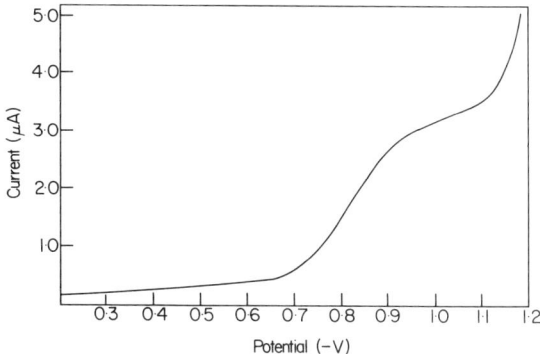

FIG. 1. Current–potential curve for 0.0002M solution of phenylarsonic acid in 0.1M HCl.

graphy of an azo dyestuff complex containing the arsonic group, the Arsenazo I dye which is used in the spectrophotometric determination of various metals. None of these authors have studied the reaction paths of the electrode processes or the accuracy of the analytical determinations. This has largely been carried out by the present author (Watson et al, 1979; Watson and Svehla, 1975a) and is discussed below.

Table I shows the range of derivatives examined by the various authors. The phenyl arsonic acids, without a nitro or azo group, all give rise to a single well-defined cathodic wave i_1, ($E_{\frac{1}{2}} = -0.8 \rightarrow -1.0$ V at pH 1) in acidic solutions below pH 3 (Fig. 1). At higher pH values the wave becomes merged with the decomposition of the supporting electrolyte and so no further cathodic behaviour is shown throughout the remainder of the pH region.

The nitro substituted phenyl arsonic acids give rise to two waves of approximately equal height (Fig. 2). The more negative wave i_1 is similar to

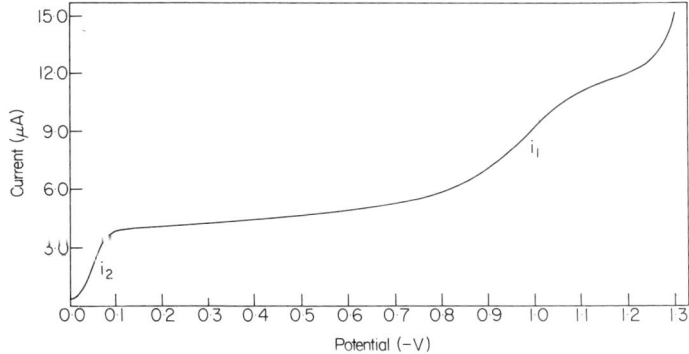

FIG. 2. Current–potential curve for 0.0002M solution of 4-nitrophenylarsonic acid in 0.1M HCl.

the wave produced by the other arsonic acids and can be observed only below pH 3. The more positive wave i_2 can be observed over the entire pH range and is due to the reduction of the nitro group to the hydroxylamine derivative. In addition the 2-nitro phenyl arsonic acid gives a third poorly defined complex reduction phenomenon i_3 between the waves i_1 and i_2 consisting of small waves and a maximum. The 5 nitro-2-4 dihydroxy derivative displays a maximum of the first kind on the wave i_2 while the 3-nitro-4-hydroxy derivative produces a maximum of the second kind on the upper plateau of the wave i_2.

Yoshidu (1970) and Wallis (1959) have shown that azo dyestuffs containing an arsonic group give three cathodic waves (or peaks in cathode ray and a.c. polarography). The two more positive waves are due to the reduction of the azo group, while the most negative wave, believed to be due to the reduction of the arsonic group, exists only below pH 3.

The arsonic acids do not display anodic phenomena throughout the pH range. Only the wave i_1 due to the reduction of the arsonic group is considered here.

The wave height is proportional to the square root of the height of the mercury column for all the arsonic acids examined, including the wave due to the arsonic group in the azo dyestuffs (Yoshidu, 1970), which indicates a diffusion controlled process.

Plots of the logarithm of the instantaneous current (recorded oscillographically) versus the logarithm of time during the drop life are somewhat curved, due to the irreversibility of the electrode process and residual depletion effects. The slope measured over the last 10% of the drop life time, where these effects are least marked approaches the theoretical value +0.19 for a diffusion controlled process (Zuman, 1969a) at the beginning of the upper plateau and increases towards the base of the wave—a characteristic of irreversible diffusion controlled processes.

Evidence of adsorption effects are only found with 2-nitro phenyl arsonic acid and the azo dyestuffs. In the former case these are associated with the process i_3 only, resulting in complex instantaneous current time curves and minima in the electrocapillary curve at these potentials. Both Wallis (1959) and the present author (Watson, 1978) have shown this phenomenon to involve the reduction of the hydroxylamine and the arsonic groups, together involving an adsorbed state(s) and that the expulsion of this adsorbed state(s) results in the maximum followed by the normal "arsonic" wave. The process i_3 is suppressed by the addition of 0.01% Lissapol N or Triton X-100. Electrocapillary curves also indicate adsorption effects with the azo dyestuffs but again not at the potential of the "arsonic" wave (Yoshidu, 1970).

The irreversibility of the electrode process was further investigated for twelve phenyl arsonic acids by logarithmic analysis of the shape of the wave

10. POLAROGRAPHY OF THE ORGANIC COMPOUNDS OF ARSENIC

using various logarithmic functions of the current which were plotted against the electrode potential corresponding to six possible current potential relationships and six different types of process (Perrin, 1965; Delahay, 1954). αn_a values of ~ 0.5 were found for all the phenyl arsonic acids, except the 4-chloro derivative ($\alpha n_a \sim 0.4$) (Watson and Svehla, 1975a). These results indicate an irreversible process described by the relationship

$$E = E_{\frac{1}{2}, \text{pH} = 0} + \frac{2.303RT}{\alpha nF} \log \frac{(i_d - i)}{i} - \frac{2.303RT}{\alpha nF} \text{pH}$$

The first attempt to examine the effect of substitution on the half wave potential and to relate it to other properties of the derivative was made by Breyer (1938, 1939). This was largely qualitative but was one of the earliest papers to make this sort of investigation. He has shown that the $E_{\frac{1}{2}}$ values of the phenyl arsonic acids become more negative with increasing pK value.

The close similarity of their polarographic behaviour fulfils the conditions necessary for a meaningful study of the effect of substitution by a graphical comparison of half wave potentials to the Hammett substituent constants (Zuman, 1969b).

A recent publication (Watson and Svehla, 1975a) has shown that for the 4-amino, the 3-amino-4-hydroxy, the 4 methyl and the unsubstituted phenyl arsonic acids a linear relationship exists between the half-wave potential and

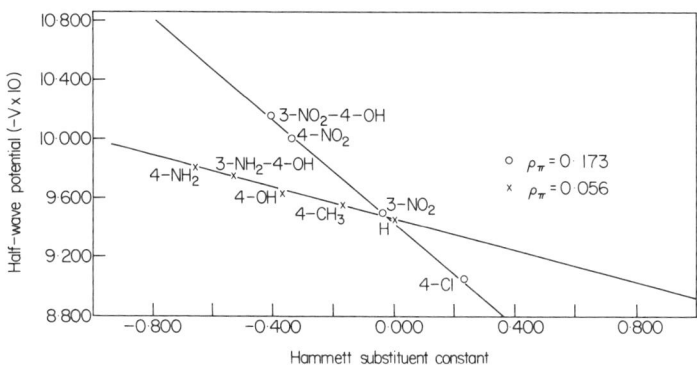

FIG. 3. Half-wave potential *versus* polar Hammett substituent constants for some phenyl-arsonic acids. ○, $\rho_\pi = 0.173$ and ×, $\rho_\pi = 0.056$.

the polar Hammett substituent constants (Fig. 3), with a coefficient of correlation of 0.9946 and a polarographic reaction constant of 0.053 ± 0.01 V. This relationship applies only to polar substituent constants which do not take mesomeric effects into account. Similar analysis of the half wave potentials given by Breyer (1938, 1939) and Maruyama and Furuya (1957a,b) for 3- and

4-substituted phenyl arsonic acids yields a similar result with, however, a higher scatter of the points due to the lower precision of the data. The reaction constant for Breyer's data is somewhat higher.

The low value of the polarographic reaction constant in comparison with those for many other systems given in the literature (Zuman 1967a) indicates the formation of a neutral transition state and so confirms that the potential determining step is the reduction of the protonated cation, as is suggested by the half wave potential pH dependency. This low value combined with the lack of mesomeric effects indicates that reduction does not directly involve the benzene ring but is confined to the arsonic group itself. The central importance of the arsenic atom is confirmed by the fact that the phosphorus analogue, phenyl phosphonic acid, is not electroactive.

As can be seen in Fig. 3, the nitro containing phenyl arsonic acids fall on a separate line and this can be explained by the reduction of the nitro compound to the hydroxylamine* occurring at the same potential as the arsonic group and thus affecting the overall electrode process. The polarographic reaction constant is 0.173 V. The 4-chloroderivative also deviates from the original line due, it is believed (Zuman, 1967b), to the highly polarizable chlorine atom acting as an "electron bridge" between the electrode and the electroactive group, thus involving a different orientation at the electrode surface.

A plot of $E_{\frac{1}{2}}$ vs pK_1† of the phenyl arsonic acids shows that 2-substituted molecules are involved in steric and hydrogen bonding effects, as would be expected.

Such information on the effect of substituents on $E_{\frac{1}{2}}$ is of importance in toxicity studies since, in general, those derivatives with the most negative half wave potentials tend to be the least toxic (Breyer, 1939). Such a trend could be expected if toxicity is connected with the reduction of As(V) to As(III), followed by the formation of stable adducts with protein thiols, as is believed.

Watson and Svehla (1975a) have proposed the following reduction path for phenyl arsonic acid

* Whose Hammett constant must be used in the construction of Fig. 3.
† Determined by potentiometric titration with $Ba(OH)_2$.

The postulated reaction path requires eight electrons for the reduction of two molecules of phenyl arsonic acid. This is in good agreement with the average of 4.05 ± 0.15 electrons per molecule obtained for 4-hydroxy and the unsubstituted phenyl arsonic acid by microcoulometry (Watson and Svehla, 1975a) that is by means of prolonged electrolysis on a small volume. The close similarity of behaviour of the other derivatives would indicate that they are also reduced along this reaction path. A value of four electrons per molecule was also obtained for the "arsonic" wave of Arsenazo I (Yoshidu, 1970).

The wave height for i_1 is linearly related to the concentration in the range 10^{-3}M–10^{-5}M using sampled d.c. polarography. The standard error of estimate for a calibration curve with ten points is 1–2% for phenyl arsonic acids and 3–5% for nitro-containing molecules.

III. PHENYL ARSENOXIDE

Phenyl arsenoxide exists in the solid form as an oligomeric oxide (also known as arsenosobenzene) and in aqueous solutions as a monomeric dihydroxyl form (also known as phenyl arsonous "acid")

and is the most stable of the mono-substituted arsenic III compounds. It is a very weak acid indeed with a pK_1 value of about 11.

The phenyl arsenoxides are believed to have been the active agents produced by oxidation in the now obsolete drugs based on arsenobenzene such as Salvarsan (the 3-amino-4-hydroxy derivative) (Doak et al., 1940). The difficulty in determining the degree of oxidation and hence the toxicity and therapeutic strength was a large factor in their replacement by other agents in the nineteen-fifties. However the arsenobenzenes were originally of central importance as antibiotic and trypanocidal agents.

The phenyl arsenoxides exist in the lower oxidation state corresponding to the phenyl arsonic acids, which find widespread use as additives to animal feeding stuffs. The arsenoxides, however, are considerably more toxic and the maximum permitted levels are an order of two lower (Morrison, 1968). The arsonic acids are believed to owe their therapeutic value to reduction in the body to the arsenoxide which again is the active agent (Doak et al., 1940). The arsonic acids are preferred today to the arsenobenzenes for medical and veterinary purposes because of their greater stability and more constant toxicity.

The arsenoxides are also formed by hydrolysis of dichloroarsines such as the Lewisite poison war gas chlorovinyl dichloroarsine.

The ability of the arsenoxides to react with protein thiol groups to form adducts stable at the pH values found in most living cells is believed to be responsible for both their high toxicity and therapeutic value (Eagle and Doak, 1951). Analytical methods for the determination of this species in the evaluation of the toxicology of the organic compounds of arsenic are clearly of great value.

The arsenoxide was one of the earliest organic compounds to be examined polarographically (Brdička, 1933). This was concerned with the polarographic determination of the product (3-amino-4-hydroxyl-phenyl arsenoxide) of the oxidation in air of samples of Salvarsan (3-amino-4-hydroxy arsenobenzene), and so was one of the very first attempts to employ polarography in pharmaceutical analysis. Unfortunately the work was carried out in unbuffered lithium chloride solutions and resulted in poorly formed waves, broad maxima and poor reproducibility. For this reason a detailed study of the polarographic behaviour of phenyl arsenoxide was published recently (Watson and Svehla, 1975b) and is discussed in the following paragraphs.

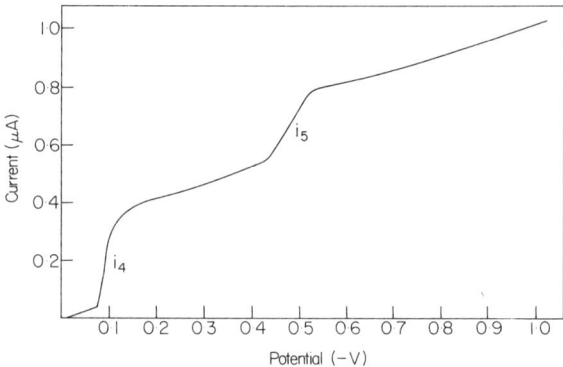

FIG. 4. Current–potential curve for 0.00005M solution of phenyl arsenoxide in 0.1M HCl.

Phenyl arsenoxide shows two main reduction phenomena, i_4 and i_5, throughout the pH region. However, in the less acidic region above pH 2–3 the waves are very poorly formed and exhibit broad maxima, as was also observed by Brdička (1933). A change of buffer at the same pH value did not alter or improve this situation. For this reason most of the studies have been carried out in 0.1M hydrochloric acid in which the waves below 1×10^{-4}M are well-defined, reproducible and exhibit no maxima (Fig. 4).

Watson and Svehla (1975b) have investigated in detail the mechanism of the polarographic reduction of phenyl arsenoxide using instantaneous current measurements during the drop life, effect of pH, wave shape analysis, coulometry etc. and proposed the following overall reaction path

$$\text{C}_6\text{H}_5\text{As(OH)}_2 \xrightarrow[4\text{H}^+]{4e} \text{C}_6\text{H}_5\text{AsH}_2$$

$$\text{C}_6\text{H}_5\text{AsH}_2 + \text{C}_6\text{H}_5\text{As(OH)}_2 \longrightarrow \frac{1}{3}(\text{C}_6\text{H}_5\text{As})_6$$

This proposed reaction path consumes an average of two electrons per molecule to form arsenobenzene, the value obtained from the microcoulometric data for electrolysis at the upper plateau of the wave i_5.

A linear relationship exists between the wave height of both waves and the concentration below 1×10^{-4}M with a standard error of estimate about 2% of the average current for both waves, behaviour characteristic for diffusion control. Within experimental error the slope of the plots do not vary from sample to sample of the solid, confirming that solid phenyl arsenoxide yields a reproducible concentration of the electroactive form. The ratio of the waves i_4 to i_5 is approximately three to two.

IV. DIPHENYL ARSINIC ACID

Diphenyl or dialkyl arsinic acids are often found as the chief impurity, formed as a by-product in the manufacture of the corresponding arsonic acid, thus causing uncertainty in the toxicity and therapeutic value of the latter compounds (Doak and Freedman, 1970b). They also find application as defoliants, particularly dimethyl arsinic acid (cacodylic acid) (Sachs et al., 1971). A mixture known as Agent Blue, largely based on dimethyl arsinic acid, was used in the Vietnamese war to defoliate the jungle for military purposes. A few rare medical applications remain, e.g. as an agent against amoebic and similar infections.

In acidic solutions diphenyl arsinic acid gives a wave at similar potentials to the arsonic acid, preceded by a prewave whose height is limited above approximately 1×10^{-1}M and proportional to the height of the mercury column (Fig. 5) (Watson and Svehla, 1974). For the prewave the instantaneous current time relationship is typical for a Brdička adsorption prewave with an exponent of approximately -0.33. The total height of the wave and prewave is proportional to the concentration and the square root of the height of the mercury column, which indicates diffusion control.

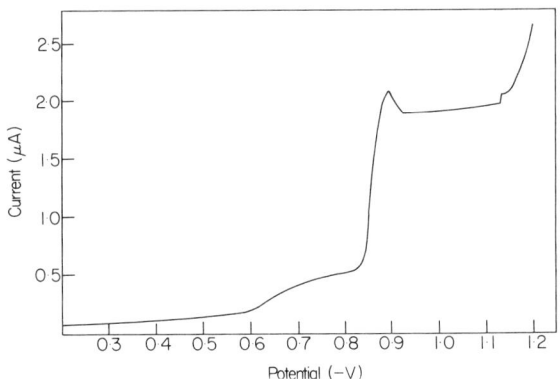

FIG. 5. Current-potential curve for 0.0002M diphenyl arsinic acid in 0.1M HCl.

The prewave, the maximum on the principal wave and all inhibition effects are completely suppressed by the addition of a surface active agent such as Triton X-100, which considerably improves the reproducibility of the wave. Analytical calibration plots are obtained with a standard error of estimate of 2–3% of the average current.

The wave height is independent of pH below pH 6; above pH 6 as the pK_1 value of the arsinic acid is reached the wave height falls rapidly as the electro-inactive anion is formed. The half wave potential moves to more negative potentials at a rate of 0.09 V pH unit^{-1} compared with about 0.12 V pH unit^{-1} for the arsonic acid. This and the higher pK_1 value of the diphenyl arsinic acid results in the arsinic acid being electroactive at pH values at which the arsonic acid is not polarographically active.

Logarithmic analysis of the shape of the wave, as for the arsonic acid, shows an irreversible process with a value of approximately 0.09 V for $2.303RT/\alpha nF$ equal to the slope of the half wave potential pH plots which indicates that protonation of the arsinic acid occurs prior to the first potential determining reduction step (Watson, 1978).

Prolonged electrolysis on a small volume gave a figure of three electrons consumed per molecule (Watson and Svehla, 1974), the value expected for the following reduction mechanism analogous to that of the phenyl arsonic acid.

$$Ph_2As(=O)OH + 2H^+ + 2e \longrightarrow Ph_2As-OH + H_2O$$

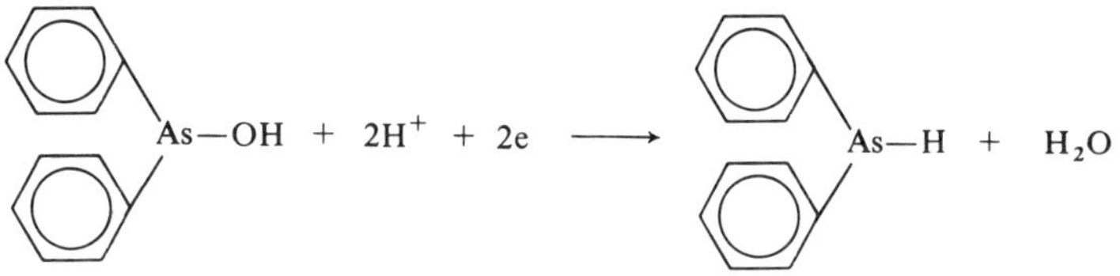

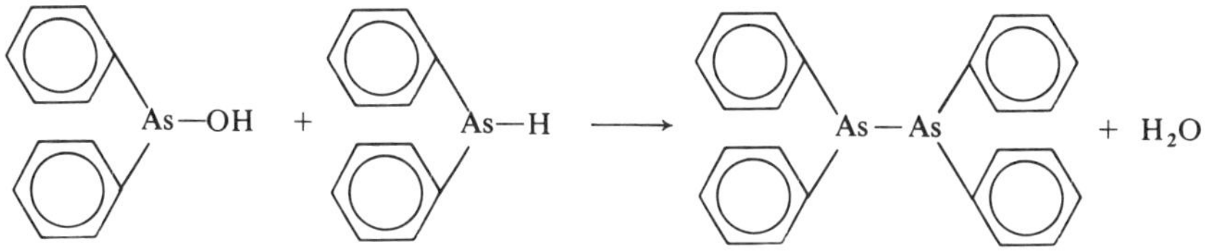

$$\text{Ph}_2\text{As-OH} + 2\text{H}^+ + 2e \longrightarrow \text{Ph}_2\text{As-H} + \text{H}_2\text{O}$$

$$\text{Ph}_2\text{As-OH} + \text{Ph}_2\text{As-H} \longrightarrow \text{Ph}_2\text{As-AsPh}_2 + \text{H}_2\text{O}$$

V. DIPHENYL ARSINE OXIDE

Diphenyl arsine oxide, the lower oxidation state of diphenyl arsinic acid, exists both as a dimeric oxide and as a monomeric hydroxide. It is chiefly of interest as the hydrolysis product of such irritants and chemical warfare agents as diphenylcyanoarsine or diphenyl chloroarsine (Kalvoda, 1954). However very little work has been done on the polarographic behaviour of this compound.

Dessy et al. (1966) have shown that in anhydrous glyme with tetrabutylammonium perchlorate as the supporting electrolyte the following reaction occurs as a single wave at -2.8 V.

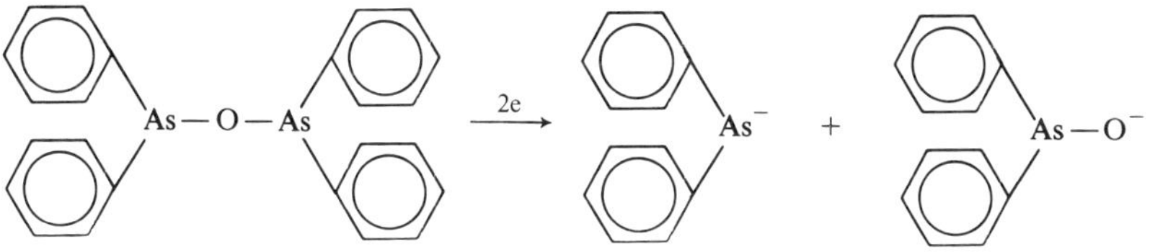

$$\text{Ph}_2\text{As-O-AsPh}_2 \xrightarrow{2e} \text{Ph}_2\text{As}^- + \text{Ph}_2\text{As-O}^-$$

In the presence of a large counter ion and non-aqueous conditions these ions were found to be stable. In aqueous conditions, one would expect these ions to protonate and combine with loss of water to give tetraphenyl diarsine, the product also of the reduction of diphenyl arsinic acid.

Suzuki and Tachi (1948) have shown that diphenyl arsine oxide produced by the hydrolysis of diphenyl cyanoarsine, gave two waves in aqueous alcoholic solutions, whose wave heights and ratio were not very reproducible. Similarly Kalvoda (1954) has found two reduction phenomena in the Heyrovsky–Forejt oscillographic polarography of the hydrolysis product of diphenyl chloroarsine, in alkaline solutions containing acetone. Neither author has examined the reaction path but it is probable that the waves would be due to the reduction of the dimeric and monomeric forms of diphenyl arsine oxide respectively and that the uncertainty in wave height arose from the uncertainty in the extent of dimerization.

VI. TRIPHENYL ARSINE OXIDE

In the reductions described in the preceding sections phenyl arsine (or diphenyl arsine) reacted with its own higher oxidation state, phenyl arsenoxide (diphenyl arsine oxide) with the formation and loss of water between the arsine hydrogen and the hydroxyl group of the higher oxidation state. Such a reaction is not possible for triphenyl arsine, and so a study of the polarographic behaviour of its oxidation product triphenyl arsine oxide makes an interesting comparison with the investigations described already. This is discussed in the preceding paragraphs (Watson and Svehla, 1975c).

Considerable interest has been shown in triphenyl arsine and its oxide as ligands in co-ordination chemistry (Booth, 1963: Doak and Freedman,

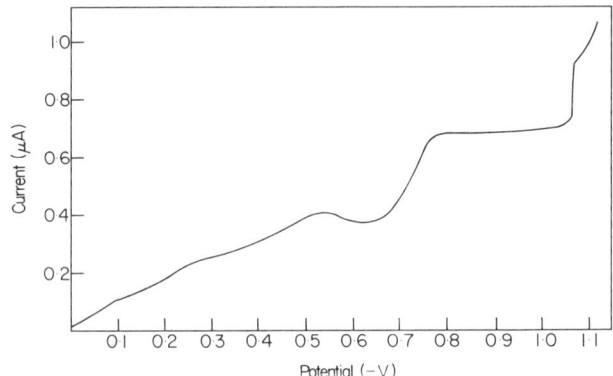

FIG. 6. Current–potential curve for 0.0001 M solution of triphenylarsine oxide.

1970b). The chief significance of triphenyl or trialkyl arsine oxides in the biological context is their presence as by-products in the manufacture of the arsonic acids used in animal feeding stuffs or the arsinic acids used as pesticides and defoliants (Doak and Freedman, 1970b).

Triphenyl arsine oxide in 0.1M hydrochloric acid gives rise to basically one main wave at −0.80 V which is complicated by adsorption effects.

Triphenyl arsine oxide is electroactive throughout the pH range as could be expected from the absence of ionizable hydrogen atoms. The investigation described in this section utilizes dilute hydrochloric acid as the supporting electrode.

Below 1×10^{-4}M some very poorly formed adsorption prewaves can be observed (Fig. 6) whose height does not increase with increasing concentration. Above 1×10^{-4}M these waves become so indistinct in shape that they

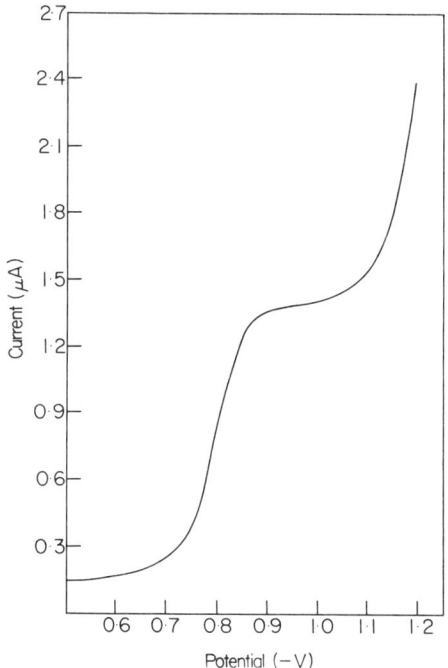

FIG. 7. Wave for 0.0002M solution of triphenylarsine oxide in 0.1M HCl containing 0.005 percent of Triton X-100.

are observed only as a general increase in residual current although marked inhibition effects appear at this concentration on the upper plateau of the main wave. These effects can be removed by addition of Triton X-100 (Fig. 7).

Following a study of the current–potential relationships, microcoulometry (which yielded a value of $n = 2.03 \pm 0.15$) and subsequent thin layer chromatographic identification of triphenyl arsine as the reduction product, Watson and Svehla (1975c) proposed the following mechanism of reduction:

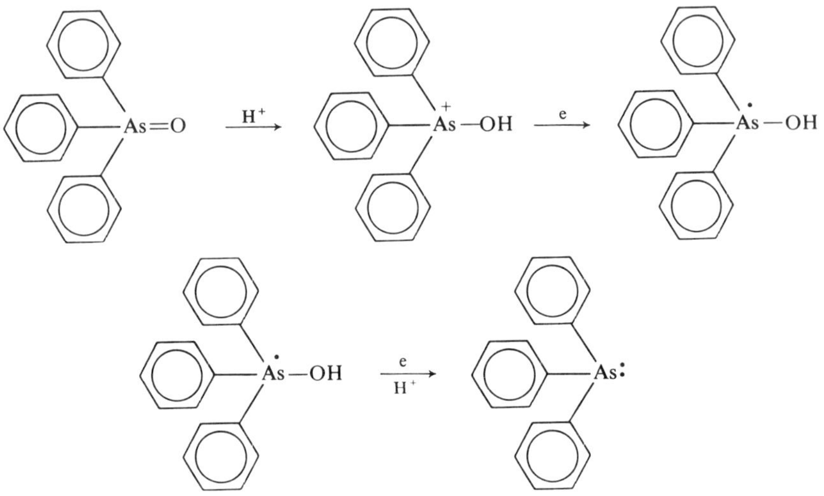

$$Ph_2As{=}O \xrightarrow{H^+} Ph_2\overset{+}{As}{-}OH \xrightarrow{e} Ph_2\overset{\cdot}{As}{-}OH$$

$$Ph_2\overset{\cdot}{As}{-}OH \xrightarrow[H^+]{e} Ph_2As{:}$$

A proportional relationship exists between the wave height and the concentration of triphenyl arsine oxide. A standard error of estimate of the order of 2–3% of the average wave height can be obtained, slightly higher than the value for the arsonic acids and phenyl arsenoxide obtained under similar conditions (Watson and Svehla, 1974; 1975a, b, c). This and the proportional relationship between the wave height and the square root of the height of the mercury column indicate a diffusion controlled process.

VII. QUATERNARY ARSONIUM SALTS

These compounds are principally reagents in analytical chemistry (Welcher, 1968b) and for a discussion of their polarographic behaviour Hörner et al. (1963) should be consulted.

VIII. ORGANIC ARSENIC HALIDES

These compounds are highly reactive towards water and hydrolyse to oxides or hydroxides. Many have been applied as irritants or poisons for military or law enforcement purposes (Kalvoda, 1954; Ludemann et al., 1969). Formation on hydrolysis of the arsenoxides which can react with protein thiols account for their corrosive toxic nature.

The hydrolysis products can be examined polarographically in aqueous solutions or the halides themselves can be examined in anhydrous aprotic organic solvents. Kalvoda (1954) has chosen the former course and has shown that diphenyl chloroarsine, diphenyl cyanoarsine and mixtures of chlorovinyl chloroarsines can be detected semiquantitatively by the characteristic curves of the Heyrovsky–Forejt oscillographic polarography of their hydrolysis products.

For the more detailed cyclic voltammetric and polarographic studies of these compounds in anhydrous glyme with tetrabutyl ammonium perchlorate as the supporting electrolyte, the work of Dessy et al. (1966) should be consulted.

IX. THE ELECTROINACTIVE COMPOUNDS

Phenyl arsine, diphenyl arsine, triphenyl arsine, arsenobenzene and tetraphenyl diarsine have been found to be electroinactive in aqueous solutions, while tetraphenyl diarsine and triphenyl arsine in anhydrous glyme have very negative half wave potentials indeed. All the other organic compounds of arsenic discussed in this chapter can thus be differentiated from these by polarography.

X. ANALYTICAL APPLICATIONS

The majority of papers published on the analysis of the organic compounds of arsenic deal exclusively with elemental analysis. In both the biological and analytical context this has several major disadvantages. The volatility of many arsenic salts particularly the chloride often results in a loss of arsenic during the decomposition procedure (Table II) (Reay, 1974). Also with oxygen flask combustion methods arsenic alloys with the platinum spiral and so a quartz spiral must be used and problems often arise due to incomplete combustion of the sample (MacDonald, 1961). While both these problems can be overcome with care and suitable techniques elemental analysis does not differentiate between the different oxidation states, which vary considerably in toxicity, and so is not a reliable measure of the toxicity or therapeutic value of a sample.

The most widespread technique for the determination of arsanilic acid (4-amino phenyl arsonic acid), the most commonly used arsonic acid in animal feeding stuffs, is diazotization of the amino group followed by coupling with an agent such as N-naphthylethylene diamine to form a highly coloured dyestuff which is then measured spectrophotometrically (Malaiyandi et al., 1969). This method has the advantage of greater sensitivity than other spectroscopic methods and can be easily automated (Kaufmann, 1966). It can also be used to selectively determine aminophenyl arsonic acids in the

TABLE II. *The effect of the method of oxidation on the recovery of arsenic in elemental analysis*

Plant material[a] (mg)	Addition	As found ($\mu g\, g^{-1}$)
Dry ashing at 450°C		
100	—	662
100	MgO (50 mg)	680
100	MgO (100 mg)	718
100	$NaHCO_3/Ag_2O$ (100 mg)	842
100	$NaHCO_3/Ag_2O$ (200 mg)	850
50	$NaHCO_3/Ag_2O$ (200 mg)	845
25	$NaHCO_3/Ag_2O$ (200 mg)	852
100	$NaHCO_3/Ag_2O$ (100 mg) Then fused with NaOH (2 g)	843
Wet ashing		
100	Nitric (2 ml)[b]	846
100	Nitric (2 ml) + perchloric (1 ml)	852

[a] Dried ground *Ceratophyllum demersum*.
[b] Digested with acid in a 22 × 150 mm test tube at 120°C and dried at 160°C.
[c] (From Reay, 1974 for dry ashing procedure).

presence of other arsonic acids. The method itself has a precision of about 2% (Weston et al., 1971) but again the disadvantage of this technique is that it does not differentiate between the oxidation states; also as the therapeutically active arsonic group is being determined indirectly with any other amino compounds present interfering, the method is again not a direct measure of the toxicity of the sample. Interferences of this type considerably lower the accuracy in the practical analysis of feeding stuffs.

An interesting method is that in which the carbon–arsenic bond of the arsonic acid is broken by reduction with for example Raney–Nickel and the organic residue is quantitatively identified and determined by gas chromatography (George and Morrison, 1971). This method has the great advantage of selective analysis of different substituted phenyl arsonic acids, but again does not differentiate between the oxidation states.

The vast majority of methods described in the literature do not differentiate between either the oxidation states or the number of organic substituents on the arsenic atom without a prior separation by chromatography or extraction methods. From the wide variation of behaviour described in the previous sections it can be seen that polarography does offer such a possibility.

The derivatives of phenyl arsonic acid give rise to a single well defined diffusion controlled wave in 0.1M hydrochloric acid (Watson and Svehla,

TABLE III. Linear regression analysis of wave height versus concentration data[a] for several phenyl arsonic acids

Derivative	Slope −A/mmol l⁻¹		Intercept on the wave height axis μA		Intercept on the concentration axis mmol l⁻¹		Coefficient of correlation	Standard error of estimate μA	Mean value of the μA
	Slope	Tolerance	Intercept	Tolerance	Intercept	Tolerance			
0.1–1.0 mmol l⁻¹									
H	16.29	0.34	0.07	0.21	−0.004	0.023	0.99967	0.14	9.03
4-OH	16.19	0.29	0.09	0.18	−0.006	0.020	0.99976	0.11	8.99
2,4-di. OH	16.42	0.66	0.00	0.41	0.000	0.044	0.99880	0.26	9.03
4-NH_2	16.61	0.20	0.10	0.13	−0.006	0.013	0.99989	0.08	9.23
2-NH_2[a]	16.89	0.24	−0.05	0.15	0.003	0.016	0.99984	0.10	9.24
3-NH_2, 4-OH	14.44	1.38	0.00	0.44	0.000	0.041	0.99864	0.12	4.19
4-CH_3	16.21	0.23	0.01	0.14	−0.001	0.016	0.99985	0.09	8.93
4-Cl	15.84	0.45	0.09	0.14	−0.005	0.012	0.99988	0.04	4.68
4-NO_2	17.59	0.81	−0.34	0.50	0.019	0.051	0.99841	0.32	9.29
2-NO_2	7.82	0.82	0.95	0.51	−0.122	0.122	0.99189	0.32	5.25
3-NO_2	16.80	0.87	−0.15	0.54	0.009	0.057	0.99800	0.34	9.09
3-NO_2, 4-OH	16.14	0.70	0.02	0.43	−0.001	0.047	0.99860	0.27	8.90
5-NO_2, 2,4-di–OH	15.29	0.84	0.23	0.53	−0.015	0.061	0.99772	0.33	8.64
n-Propane–AsO_3H_2	7.03	0.28	0.13	0.17	−0.018	0.044	0.99884	0.11	3.99
(0.01–0.1 mmol l⁻¹)									
H	16.00	0.30	0.017	0.019	−0.0011	0.0021	0.9997	0.012	0.897
4-OH	16.05	0.19	−0.002	0.012	0.0001	0.0013	0.9999	0.008	0.881
4-NH_2	16.56	0.27	0.007	0.017	−0.0004	0.0018	0.9998	0.011	0.918
2-NH_2	16.60	0.39	0.011	0.024	−0.0007	0.0026	0.9996	0.015	0.924
3-NH_2,4-OH	15.61	0.21	0.001	0.013	−0.0001	0.0015	0.9999	0.008	0.860
4-Cl	15.84	0.43	0.002	0.026	−0.0001	0.0030	0.9995	0.017	0.873

[a] From ten measurements per plot, in 0.1 M hydrochloric acid. (Watson and Svehla, 1975a).

1975a). The wave height is reproducible, proportional to concentration in the range 1×10^{-5}M to 1×10^{-3}M and is independent of pH. These are conditions suitable for analytical application. Table III gives the results of linear regression analysis on some typical calibration plots for several phenyl arsonic acids (Watson and Svehla, 1975a).

In analyses of known solutions of phenyl arsonic acids (in the range 1×10^{-5}M to 1×10^{-3}M) by the standard addition technique with five additions of standard solution, errors of about 2% have been obtained with sampled d.c. polarography and 0.1 M hydrochloric acid as the supporting electrolyte (Watson and Svehla, 1975a). Chloride, sulphate, nitrate, acetate and phosphate, and constant amounts of ethanol, methanol or acetone have been found not to interfere.

The close values of the half-wave potential, within 50 mV (Table I), result in a single wave from mixtures of different substituted phenyl arsonic acids, preventing differentiation of the derivatives, except when the substituent itself is electroactive. However, as the slope of their calibration plots are equal within experimental error (Table III), this wave height can be used to determine the total concentration of the arsonic group in such mixtures (Watson and Svehla, 1975a).

The nitro phenyl arsonic acids have higher standard errors of estimate (3–5% of the average wave height) of the calibration line for the "arsonic" wave i_1 (Table III). For the determination of these compounds it is better to use the more positive "nitro" wave i_2 for which a value of about 2% was obtained (Watson and Svehla, 1975a).

The most important applications of the arsonic acids are as additives in animal feeding stuffs, as growth promoters or against diseases such as blackhead in turkeys. Many methods of extraction have been recommended in the literature for various types of sample (Malaiyandi et al., 1969; Kaufmann, 1966; Weston et al., 1971; George and Morrison, 1970, 1971; Cenci and Cremonini, 1969). These result in solutions 10^{-5}–10^{-3}M of the arsonic acid, which is suitable for polarography, and most of the extraction agents, such as phosphate, methanol, etc., have been shown (Watson and Svehla, 1975a) not to interfere with the wave. Care would have to be taken, however, to remove protein and surface active agents; several of the published extraction methods also deal with this problem (Malaiyandi, 1969; George and Morrison, 1970). Cenci and Cremonini (1969) claim that by extraction with phosphate only the simple organic compounds of arsenic artificially added to the animal feeding stuff are extracted, while naturally occurring arsenic bound to protein thiols are not extracted.

The most important derivatives for these purposes are the 4-amino (arsanilic acid) and the 3-nitro-4-hydroxy (roxarsone) phenyl arsonic acids (Morrison, 1968). The use of the wave i_2 to determine the nitro derivative and the wave i_1 to determine the total arsonic acid content enables the specific analysis of these compounds in mutual mixtures with an error of 3–5%. Again a 0.1M solution of hydrochloric or sulphuric acid is a suitable supporting electrolyte.

The lower oxidation state, phenyl arsenoxide, gives rise to two well-defined diffusion controlled waves below 1×10^{-4}M in 0.1M hydrochloric acid (Watson and Svehla, 1975b). The wave heights are reproducible, proportional to concentration in the range 1×10^{-5} to 1×10^{-4}M and are independent of pH. Again these are conditions suitable for analytical application. Table IV gives linear regression analysis of two typical calibration plots (Watson and Svehla 1975b).

TABLE IV. *Linear regression analysis of two calibration plots[a] each for the waves i_4 and i_5 of phenyl arsenoxide (wave height in microamperes and concentration in millimoles per litre)*

	Wave i_4	Wave i_5	Wave i_4	Wave i_5
Slope	6.29	4.12	6.37	4.13
Tolerance	0.15	0.13	0.17	0.11
Intercept on the wave height axis	0.003	0.002	0.002	−0.001
Tolerance	0.009	0.008	0.011	0.007
Intercept on the concentration axis	−0.0005	−0.0006	−0.0003	−0.0003
Tolerance	0.0026	0.0036	0.0030	0.0029
Coefficient of correlation	0.9996	0.9992	0.9995	0.9995
Standard error of estimate	0.006	0.005	0.007	0.004
Mean wave height	0.349	0.229	0.353	0.226

[a] Ten measurements per plot (0.01–0.1 mmol l^{-1}), in 0.1M hydrochloric acid. (Watson and Svehla, 1975b).

In analyses, with 0.1M hydrochloric acid as the supporting electrolyte, of known solutions of phenyl arsenoxide (in the range 1×10^{-5} to 1×10^{-4} M) by the standard addition technique with five additions of standard solution errors of about 2% were obtained. Chloride, sulphate, nitrate, acetate and phosphate, and constant amounts of ethanol, methanol and acetone were found not to interfere (Watson and Svehla, 1975b)

As the higher oxidation state phenyl arsonic acid gives rise to a single well formed diffusion controlled wave in 0.1M hydrochloric acid at more negative potentials (−0.8 to −1.0 V) than the waves due to phenyl arsenoxide and as all three waves are separated by 300–400 mV below 1×10^{-4}M and have been found not to mutually interfere (Watson and Svehla, 1975b), polarography in 0.1M hydrochloric acid offers a simple and reliable method for the specific analysis of the two compounds in mutual mixtures in the range 1×10^{-5} to 1×10^{-4} M of phenyl arsenoxide and 1×10^{-5}–1×10^{-3} M of phenyl arsonic acid. Above 1×10^{-4} M of phenyl arsenoxide inhibition interferes with the "arsenoxide" waves and the shift of the wave i_5 to more negative potentials interferes with the wave i_1 of phenyl arsonic acid. At higher concentrations of phenyl arsonic acid the base of the "arsonic" wave i_1 interferes with the wave i_5 of phenyl arsenoxide but not with the wave i_4. The use of the standard addition technique with five additions has been found to give an error for both compounds of about 2% (Watson and Svehla, 1975b).

Phenyl arsenoxide as an impurity or as an intentional additive in arsenic-containing animal feeding stuffs will be found at lower levels than the arsonic acid (Morrison, 1968). Up to at least a 500-fold excess of arsonic acid has been found not to interfere with analysis by means of the more positive wave of phenyl arsenoxide (Watson and Svehla, 1975b).

The problem of differentiating by non-polarographic means the oxidation states is even greater for triphenyl arsine and its oxide as neither has acidic properties and they both have similar molecular weights. Triphenyl arsine oxide (Watson and Svehla, 1975c) gives rise to a single well-defined diffusion controlled wave in 0.1 M hydrochloric acid containing 0.01% Triton X-100. The wave height is reproducible, proportional to concentration in the range 2×10^{-5} to 1×10^{-3} M and is independent of pH. Triphenyl arsine is not electroactive in aqueous solutions. These are conditions suitable for the determination of triphenyl arsine oxide in the presence of triphenyl arsine by polarography. An error of 3–4% can be expected. Linear regression analysis of two typical calibration plots is given in Table V.

TABLE V. Linear regression analysis of two calibration plots[a] for triphenyl arsine oxide (wave height in micro amperes and concentration in millimoles per litre.

Slope	5.38	5.55
Tolerance	0.18	0.24
Intercept on the wave height axis	0.06	−0.03
Tolerance	0.11	0.15
Intercept on the concentration axis	−0.011	0.005
Tolerance	0.038	0.048
Coefficient of correlation	0.9991	0.9986
Standard error of estimate	0.07	0.10
Mean wave height	3.02	3.03

[a] Ten measurements per plot (0.1–1.0 mmol l^{-1}), in 0.1 M hydrochloric acid containing 0.01% Triton X-100. (Watson and Svehla, 1975c).

During formation of the carbon arsenic bond in the manufacture of an arsonic acid from inorganic arsenic a certain percentage of di- and tri-substitution occurs and so the diphenyl arsinic acid and triphenyl arsine oxide are often found as impurities in the phenyl arsonic acid (Doak and Freedman, 1970b). Quantitative detection of these in the arsonic acid is often difficult. Again polarography offers a possibility. The phenyl arsonic acids (Watson et al., 1970; Watson and Svehla, 1975a) are not electroactive above pH 3 while the diphenyl arsenic acids (Watson and Svehla, 1974; Maruyama and Furuya, 1957) remain electroactive until their pK_1 value about pH 6 is reached, above which triphenyl arsine oxide (Watson and Svehla, 1975c) remains active in the remainder of the pH range. Table VI gives an example of analyses by Maruyama and Furuya (1957b) of bis(aminophenyl)arsinic acid in 4-amino phenyl arsonic acid. Their error of 4–5% could be improved by the use of the sampled d.c. technique.

TABLE VI. *Determination of bis(-aminophenyl)arsinic acid in 4-arsanilic acid (4-aminophenyl arsonic acid)*

No.	Calculated Value (%)		Observed Value (%)
	4-Arsanilic acid	Bis(aminophenyl) arsinic acid	Bis(aminophenyl) arsinic acid
1	85.0	15.0	14.0
2	70.0	30.0	30.1
3	60.0	40.0	41.7
4	40.0	60.0	56.0
5	20.0	80.0	76.0

The tetraphenyl arsonium salts (Hörner et al., 1963) are electroactive at much more negative potentials and as the half wave potentials of the arsonium salts vary over a 1 V range (Hörner et al., 1963) it should prove possible to selectively determine mixtures of various arsonium salts as well as in the presence of other classes of organic compounds of arsenic. Precipitation reactions with the tetraphenyl arsonium salts have been followed as amperometric titrations (Shinagawa et al., 1957).

Kalvoda (1954) has suggested the use of Heyrovsky–Forejt oscillographic polarography of their hydrolysis products for the selective detection of such irritants and poison war gases as diphenyl cyanoarsine and diphenyl chloroarsine (Clark I and II), $HN(C_6H_4)_2AsCl$ (Adamsite) and the chlorovinyl chloroarsines (Lewisites). The method is rapid but essentially semiquantitative. Suzuki and Tachi (1948) found that while analysis of diphenyl cyanoarsine by the classical polarography of its hydrolysis product diphenyl arsine oxide did not give good reproducibility, oxidation with hydrogen peroxide to the diphenyl arsinic acid and polarographic analysis of the latter at pH 3 (HCl/KCl buffer with 10% acetone) proved entirely suitable with an error of about 3%.

Thus the versatility of the polarographic technique in this context can be seen; selective analysis is possible for each class of organic arsenical stable in aqueous solution and also of those compounds from which they can be quantitatively formed, enabling a more precise knowledge of the toxicity of the sample than is offered by most other methods. The technique is generally applicable to the levels of organo-arsenicals to be found in animal feeding stuffs and other agricultural and medical preparations or as significant impurities in other organo-arsenicals, that is in the lower microgram range and above. Analysis of these compounds in the nanogram range requires ultra trace analysis of elemental arsenic (Tölg, 1970) or the use of sensitive polarographic equipment, which was not available at the time of these studies.

REFERENCES

Booth, G. (1963). *Adv. Inorg. Chem. Radiochem.* **6**, 1.
Brdička, R. (1933). *Casopis Ceskoslovensko Lekarnistva*, **13**, 51.
Breyer, B. (1938). *Chem. Ber.* **71B**, 163.
Breyer, B. (1939). *Biochem. Z.* **301**, 65.
Cenci, P. and Cremonini, B. (1969). *Aliment. Animal.* **13**, 31.
Delahay, P. (1954). *In* "New Instrumental Methods in Electrochemistry", p. 57. Interscience, New York.
Dessy, R. E., Chivers, T. and Kitching, T. (1966). *J. Amer. Chem. Soc.* **88**, 467.
Doak, G. O., Eagle, H. and Steinman, G. H. (1940). *J. Amer. Chem. Soc.* **62**, 168.
Doak, G. O. and Freedman, L. D. (1970a). *In* "Organomentallic Compounds of Arsenic, Antimony and Bismuth", pp. 17–274. Interscience, New York.
Doak, G. O. and Freedman, L. D. (1970b). *In* "Organomentallic Compounds of Arsenic Antimony and Bismuth", p. 189. Interscience, New York.
Eagle, H. and Doak, G. O. (1951). *Pharmacol. Rev.* **3**, 107.
Fichter, G. and Elkind, E. (1916), *Chem. Ber.* **49**, 239.
George, G. M. and Morrison, J. L. (1970). *J. Assoc. Off. Agr. Chem.* **53**, 875.
George, G. M. and Morrison, J. L. (1971). *J. Assoc. Off. Agr. Chem.* **54**, 80.
Hörner, L., Röttger, F. and Fuchs, H. (1963). *Chem. Ber.* **96**, 3141.
Kalvoda, R. (1954). *Stat. Zdravot. Naklad.* 58.
Kaufmann, P. (1966). *In* "Technicon Symposium on Automation in Analytical Chemistry", Vol. 1, 152–154. New York.
Ludemann, W. D., Stutz, M. H. and Sass, S. (1969). *Anal. Chem.* **41**, 679.
MacDonald, A. M. G. (1961). *Analyst*, **86**, 3.
Malaiyandi, M., MacDonald, S. A. and Barrette, J. P. (1969). *J. Agr. Food Chem.* **17**, 51–55.
Maruyama, M. and Furuya, T. (1957a). *Bull. Chem. Soc. Japan*, **30**, 650.

Maruyama, M. and Furuya, T. (1957b). *Bull. Chem. Soc. Japan*, **30**, 659.
Morrison, J. L. (1968). *J. Agr. Food Chem.* **16**, 704.
Perrin, C. L. (1965). "Progress in Physical Organic Chemistry" (Cohen, S. G., Streitwieser, A. and Taft, R. W., eds), pp. 177–184. Wiley, Chichester and New York.
Reay, P. F. (1974). *Anal. Chim. Acta*, **72**, 145.
Sachs, R. M., Michael, J. L., Anastasia, F. B. and Wells, W. A. (1971). *Weed Sci.* **19**, 412.
Shinagawa, M., Matsuo, H. and Nakashima, F. (1957). *Kogyo Kagaku Zasshi.* **60**, 1409.
Suzuki, M. and Tachi, I. (1948). *J. Electrochem. Soc. Japan*, **16**, 152.
Tech. Rep. Ser. Wld. Hlth, Org. (1969). No. 434.
Tölg, G. (1970). *In* "Ultramicro Elemental Analysis", pp. 193–195. Wiley–Interscience, Chichester and New York.
Wallis, C. P. (1959). *J. Electroanal. Chem.* **1**, 307.
Watson, A., Smyth, W. F., Svehla, G. and Vadasdi, K. (1970). *Z. Anal. Chem.* **253**, 106.
Watson, A. (1978). *Analyst*, **103**, 332.
Watson, A. and Svehla, G. (1974). *Proc. Soc. Anal. Chem.* **11**, 163.
Watson, A. and Svehla, G. (1975a). *Analyst*, **100**, 489.
Watson, A. and Svehla, G. (1975b). *Analyst*, **100**, 573.
Watson, A. and Svehla, G. (1975c). *Analyst*, **100**, 584.
Watson, A. *et al.* (1979) To be published.
Welcher, R. J. (1968a). *In* "Organic Analytical Reagents", Vol. VI, p. 49. Van Nostrand Reinhold Co., London.
Welcher, R. J. (1968b). *In* "Organic Analytical Reagents", Vol. VI, pp. 326–330. Van Nostrand Reinhold Co., London.
Weston, R. E., Wheals, B. B. and Kensett, M. J. (1971). *Analyst*, **96**, 601.
Yoshidu, T. (1970). *J. Chem. Soc. Japan Pure Chem. Sect.* **91**, 243.
Zuman, P. (1967a). "Subsequent Effects on Organic Polarography", p. 58. Plenum Press, New York and London.
Zuman, P. (1967b). "Substituent Effects in Organic Polarography", p. 68. Plenum Press, New York and London.
Zuman, P. (1969a). "Elucidation of Organic Electrode Reactions", p. 16. Academic Press, London and New York.
Zuman, P. (1969b). "Elucidation of Organic Electrode Reactions", p. 123. Academic Press, London and New York.

SUBJECT INDEX

A

Accuracy, of voltammetric measurements, 9–11.
Acids, aliphatic, 223.
A.c. polarography, 61–63.
Acridines, 178.
Adenine, 143.
Adsorption processes, 9, 12, 50–51, 142–144, 187, 244–248, 254, 262–264, 269–276.
Agrochemicals, 210–214, 229–257, 261–290, 295–318.
Alcohols, oxidation of, 159.
Aldehydes, oxidation of, 159.
Alkaloids, 24, 29.
  rauwolfia, 84–85.
Alkylbenzene sulphonates, see LAS.
Amidopyrine, 137–138.
Amines, 150–151.
Amino acids, 19, 28.
Amitriptyline, 98, 141.
Amphetamine, 15, 140.
Analgesics, 135–139.
Anodic stripping voltammetry, 11, 12, 63–66, 254–255.
  in continuous flow analysis, 56.
Antidepressant drugs, 96–99.
Antituberculins, 142.
Aqueous environment, electroanalysis of, 203–224.
Arsonic acids, 297–303.
Ascorbic acid, 19, 25, 39, 128–131, 137, 149.
Atropine, 79, 136.
Automatic Analysis, 5, 10, 89, 91, 130–131, 139, 140.
Azo compounds, 7, 15, 221–223.
Azo group reduction, 15.

B

Barbituric acids, 24, 38.
  1-benzoyl 5, 5′diethyl-, 7
  nitration of, 29, 101–103.
  a.c. polarography of, 102
  oxidation of, 102, 153.
Benzhydrylpiperazine derivatives, 7, 14.
Benzimidazoles, 153.
1,4-Benzodiazepines, 7, 13, 17, 18, 25, 54, 85–96, 103–106, 141.
Benzpyrenes, 178.
γ-BHC, 213, 231–232.
Bilirubin, 19.
Biotin, 31.
Blank subtraction, 5.
Blood flow, 176.
Body fluids, analysis of, 17.
Bridicka waves, 29, 169–172.
Bromazepam, 12, 16, 92.
Butyrophenones, 83–84.

C

Capacitive currents, 5, 49, 56.
Caprolactam, 214.
Captan, 256.
Carbamates, 249–251.
Carbazoles, 154.
Carbonyl compounds, 206–207.
Carbon electrodes, 44–45.
Carcinogenesis, 177–179.
Catalytic processes, 9, 19, 50, 147.
Catecholamines, 60, 148–150.
Catechols, 148–150.
Cathode ray polarography (see linear sweep voltammetry).

SUBJECT INDEX

Cathodic stripping voltammetry, 11, 12, 63–66, 155–158, 244–248.
Cells, selection of electrochemical, 41–43.
Central nervous system agents, 139–141.
Centrifugation, 22.
Cephalosporins, 6, 22.
Chloral hydrate, 99–100.
Chloramphenicol, 113–118.
Chlordiazepoxide, 85–89.
Chlorpromazine, 15, 20–22, 23, 80–83, 140.
Chlortetracycline, 38.
Choice of method, 4–17.
Choice of solvent/supporting electrolyte, 37–41.
Cholinesterase activity, determination of, 147.
Clinical diagnosis, 169–173.
Complexation, in derivatization, 28–29.
Computers, application of, 68–72.
  optimization of experimental parameters in, 69–71.
Conjugation reactions, 16, 18, 22.
Convenience of electroanalytical methods, 9–11.
Condensation, in derivatization, 31–32.
Coulometry, 66–67.
  controlled current, 66.
  controlled potential, 66–67.
Creatine, 16, 31.
Creatinine, 16, 20, 31.
Cyclamate, 32.
Cyclic voltammetry, 51–52.
Cystine/cysteine, 19, 146, 172.
Cystinuria, 172.

D

Data processing, 5.
D.c. polarography, 48–51.
DDT, 213, 230–231.
Dealkylation metabolic reactions, application of electroanalysis to, 14.
Deamination metabolic reactions, applications of electroanalysis to, 15.
Derivitization, 29–33.
Desulphuration metabolic reactions, application of electroanalysis to, 15.
Diazepam, 85, 89–90.

Differential pulse polarography, 11, 58–59.
Digitoxin, 174.
Digoxin, 174.
Diphenylarsine oxide, 307–308.
Diphenylarsinic acid, 305–307.
Diphenylhydantoin, 29.
Dissolution rate measurements, 10, 139.
Disulfiram, 29, 32.
Dithiocarbamates, 65, 158, 246–248.
l-Dopa, 139, 148, 184.
Dopamines, 148, 180–185.
Dropping mercury electrode, 43–44.

E

Electroactive molecules, 5–9.
Electrochemical detectors (for h.p.l.c.), 5, 11, 24, 56, 148, 183–184.
Electrodes, selection of, 43–46.
  3 electrode operation, 42
Electrophoresis, 170–172, 186.
Electrotechnology, 5
Endosulphan, 232–233.
Epinephrine, 148.
Ethionamide, 12, 154–155.
Ethylenethioureas,
  nitrosation of, 248
Explosives, 219–220.

F

Fenitrothion, 211, 239.
Ferbam, 253–254.
Filtration, as separation method, 22
Fluphenazine, 27.
Flurazepam, 24, 26, 28, 106.
Folic acid, 133.
Formulation analysis, 5, 9–10.
  single tablet assays, 11.
  degradation product analysis, 11, 12, 13, 115–116.
  crop formulations, 18.
  artificial steroids, 174–175.
Fructose, 173.
Fulvic acids, 20.
Furazolidone, 119–121, 241.
Furfural, 206.

SUBJECT INDEX

G

Gel filtration, as separation method, 22.
Glutathione, 19.
Glutethimide, 101.
Glyphosphate, 31, 210–211.
Guanine, 143–144.

H

Halides, determination by c.s.v., 159–160.
Haloperidol, 17, 83–84.
Hanging mercury drop electrode, 44.
Herbicides, see Agrochemicals.
Hexachlorophane, 6.
High performance liquid chromatography, 5, 24–25.
 of biogenic amines, 183–184.
Historical background, of electroanalysis, 4–5.
Hormones, 173–175.
Humic acids, 20.
Hydrazine, 151.
Hydrolysis, in derivitization, 16, 18, 32.
Hydroxylamines, 151.
 aliphatic, 39.
 aromatic, 8.
Hydroxylation metabolic reactions, application of electroanalysis to, 14.
Hypnotic drugs (*see* Sedative drugs)

I

Identification, of unknown molecules, 5, 13.
Ilkovic equation, 48
Immunology, applications in, 186–188.
Insecticides, see Agrochemicals.
Ion Exchange Chromatography, as separation method, 23, 148.
 of glyphosphate residues, 31.
Ion Pairing, as separation method, 28.
Iproniazid, 98.
Irreversible reactions, 50.

L

LAS, 23, 216–219.
Lignin sulphonic acids, 29, 220–221.
Linear sweep voltammetry, 4, 5, 11, 51.
Lorazepam, 91.
Lynestrenol, 6, 32.

M

Malathion, 234–235.
Malonic acid, 19.
Maximum supression, 20, 39, 114, 117, 215.
Medazepam, 91.
Membranes, 189–190.
Meprobamate, 22, 24, 85.
Mercaptans, 148.
2-Mercaptopyridine-N-oxide, 158.
Metabolic reactions, 13–16, 17, 116–117.
Metation, 23.
Methaqualone, 17, 101.
Methyldopa, 23.
Methylene blue, 12.
Methylmercury, 266–267.
Metronidazole, 18, 29, 119, 121–124.
Microprocessor controlled polarographs, 5, 10, 71–72, 114.
Mitomycins, 179.
Morphine, 23, 24, 135–136.

N

NAD/NADH, 20, 144–145.
Neurophysiology, applications in, 179–184.
Neurotransmitters, 20, 179–184.
Nialamide, 98.
Nicarbazin, 38, 241.
Nitration, 29–30, 249–251.
 of alkylbenzene sulphonates, 23.
 of benzene, 209.
 of phenol, 209.
Nitrazepam, 20, 103–104.
Nitrilotriacetic acid, 29, 207–209.
Nitro-compounds, 111–125, 239–244.
 aliphatic, 7, 15.
 aromatic, 7, 13, 14, 15, 209.
Nitrofurantoin, 18, 54, 119–210.

Nitrofurans, 118–121.
Nitrogroup reduction, in metabolic reactions
  application of electroanalysis to, 15.
Nitroimidazoles, 121–125.
4-Nitroquinolines, 177.
N-Nitrosamines, 22, 39, 178–179.
Nitrogen, determination in water, 223.
Nitrosation, in derivitization, 30–31, 249–251.
  of ethylenethiourea residues, 24, 248.
  C-, metabolic reaction, 16.
  N-, metabolic reaction, 16.
  of Intration, 238.
Nitroso compounds
  aliphatic, 7.
  aromatic, 7.
Norepinephrine, 148.
Normal pulse polarography, 56–58.
N-oxidation, in metabolic reactions.
  application of electroanalysis to, 15, 31.
N-oxides
  aliphatic, 8.
  aromatic, 8.
  of chlorpromazine, 15, 20–22.

O

Organoarsenic compounds, 295–318.
Organochlorine pesticides, 230–234.
Organogermanium compounds, 286–290.
Organolead compounds, 280–286.
Organomercury compounds, 262–268.
Organometallic compounds, 253–255, 261–290.
  oxidation states, application of electroanalysis to the determination of, 13, 295–296.
  reduction of, 9.
Organophosphorus pesticides, 234–238.
Organotin compounds, 268–279.
Oxazepam, 85, 90–91.
Oxidation processes, 9.
Oximes, aliphatic, 8.
Oxygen, 173, 175–176.
Oxyphencyclimine, 79–80.
Oxytetracycline, 38.

P

Paper chromatography, as separation process, 24.
Paracetamol, 137.
Paraquat, 252.
Parathion, 239–240, 244.
Penicillins, 22.
Persedon, 17.
Pesticides (see Agrochemicals).
Pethidine, 136.
Phenylarsenoxide, 303–305.
pH, choice of, in electroanalysis, 37–41.
Phenacetin, 137.
Phenobarbital, 101–102.
Phenols, oxidation of, 158, 159.
Phenothiazines, 80–83, 140–141.
Picloram, 212, 252.
Piperidine, 153.
Plasticisers, 219–220.
Polychlorinated aromatics, 233–234.
Polymers, 214–215.
Polysaccharides, 20.
Potassium chlorazepate, 95.
Potentiostats, 46–48.
Prazepam, 92.
Precision, of electroanalytical methods, 9–11.
Progestrogens, 175.
Proteins, 19, 20, 22, 28, 114.
Protein binding, 18, 20, 124.
Prothionamide, 12.
Psychosedative drugs, 80–96.
Psychotomimetic drugs, 79–80.
Pulse polarography, 4–5, 11.
  alternate drop, 5, 59.
  normal, 56–58.
  differential, 58–59.
  differential double pulse voltammetry, 60.
Purines, 142–147.
Pyridines, 153.
Pyridoxine, 134.
Pyrimidines, 143.

Q

Quinolines, 153.

R

Radicals, electroanalysis of, 262–264, 271, 274–275.
Rapidity, of electroanalytical methods, 9–11.
Reduction processes, 5–9.
Reference electrodes, 46.
Reserpine, 84–85.
Respiratory physiology, applications in, 175–176.
Reversible reactions, 49–50.
Riboflavin, 133–134.

S

Saccharin, 22, 23.
Salicylic acid, 138.
Sample solution, preparation of, for electroanalysis, 37–41.
Sampling and initial treatment of sample, 17–18.
Sedative drugs, 99–107, 140–141.
Selectivity, of electroanalytical methods, 12–13.
Semicarbazones, 31–32, 206–207.
Sensitivity, of electroanalytical methods, 11–12, 64.
Separation techniques, 18–28.
Sequestering agents, 207–209.
Solvent, choice of, for electroanalysis, 37–41.
Solvent extraction, as separation technique, 25–28, 30.
S-oxidation, in metabolic reactions, application of electroanalysis in, 15, 31.
S-oxides,
 aromatic, 8.
 of chlorpromazine, 15, 20–22.
Specificity, of electroanalytical methods, 12–13.
Steroids, 173–175.
 3-keto-$\Delta_{4,5}$-, 6, 39, 174–175.
 cortico-, 24.
 formation of hydrazone derivatives, 32.
Structure Elucidation, by electroanalytical methods, 13–17.

Substituent Effects, 13, 96, 99, 138, 158, 265, 266, 301, 302.
Sugars, 6, 19, 30.
Supporting Electrolyte, 41.
 choice of, 37–41.
 purification of, 40–41.
Surfactants, 23, 215–219.

T

Tannins, 20.
Temazepham, 93.
Tensammetry, 9, 62–63, 216.
Tetraalkyllead compounds, 286.
Thalidomide, 101.
Thiamine, 134.
 determination using a Co(II) catalytic prewave, 9.
Thin layer chromatography, 24.
Thioamides, 10, 12, 13, 154–156.
2-Thiobarbiturates, 14, 15, 102.
 5,5-disubstituted, 8, 158.
Thiols, 221.
6-Thiopurine, 146.
Thioureas, 154–157.
Tocopherols, 131–133, 174.
Thiram, 246–247.
Tranquilizers (see sedative drugs).
Triazines, 210, 251–252.
Trifluralin, 243.
Trimethoprim, 24.
Triorganogermanium compounds, 289–290.
Triphenylarsine oxide, 308–310.
Triphenyl lead compounds, 284–285.
Triphenyltin compounds, 52, 254–255, 276–279.

U

Unit Processes, in organic voltammetric analysis, 3–72.
Uracil, 145.
Ureas, 249–251.
Uric acid, 20, 143–145.

V

Vacor, 241.
Vasopressin, 175.
Virology, applications in, 185–186.
Vitamins, 20, 128–135, 175.
Voltammetry
   at solid electrodes, 52–54.
   hydrodynamic, 54–56.

Y

Yohimbine, 85.

Z

Zineb, 253–254.
Ziram, 253–254.